D1829368

APPLICATIONS OF FIBER OPTIC SENSORS IN ENGINEERING MECHANICS

A collection of state-of-the-art papers in the applications of fiber optic technologies to civil structures

Approved for publication by the
Engineering Mechanics Division
of the American Society of Civil Engineers

Edited by Farhad Ansari

Published by the
American Society of Civil Engineers
345 East 47th Street
New York, New York 10017-2398

ABSTRACT

Recent advances in fiber optic technology, and the possibility of their use in civil structures has instigated the development of a number of research activities in the civil engineering community. This book is intended for rapid dissemination of current state of the art on this emerging technology. Fiber optic sensors, and smart structure technology have been successfully employed in space and aeronautics applications. With minimal research, these technologies can be applied to civil structures. Fiber optic sensors can help protect our infrastructure. This book contains information regarding: 1) Techniques for measurement of strain, temperature, pressure and magnetic fields; 2) smart structure technology; 3) applications in fracture mechanics; and 4) embedment in concrete.

Library of Congress Cataloging-in-Publication Data

Application of fiber optic sensors in engineering mechanics : a collection of state-of-the-art papers in the application of fiber optic technologies to civil structures / edited by Farhad Ansari.

p.cm.
Includes indexes.
ISBN 0-87262-895-7
1. Optical fiber detectors. 2. Smart structures. 3. Mechanics, Applied. I. Ansari, Farhad.
TA1815.A67 1993
621.382'75—dc20   93-604
CIP

The Society is not responsible for any statements made or opinions expressed in its publications.

Authorization to photocopy material for internal or personal use under circumstances not falling within the fair use provisions of the Copyright Act is granted by ASCE to libraries and other users registered with the Copyright Clearance Center (CCC) Transactional Reporting Service, provided that the base fee of $1.00 per article plus $.15 per page is paid directly to CCC, 27 Congress Street, Salem, MA 01970. The identification for ASCE Books is 0-87262/93.$1+.15. Requests for special permission or bulk copying should be addressed to Reprinting/Permissions Department.

Copyright © 1993 by the American Society of Civil Engineers,
All Rights Reserved.
Library of Congress Catalog Card No: 93-604
ISBN 0-87262-895-7
Manufactured in the United States of America.

PREFACE

One of the principal objectives of the American Society of Civil Engineers has been the disclosure of information in the broad area of Civil Engineering. This has been accomplished by direct interchange of ideas at many of the yearly meetings or through societies publications, such as journal of the Engineering Mechanics, and conference proceedings. These publications tend to emphasize research papers on refined topics or application papers on specific subjects. A number of excellent text books are also available for general treatment of Civil Engineering related topics. At times, however, it is felt that some of the emerging technologies require special treatment for rapid dissemination. The present special publication seems to accomplish such a task.

Recent advances in fiber optic sensor technology, and the possibility of their use in civil structures has instigated the development of a number of research activities in the civil engineering community. However, the civil engineering related research in fiber optic sensor technology is still at a stage of infancy. Fiber optic sensors, and smart structure technology have been successfully applied in other disciplines. In contrast to existing nondestructive evaluation techniques, optical fibers will be able to detect minute variations in structural conditions through remote measurements. The fiber optic sensor technologies already developed for aeronautics and space applications could, with minimal research be applied for use in civil structures. For instance, optical fibers can be embedded in concrete elements. They can be put to use immediately to determine, for example, concrete strength in form removal operations. The technology can help protect our infrastructure and, therefore current tax expenditures. Hence, there is a need for spreading the word about the fiber optic sensor technology among the civil engineers, especially to those in government agencies.

The experimental analysis and instrumentation (EA&I) committee of the American Society of Civil Engineers has felt the need for introducing some of the recent activities in the applications of fiber optic sensor technology to practicing engineers and academic researchers. Due to highly interdisciplinary nature of the topic, it was decided to create a special forum by inviting distinguished researchers from different disciplines to contribute a paper for inclusion in this book. This book is intended to encompass the state of the art in fiber optic sensor technology, and its contents include basic research, applications, and use in smart structures.

The organization of papers contained in this book are based on the broad application category to which they pertain. Accordingly, papers selected for this publications are arranged within the following chapters:

1. *Overview* of applications, smart structure technology, potential for civil engineering use, and international achievements in the discipline.

2. *Measurement of Strain* and interpretation of results by employment of a variety of sensor configurations is described. In particular, use of

interferometric, and novel intensity modulated techniques are emphasized.

3. *Embedment in Concrete Structures* for condition monitoring, measurement of rebar strains, and detection of cracks.

4. *Fracture Mechanics* applications to a variety of materials for measurement of Crack Opening Displacements, or the Stress Intensity Factors.

5. *Strain and Temperature* for measurement of strains at elevated temperature by specially coated fibers, or for use as a Temperature measuring device.

6. *Measurement of Pressure* for a variety of applications including the measurement of hydrostatic pressure distribution.

7. *Measurement of Magnetic Field* and its applications in civil engineering.

8. *Localized Detection and Smart Structures* pertaining to a major goal for application to control of civil structures.

Each one of the papers included in this book has received two positive reviews. All papers are eligible for discussion in the Journal of Engineering Mechanics. All papers are eligible for ASCE awards.

Farhad Ansari,
February 1993
New Jersey Institute of Technology

CONTENTS

V - STRAIN AND TEMPERATURE

VI - MEASUREMENT OF PRESSURE

VII - MEASUREMENT OF MAGNETIC FIELD

VIII - LOCALIZED DETECTION AND SMART STRUCTURES

I - OVERVIEW

Engineering Applications
For Fibre Optic Sensors

David A.Jackson[1]

Abstract

The mode of operation of a range of fibre optic sensors developed over the last decade is described. These new devices offer comparable or superior performance to conventional electronic sensors which together with their electrically passive nature and capacity for remote operation, makes them very attractive for a diverse range of engineering applications. Measurands addressable by these sensors include vibration, flow, temperature, strain, displacement and pressure; where contact or non-contact measurements are possible. A Distributed Temperature Sensor is also described which has a demonstrated spatial resolution of 10m over a 22.5Km range.

Introduction

It is well known that optical techniques offer extremely high precision in metrological applications, particularly if the technique is based upon optical interferometry. Unfortunately, it has been very difficult to apply these high precision techniques outside the laboratory as environmental perturbations tend to cause rapid misalignment of the optical components. Another factor that has also limited the application of optics outside the laboratory is the requirement of a 'line of sight' between the measurement volume and the optics. This situation has changed dramatically during recent years with the incorporation of fibre optic waveguides into sensors, where the fibre optic waveguide can act as the sensor or simply serve as a transceiver link to transfer light to and from a remote sensor; the remote sensor being fabricated either from the fibre or from conventional optical

[1] Applied Optics Group,University of Kent, CANTERBURY, Kent. CT2 7NR. Tel.(0227) 764000 Ext.3288

components. At the University of Kent we have been researching into a large range of fibre optic based sensors with a particular emphasis on interferometric configurations. In this paper an overview of fibre optic sensor developments at Kent will be given together with their potential for mechanical engineering applications.

Fibre Optic Interferometers (FOI)

Fibre optic variants of most of the classical optical interferometers have been demonstrated; of these configurations,the Mach Zehnder, Michelson and Fabry-Perot, figure 1, show the most promise as basic sensing elements. The transduction mechanism common to all these FOI when used as sensors is that the measurand induces an optical phase change which is detected as a change in intensity in the interferometer's output signal.Of the possible measurands that can induce a phase change in the fibre, strain and temperature produce by far the largest effects. For a monomode fibre illuminated with a source at ~800nm - typical of a laser diode - the strain and temperature sensitivities are 10^7 rad m^{-1} and 10^2 rad K^{-1} m^{-1} respectively. With appropriate signal processing, phase resolutions in excess of 10^{-6} rads/$\sqrt{Hz}$ can be detected, (Jackson, 1985) Lower sensitivity differential interferometers based upon the bimodal properties of the optical waveguide have also been developed, the sensitivity of these interferometers is typically two orders of magnitude less.

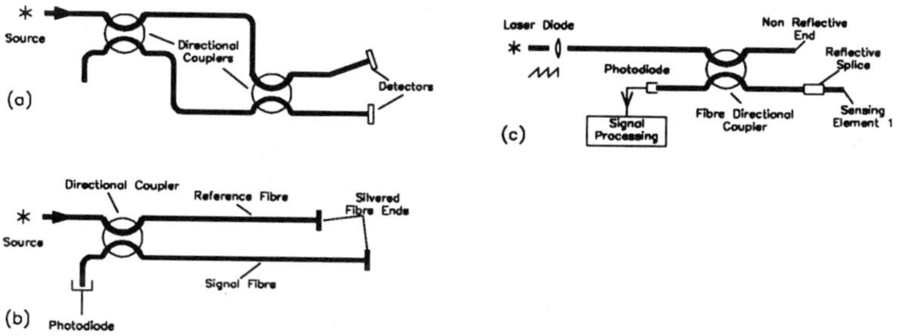

Figure 1. Fibre Optic variants of the classical optical interferometers:
a. Mach Zehnder
b. Michelson
c. Remote Fabry-Perot

Signal Processing

The output signal I from a 2 beam interferometer can be written

$$I = C(1 + V \cos \phi(t)) \quad (1)$$

where C is a constant including such factors as the conversion efficiency of the photodetector, V is the fringe visibility and $\phi(t)$ is the time dependent differential phase between the two arms of the interferometer. The periodic nature of this transfer function tends to make signal recovery more complex than from a conventional sensor with a linear transfer function, such as a strain gauge for example. For displacement resolutions of the order of the wavelength of the source, (i.e 0.3μm for a source wavelength of 0.6μm) simple 'fringe counting' will suffice - providing the fringe direction is known - if this is not the case then additional optical components are required in order to generate a second output from the interferometer which is out of phase by 90° (i.e. a quadrature signal). To obtain higher sensitivity it is necessary to determine the instantaneous phase of the interferometer. In a conventional interferometer this is usually accomplished by incorporating some form of high bandwidth frequency shifter such as an acousto-optic modulator in one of the arms of the interferometer, such that heterodyne processing may be used. Phase resolutions of ~ 10^{-6}rads/$\sqrt{\text{Hz}}$ can be achieved, with the additional advantage that the 'fringe direction' can also be determined, eliminating the directional ambiguity associated with equation (1).

In many applications for FOIS, the sensor must be miniature, remote and passive such that it is not possible to incorporate a conventional acousto optic modulator. To overcome this problem a form of heterodyning processing has been developed (often called pseudo-heterodyne), in which the absolute optical frequency (Jackson et al 1982 and Dandridge et.al 1982)of the illuminating source is modulated - either linearly or sinusoidally such that the output of the interferometer is swept over ~1 cycle of its transfer function. This requires that the optical path difference (OPD) of the interferometer must be finite; for a typical laser diode the maximum continuous frequency excursion is ~30GHz which results in a minimum OPD of 2-3cms.This method of processing has proved highly effective for periodic signals where only the relative amplitudes of the spectral components are required, however the technique cannot be easily applied to quasi-steady state measurands such as temperature, pressure or slowly varying strain as the absolute phase of the interferometer cannot be determined. A technique known as low coherence interferometry (LCI) (or White Light interferometry) has been developed for such low frequency measurands (Jackson, 1985). A typical sensing system exploiting LCI is shown in figure 2.In principle LCI allows the absolute value of the measurand to be determined when the sensor network is 'powered up' (Gerges et al, 1988).

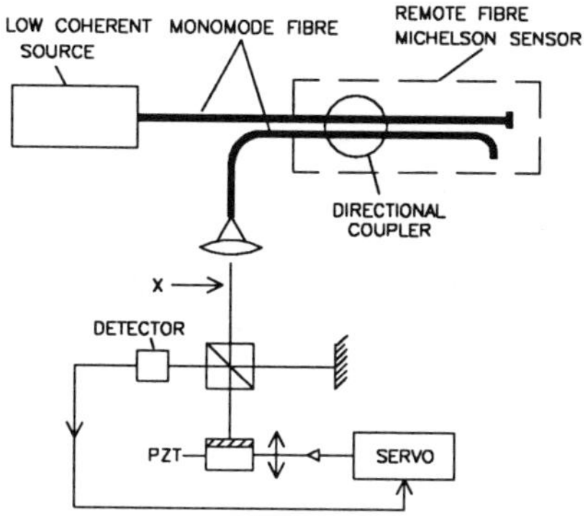

Figure 2. Basic low coherence interferometric system. The sensor may be one of the configurations shown in figure 1. The coherence length of the source must be significantly less than the OPD of the sensor such that no interference fringes are observed at point 'x'.

SYSTEMS AND APPLICATIONS

VIBRATION ANALYSIS

Non-contact - Point Measurement

We have developed several systems to detect vibration at a remote location. These systems are shown schematically in figure 3. The vibration sensor shown in figure 3(a)(Lewin,et al, 1985;Meggit et al, 1989) is basically an all fibre Michelson interferometer fabricated from monomode fibre. The output lens acts as an autocollimator, the light back reflected from the surface being re-injected into the fibre. The piezo-electric cylinder serves to phase modulate the interferometer to facilitate the signal processing. The operational range is primarily governed by the lens aperture and can be up to 20m. A rather more sophisticated vibrometer is shown in figure 3(b), here a polarization selective bulk Mach Zehnder interferometer (including an acousto optic modulator) is combined with a (highly birefringent) fibre optic transceiver link. This system offers very high performance and can be easily built in the laboratory, a commercial version is however available.

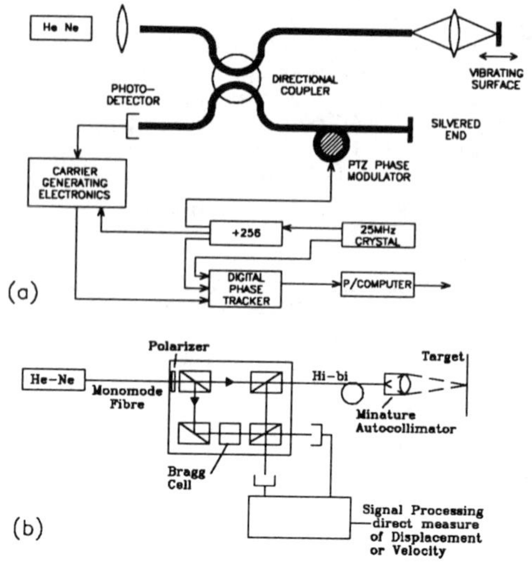

Figure 3. Non-contact vibration sensors

a. Low cost system based on an all fibre Michelson - using pseudo heterodyne processing.

b. Conventional non-contact vibration system incorporating fibre optic links to enable remote operation.Both system have displacement resolutions of 10^{-9}m - the velocity range of a) is restricted by the signal processing to $3 \times 10^{-2}ms^{-1}$ whilst b) operates up to $10ms^{-1}$

Non-contact Field measurements

Holography

In order to study the vibrational dynamics of large structures it is necessary to determine the motion of the whole surface rather than an isolated point. Optical holography has been used for this purpose for many years however it is essential that the relative positions of all the optical components remain constant to within the wavelength of the light during the exposure time tending to restrict the technique to the laboratory. The optical topology of most holographic systems is very similar to a Michelson interferometer and we have shown that excellent holograms may be obtained using the arrangement shown in figure 4(a)(Jones et al, 1984), which is very similar to the system shown in figure 3(a) except now the role of the PZT has changed. Here it is driven by a closed loop servo to eliminate the effects of drift in the system caused by environmental perturbations.

T.V. Holography

The main problem encountered in applying holography in practical measurements is the relatively long time required to process the photographic plate in order to obtain the hologram. As a result of this a lower resolution technique has been developed where the photographic plate is replaced by a T.V.camera(Jones and Wykes, 1989). Commercial systems are available for T.V.Holography however they tend to be very large (and very expensive) hence their use tends to be restricted. Again, virtually all the conventional optics can be replaced with optical fibres making a robust inexpensive system as shown in figure 4(b). The PZT's can be used not only to stabilize the drift in the fibre optic leads but also to 'phase step' and to impose periodic phase modulation as required in order to obtain the dynamic behaviour of the surface.(Santos,et al, 1990)

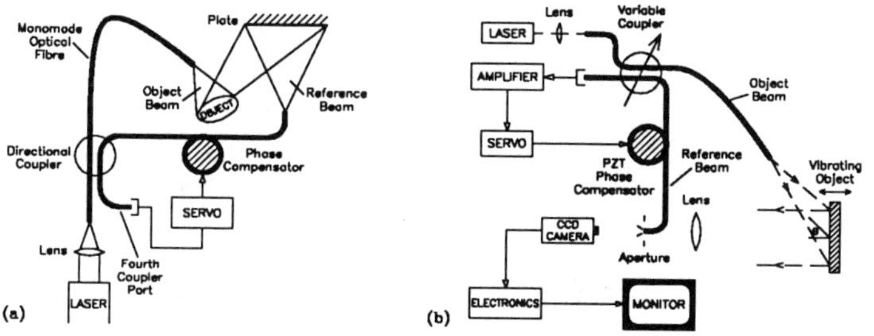

Figure 4 (a) Fibre optic based holographic system
(b) Fibre optic based T.V.Holography

Systems of this type have been successfully used to measure the vibrational profiles of auto engines in situ (Davies and Buckberry, 1986).

If a holographic system is illuminated with two sources of different wavelengths then it is possible to obtain measurements of the 'surface roughness' or 'profile' with high accuracy. For example if two laser diodes with a wavelength difference of ~100μm are used, surface profile resolutions of better than 1μm can be obtained(Atcha and Tatam, 1992).

Contact Point Measurements

Compliant Cylinder Accelerometer

As the fibre exhibits very high strain sensitivity there has been considerable interest in designing new forms of accelerometers based upon a fibre sensing element, where the fibre forms part of the spring in a 'mass loaded spring' accelerometer. In the first accelerometer of this type the mass was simple supported by the fibre which also formed part of a fibre interferometer. Although the sensitivity of this arrangement was high it is not very practical because of the high sensitivity to orthogonal accelerations. A more practical approach is shown in figure 5(a) where the fibre is wrapped under tension around a compliant mandrel loaded by the seismic mass. In the original design only 1 mandrel was used(Kersey et al, 1982), an improved design (Gardner et al, 1987), used two mandrels in a Push-Pull arrangement. A major advantage of the compliant cylinder fibre combination is that it effectively operates as a mechanical amplifier with the amplification factor being directly dependent on the number of fibre turns. Sensitivities of > 10^{-9}g/$\sqrt{}$Hz have been demonstrated by this type of optical accelerometer.

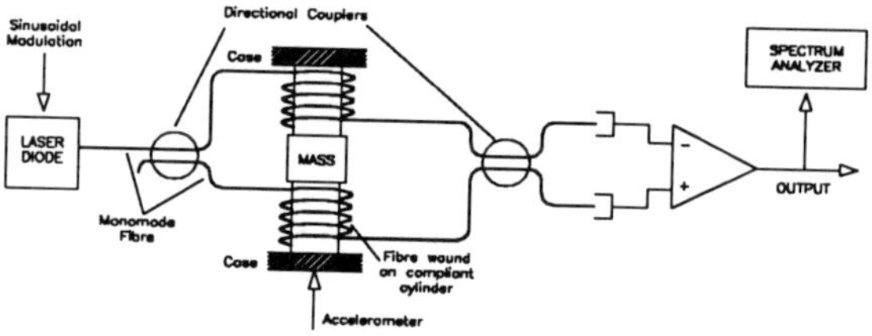

Figure 5(a) All fibre Mach Zehnder Accelerometer.

Diaphragm Accelerometer

We have also introduced (Gerges et al, 1989) a rather different design for an optical, 'mass loaded spring' accelerometer where the spring is a taut diaphragm. This accelerometer is shown in figure 5(b,c) and is based on a hemispherical Fabry-Perot cavity formed between the distal end of the fibre transceiver link and a very low cost metal mirror mounted centrally on the diaphragm acting as the mass. The sensitivity of this accelerometer is ~10^{-7}g/√Hz. One of the major advantages of this accelerometer is that the all metal construction makes it both extremely rugged and well suited for high temperature operations.

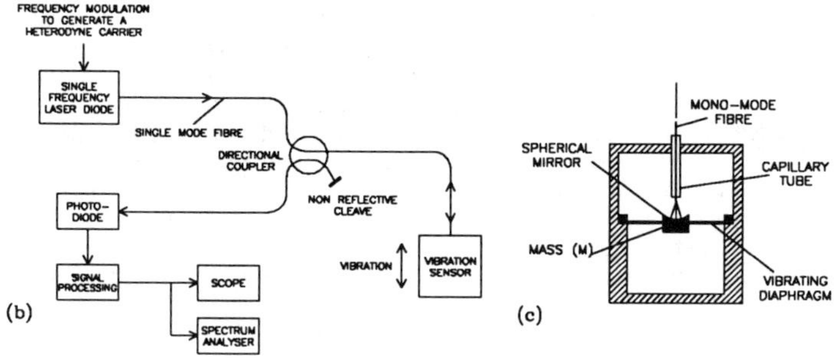

Figure 5(b) Optical arrangement used to interrogate the all metal optical accelerometer; (c) shows the sensor in detail.

This accelerometer shows very low cross-sensitivity to orthogonal accelerometers and the performance is unaffected by any environmental noise induced in the optical fibre transceiver link.

Temperature Sensors

The very large thermo-optic coefficient of a typical optic fibre (included phase change/°C) makes it an ideal thermometer or heat flux sensor when the induced phase change is measured interferometrically. Given a modest optical resolution of only 10^{-9}m, temperature changes of less than a millidegree can be determined using a 5mm length of fibre(Corke et al 1983). Most applications for temperature sensors require that the output is always unambiguous which requires special signal processing techniques. The classical method to remove ambiguities in optical interferometry is to illuminate the interferometer with two sources of different wavelengths, λ_1 and λ_2. This gives an unambiguous range of $\lambda_\mu = \lambda_1\lambda_2/(\lambda_1 - \lambda_2)$, for example with λ_1 = 800nm and λ_2 = 820nm λ_μ equals 32 microns.

Although this approach has been demonstrated by several authors(Kersey and Dandridge, 1987), the requirement to use two tunable stabilized lasers to implement the technique has not proved attractive, particularly as the output power from each laser must be coupled into the fibre via an (expensive) optical isolator, to prevent optical feedback induced instabilities in the absolute frequency of the laser.

Low coherence techniques are currently being researched by several groups to overcome the ambiguity problem. Low coherence interrogation operates as follows: referring to figure 2, the OPD of the remote interferometric sensor is set such that it exceeds the coherence length of the source, hence no interference effects will be observed at point x in the figure. The second or local receiver interferometer (LRI) (in this case a Michelson although other configurations are possible) is initially adjusted such that its OPD is within a few microns of that of the sensor and a servo unit is then used to 'lock' the tandem interferometers to a fiducial point in the transfer function. Hence any change in the OPD of the sensor can be monitored from the servo's voltage to displacement coefficient.

In order to obtain nanometer accuracy the coherence length of the source must be extremely short. Optical sources which exhibit the ideal coherence properties are LED's or incandescent lamps - unfortunately the radiation profiles of these sources are not well matched to optical fibres. The super radiant diode specially developed for fibre optic gyroscopes is well suited for LCI however as they tend to be rather expensive we have exploited the unique coherence properties of multi-mode lasers(Gerges, et al, 1990) to achieve a source with suitable coherence properties.

For many practical applications it is desirable to implement a sensor network capable of interrogating several sensors. This is possible with LCI as the LRI can be sequentially coherence tuned to each sensor in the network. Figure 6 shows such a network with two fibre Fabry-Perot sensors(Farahi et al, 1988).

For applications where the temperature may exceed 1000°C the fibre cannot be used as the sensor. It is possible, using the type of probe shown in the insert in figure 6, to make precise high temperature measurements (Gerges and Jackson, 1991). The probe shown in the figure was fabricated from fused quartz and operated up to 1000°C with a resolution 0.01°C If the quartz tube and quartz sensing element were replaced with an alumina tube with a sapphire sensing element temperature measurements above 1500°C could be made.

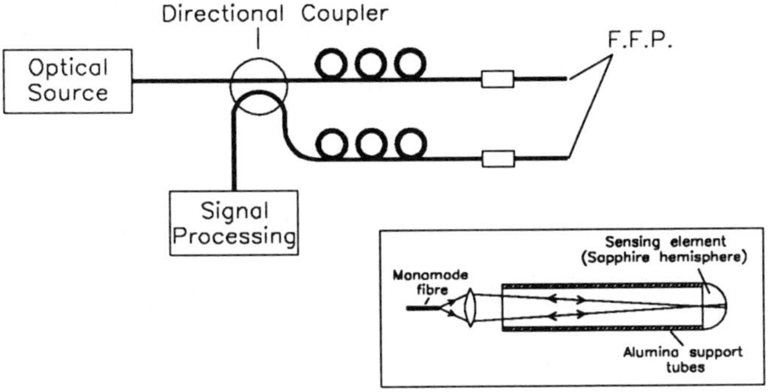

Figure 6. Two channel Fibre Fabry-Perot temperature sensor Insert indicates modifications for operation up to 1500°C

One of the unique features of the miniature FFP is their very high bandwidth >200KHz which makes them ideally suited for the study of fast phenomena such as very fast chemical reactions and heat diffusion studies. In this type of application it is unnecessary to determine the absolute value of the induced phase change; and a single laser diode can be used to illuminate the sensors(Farahi et al, 1991).

Strain

In principle any of the fibre optic interferometer can also be used as strain sensors, however the very high temperature sensitivity of the fibre means that either the measurements must be made at constant temperature or an ancillary measurement of temperature must be made. One approach to solve the cross sensitivity problem is to exploit the bi-modal nature of the optical fibre. For example if we implement a Fabry-Perot on bi-modal fibre then two sensors are formed with the same physical length but with different optical lengths due to the different propagation constants of each mode(Akhaven Leilabady et al, 1986) Hence it is feasible to simultaneously determine both the temperature and strain at a specific location.(Farahi et al, 1990).

Major applications for fibre optic strain sensors are in composite structures as conventional methods such as eddy current testing cannot be used, an early example of such an application was the investigation of a carbon fibre laminate using a polarimetric (differential interferometer) technique(Waite,et al, 1988)

Other Measurands related to displacement

The techniques described above allow either the absolute or relative displacement to be measured hence they can be applied to measurement problems such as pressure where the movement of a pressure sensitive diaphragm would be determined or the dimensional change in a gauge block for a force sensor. As the sensors would be passive and capable of operating at considerable ranges >> 1Km by virtue of the optical fibre link, many applications are likely to be found in industrial measurements.

Non-contact displacement sensors are also possible where some form of interferometer is formed between the distal end of the fibre and the surface of interest. Applications of such systems are very diverse including accurate surface profiling(Farahi and Jackson, 1990).

Flow
Contact - Vortex Shedding

The measurement of flow is a very important parameter in many industries and there are a large number of ways in which it can be measured. Of these the so called Vortex Shedding Flow meter has become an industrial standard. Vortex shedding occurs in obstructed turbulent flows; the well know phenomenon of 'whining'telephone lines in strong winds is an example of vortex shedding - the frequency of vibration of the obstructing body is directly related to the flow velocity. In a conventional vortex shedding flow meter, the movement of the shedder is measured by a piezo-electric element. As the movement of the body is relatively small, the resulting electrical signal is similarly small. It is possible to use a fibre optic strain sensor to measure the vortex shedding frequency - either directly by using the fibre as the obstruction (in gas flows) (Akhaven Leilabady et al, 1984)or indirectly by interrogating the movement of the vortex shedder externally to the flow. The performance obtained with an optical system is comparable to that achievable with a conventional system, however, the optical system has the advantage that it is not affected by electromagnetic interference and can also operate at much higher temperatures than is possible with vortex shedders using a piezoelectric element(Chu et al, 1991).

Figure 7 shows how a commercial vortex shedding flow meter has been modified such that the movement of the secondary body can be monitored by a Fibre Fabry-Perot.

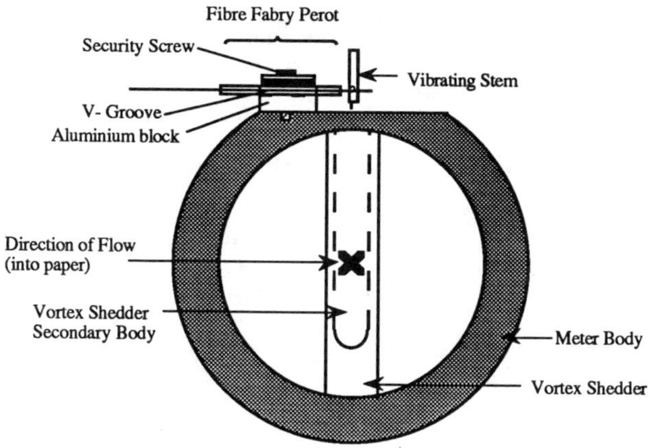

Non-Invasive Flow Measurement

Optical interferometers have been developed during the last twenty-five years for non-invasive flow measurement. This technique is generally referred to as Laser Doppler Anemometry (LDA). In this technique laser light is scattered from small particles which are entrained in the flow. When the light is scattered it suffers a frequency shift known as a Doppler Shift - which is directly proportional to the particle velocity. LDA is an absolute measurement technique with an extremely large measurement range, being capable of determining particle velocities ranging from microns/sec - the velocity at which molecules move in solutions, to more than 1000m/sec - the velocity of experimental hypersonic aircraft. In a conventional LDA system it is necessary that there is a 'direct line of sight' between the laser, the optical components generating the probe beams and the measurement volume. This requirement can make it very difficult to apply the LDA technique when there is restricted access to the flow. In wind tunnel studies, this can be a major limitation as it is often necessary to measure velocity profiles around complex aerodynamic structures.

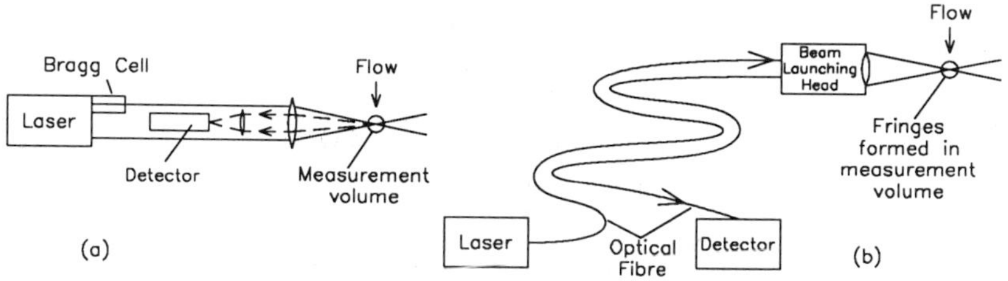

Figure 8 (a) Conventional LDA system,
(b) Fibre optic based LDA system.

The logistics become even worse when all three orthogonal velocity components must be measured in order to fully characterize the flow and its level of turbulence. These difficulties have been greatly ameliorated by incorporating optical fibre links into LDA systems(Jackson et al, 1984, Jackson and Jones, 1986). With this relatively straight forward modification, the applicability of LDA has greatly increased, for example it has enabled LDA to be applied to a range of flows as diverse as blood flow in vitro to under water studies of turbulent flow around ships' propellers.Figure 8 shows the optical configuration for conventional and 'fibre-ized' LDA systems.

Other Fibre Optic Based Sensors

Non-Contact displacement measurement via sub-carrier modulation

Optical methods for the measurement of modest distances 1-100m offer advantages over methods based upon microwave radar as it is very straight forward to produce a well collimated optical beam with a small cross-section. Although, in principle, optical interferometry could be used to make these measurements, the extremely high displacement resolution of less than 1 micron is clearly unnecessary and could be a disadvantage, especially as resolutions of 1mm would generally be more than adequate. The absolute resolution of an interferometric device can be greatly reduced if the source is amplitude modulated;for example modulation at 100MHz gives the source an effective wavelength of 3m, i.e. over a million times larger than the 6.3×10^{-7}m wavelength of the He-Ne gas laser. This technique is known as sub-carrier modulation and can be used with or without optical fibres. One application in which fibre transceiver links are used, is for the measurement of liquid levels in storage tanks, the accuracy of the measurement being ~1mm(Taylor and Jackson, 1986).

Distributed Temperature Sensors

When light propagates through an optical medium it is scattered by a variety of fundamental processes such as Raman and Rayleigh-Brillouin scattering. Raman scattering occurs due to the light scattering from molecular vibrations, and Rayleigh-Brillouin scattering due to the light scattering from spontaneous density fluctuations. As the scattering cross-sections of these fundamental processes are affected by temperature and strain it is possible by combining optical time domain techniques with suitable signal processing to realize a unique type of sensor which has been classified as a distributed sensor. To date Raman scattering is the most exploited of these scattering mechanisms. By determining the ratio of the Stokes to Anti-Stokes Raman signals, distributed temperature sensors with a continuous sensing range up to 8km with a spatial resolution of 1m and $1^{o}C$ accuracy have been achieved. Clearly such a sensor will find many applications in the construction industry (fire warning), power industry (monitoring temperature of the bus bars), and the oil and gas industries (pipe line monitoring).

Recently we have developed a distributed temperature sensor with a working range of 22km(Bao et al, 1992) by exploiting the variation of the Brillouin gain with temperature. It appears technically possible to achieve ranges of up to 100 km. Unlike Raman scattering Brillouin scattering depends on both temperature and strain(Horiguchi et al, 1990) hence it may be feasible to monitor the structural integrity of major facilities such as dams, nuclear power plants, buildings in earthquake zones, et.

Summary

A brief review has been presented of the research undertaken by the Applied Optics Group at the University of Kent (UK) into fibre optic sensors. It has been shown that a range of different configurations for the sensors is possible and that these configurations can be adapted for many mechanical engineering problems. Although not discussed here techniques have been developed which enable the outputs of the sensors to be multiplexed.

APPENDIX 1

References:

P. Akhavan Leilabady, J.D.C. Jones, A.D. Kersey, M. Corke and D.A. Jackson, Electron.Letts. **20**, 664, 1984.
P. Akhavan Leilabady, J.D.C. Jones, M. Corke and D.A.Jackson, J. Phys. E: Sci. Instrum. **19**, 143, 1986.
H.Atcha and R.P.Tatam, 8th Optical Fiber Sensor Conference, Monterey, Ca.217, 1992.
X.Bao, D.J.Webb and D.A.Jackson. submitted to Optics Letters, 1992
B.C.Chu, T.P.Newson, A.S.Gerges and D.A.Jackson Int.Journal of OptoElectronics.**6**,319, 1991.
M. Corke, A.D. Kersey, D.A. Jackson and J.D.C. Jones, Electron. Letts. **19**, 471, 1983.
A,Dandridge, A.B.Tveten and T.G.Giallorenzi IEEE J.Quant. Electron. **QE-18**, 1647, 1982.
J.C.Davies and C.H.Buckberry,VDI Berichte,**617**,279 1986.
F.Farahi, T.P.Newson, J.D.C.Jones and D.A.Jackson. Opt. Comm. **65**, 319, 1988.
F.Farahi, D.J.Webb, J.D.C.Jones and D.A.Jackson Journal of Lightwave Technology. JLWT **8**, 138, 1990.
F.Farahi and D.A.Jackson. Rev.Sci.Instrum. **61**,753, 1990.
F.Farahi, J.D.C.Jones and D.A.Jackson.

Optics Letters, Vol.**16**, 1800, 1991.
D.L.Gardner, T.Hofler, S.R.Baker, R.K.Yarber and S.L.Garrett. J.Lighwave Tech. **5**, 953, 1987.
A.S.Gerges, F.Farahi, T.P.Newson, J.D.C.Jones and D.A.Jackson. Electron.Letts. **24**, 472, 1988.
A.S.Gerges, T.P.Newson and D.A.Jackson. Optics Letters. **14**, 1155, 1989
A.S.Gerges,T.P.Newson and D.A.Jackson. Applied Optics. **29** 4473, 1990.
A.S.Gerges and D.A.Jackson. Optics Comms.**80**,210, 1991
T.Horiguchi, T.Kurashima and M.Tateda. IEE Photonics Tech. Letts. **2**, 352, 1990.
D.A. Jackson, A.D. Kersey, M. Corke and J.D.C. Jones, Electron. Letts. **18**, 1081, 1982.
D.A.Jackson, J.D.C.Jones and R.K.Y.Chan. J.Phys.E.**17**,977, 1984.
D.A. Jackson, J.Phys.E:Sci.Instrum., Instrument Science and Technology, **18**, 981, 1985.
D.A.Jackson and J.D.C.Jones, Optics and Laser Technology, **18**, 243 and 299, 1986
J.D.C. Jones, M. Corke, A.D. Kersey and D.A. Jackson, J. Phys. E: Sci. Instrum. **17**, 271, 1984.
R.Jones and C.Wykes, Holographic and Speckle Interferometry, Cambridge University Press, 1989
A.D.Kersey, D.A.Jackson and M.Corke, Electron Letts. **18**, 39, 1982
A.D.Kersey and A.Dandridge, SPIE vol. 798, Fibre Optic Sensors **11**,176, 1987.
A.C. Lewin, A.D. Kersey and D.A. Jackson, J. Phys. E: Sci. Instrum. **18**, 604, 1985.
B.T.Meggitt, A.C.Lewin and D.A.Jackson. 7th Int. Conf. Fibre Optics and Opto-Electronics, London, SPIE Proceedings, **1120**,307,1989.
J.L.Santos, T.P.Newson and D.A.Jackson. Optics. Letts. **15**, 573, 1990
M.Taylor and D.A.Jackson,Optica Acta,**33**,1571,1986.
S.Waite, R.P.Tatam and D.A.Jackson. J.Composites Technology and Research. Vol.**19** 435, 1988.

Smart Structure Technology

..................*and its Potential for Civil Engineering*

By

Raymond M. Measures

University of Toronto Institute for Aerospace Studies
4925 Dufferin St., Downsview, Ontario, M3H-5T6, CANADA

Abstract

Smart Structure technology could revolutionize civil engineering by the 21st Century. New bridges, highways, dams, pipelines, tunnls and buildings could be constructed with built-in optical fiber sensing systems that would make the civil infrastructure self monitoring and even self scheduling of its maintenance, repair and replacement. Fiber optic sensing systems could be integrated directly within concrete structures. They could also accelerate the replacement of steel reinforcing members by composite materials by providing them with a resident self diagnostic capability in the form of embedded fiber optic sensing systems. Improvements in safety, reductions in cost, and enhancement of the environment are some of the expected benefits.

Introduction

Smart Structure Definitions

"Smart Materials or Structures" possess a structurally integrated sensing system for determining their state [**Measures, 1989**]. A Smart Structure should contain a resident sensing system that is capable of maintain a constant vigilance over its structural integrity and indicate if damage is inflicted or fatigue problems developing. This same built-in sensing system should also be able to evaluate the loads imposed on the structure and monitor the static and dynamic response of the structure to these loads. The quality of fabrication of the structure may also be ascertained with this sensing system. More advanced "Adaptive Smart Structures" would control their: stiffness, shape, position, or orientation, based on the information provided by the resident sensing system. In is expected that eventually experienced based self learning systems will become feasible and the concept of self healing structures may prove to be more than a pipe dream.

Need for Smart Structures in Civil Engineering

Buildings, bridges, dams, tunnels, seaports, highways, railways, pipelines and airports represent an enormous financial investment. There is a growing need for a built-in sensing system for monitoring these structures due to increasing: traffic, air pollution, corrosive precipitation (acid rain) and the rising costs of labour intensive inspection, maintenance and repairs. If resident optical fiber sensors can improve concrete evaluation, the ramifications could have enormous value to the multi--billion dollar annual construction business as aging and deterioration of many highway bridges is recognized as one of the major problems facing structural engineers in the US, EC and Japan. In cold-climate countries this deterioration has been greatly accelerated by the use of de-icing salts. In Europe, a large scale research effort has been directed at using advanced composite materials (ACM) in bridges. This research has addressed issues of corrosion, rehabilitation and monitoring and has led to new concepts and designs in bridge repair and construction. One of the most signicant advances is the replacement of steel prestressed tendons with ones made of ACM. These fiber reinforced polymers are practically immune to corrosion and also lend themselves to internal monitoring by means of embedded fiber optic sensors.

A structurally integrated sensing system could monitor the state of a structure throughout its working life. It could determine the: strain, deformation or deflection, load distribution, temperature or environmental degadation, experienced by this valuable asset and warn if any excursions deviate from accepted values. Such an integrated sensing system could also monitor the various structural components during construction, possibly leading to improved quality control. In certain instances this resident sensing system could also provide valuable information on the usage of the structure. For example, in the case of a bridge the sensing system could be used to generate information on the traffic itself. It is quite conceivable that a suitably instrumented bridge could monitor not only the number of vehicles per minute, but their weight (even the distribution of this weight) size, direction of travel and the speed of each vehicle. In terms of degradation, especially chemical, it may be possible to have sufficient warning of its onset and location that a low cost repair is possible. The impact on civil engineering of structurally integrated sensing could be profound, leading to a revitalization of the field. For optimum progress civil engineers should work closely with those developing fiber optic sensors in order to ensure that full use is made of the wealth of knowledge and experience currently available.

Fiber Optic Based Sensing Systems for Smart Structures

Merits and Types of Fiber Optic Sensors

Fiber optic sensors are: extremely small and light weight, resistant to corrosion and fatigue, immune to electrical interference, safe and practically incapable of initiating fires or explosions. They are also particularly attractive as they: serve the dual role of sensor and pathway for the signal, are dielectric in nature and are therefore compatible with ACM, and avoid the creation of electrical pathways within the structure. They also have a high sensitivity and very large bandwith and are consequently well suited as sensors for embedding within ACM and concrete. The fact that no protection is required against lightning and various forms of electromagnetic interference also figure prominently in their choice for a sensing system in large civil structures. Even in the absence of a direct lightning strike, magnetic field coupling can generate voltages that can burnout various components and fragile circuits used in many electronic systems, especially if distributed over a large structure. It is expected that the increased use of fiber optics for communications will help reduce the cost of this technology and if it is also used substantially in the construction business this, in itself, could bring down the costs due to the potential scale of this market.

In general we can characterize optical fibers sensors as either "distributed" or "localized." Distributed sensors are, in principle, very attractive for they would permit the use of fewer sensors and represent a more effective use of the optical fibers, in that *each element of the optical fiber is used for both measurement and data transmission purposes.* In essence by making a time of flight measurment along the optical fiber the value of the measurand is determined along the entire length of the optical fiber. In practice,however, the spatial resolution attainable is somewhat limited. This is particularly true if the measurements have to be made in real time. Raman backscattering has been used to make a "distributed" fiber optic temperature sensor, [**Dakin, 1989**] with a few degrees resolution. However, its spatial resolution is limited to about one meter. In principle, Raman backscattering might also be used for distributed strain measurements. Our recent development of a simple, compact, rugged and low cost method of evaluating the wavelength of any spectrally narrow source of radiation [Melle et al., 1992] enhances such a prospect. Unfortunately, the need for high resolution, flexible sensing architecture and graceful degradation is not satisfied with distributed sensors. Often a single break in the optical fiber effectively takes out of commission a large proportion of the sensing capacity.

Localized fiber optic sensors *determine the measurand over a specific segment of the optical fiber* and are similar in that sense to conventional strain or temperature gauges. If a scalar field, like pressure or temperature, is to be mapped a grid of such localized line integrated sensors can be used in conjunction with an inverse Radon transform. Exceptions can arise where the field is constrained in such a way that it is not truly two dimensional in nature. Where the field is two dimensional, three independent measurements are required at each point to uniquely specify the field. This can be accomplished, in certain situations, by using an optical strain rosette sensor, comprising three small identical fiber optic sensors set at different orientations, **figure 1, [Valis et al., 1990]**. Under these conditions high spatial resolution is important as each sensor must have a gauge length that is small compared to the scale length for the field gradient.

Figure 1. (a) Fiber optic strain rosette based on three Fabry-Perot fiber optic sensors: (b) Comparison between a conventional resistive foil strain rosette [shown on the left] with its new fiber optic counterpart [shown on the right]. Note that overexposure makes the fiber optic rosette seem thicker than it is in reality {lettering on pen is about 0.1 inch high}.

There are two basic types of localized fiber optic sensor that are well suited for use with Smart Structures: the Fabry-Perot and the intracore Bragg grating. In the Fabry-Perot fiber optic sensor two parallel mirrors that are perpendicular to the axis of the optical fiber form a cavity which constitutes the localized sensing region. The Fabry-Perot sensor is normally configured to be single ended, and if built with low reflectivity mirrors will provide a sinusoidal phase strain relationship. Fabry-Perot sensors can be configured as either "intrinsic", **figure 2 [Hogg et al., 1991]** or "extrinsic",**[Kruschwitz et al., 1992]**. The former has the advantage of being the least perturbative, capable of being used in the form of a strain rosette, and possibly the more robust. As a consequence of the separation between gauge length and cavity length in the latter, it can be built with long gauge lengths yet work with lasers of modest coherence length. The extrinsic sensor also has less transverse coupling and can therefore evaluate more directly the axial component of strain in a host material **[Sirkis and Haslach, 1991]**. The intrinsic form of this sensor with a demodulation system giving a performance comparable to conventional resistive foil strain gauges has become commercially available from **FiberMetrics Corporation**. These optical strain gauges can have a gauge length as short as 5 mm, and have a submicrostrain resolution. It should be noted, however, that unlike foil strain gauges.....optical strain sensors can be used to very high frequencies and can therefore used to detect acoustic emission signals **[Liu et al 1990]**.

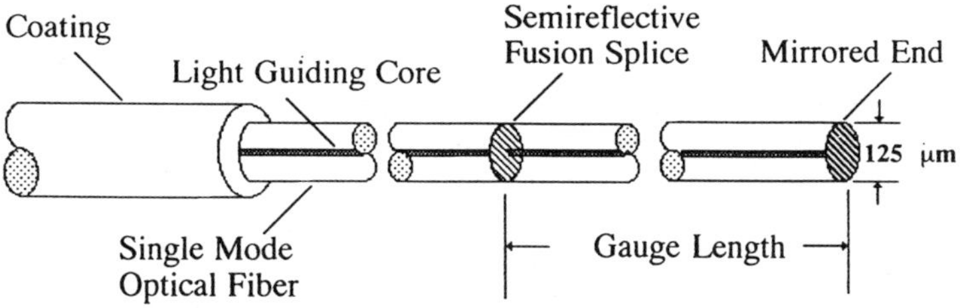

Figure 2. Mirror cavity formed in an optical fiber serves as the localized sensing region of a Fabry-Perot fiber optic sensor.

The intracore Bragg grating fiber optic sensor [**Morey et al., 1989**] relies on the narrow band reflection from a region of periodic variation in the core index of refraction of a single mode optical fiber. In this sensor the centre (Bragg) wavelength of the reflected signal is linearly dependent upon the product of the scale length of the periodic index variation and the mean core index of refraction. Changes in strain or temperature, to which the optical fiber is subjected, will consequently shift this Bragg wavelength leading to a spectrally encoded optical measurement that can be determined by the simple wavelength dependent ratiometric technique that we have recently developed, **figure 3**, [**Melle et al., 1992 and Measures et al., 1992**]. Although the method of fabrication of the Bragg grating sensor is not at present as mature as the Fabry-Perot there is recent evidence which suggests that the Bragg grating sensor could be manufactured in an automated manner during the fiber pulling process. This would greatly reduce its cost and ensure full strength. Other important advantages of this sensor are: its signal is spectrally encoded, it provides an absolute measurement that makes it good for tracking creep and it is immune to electrical interruption.

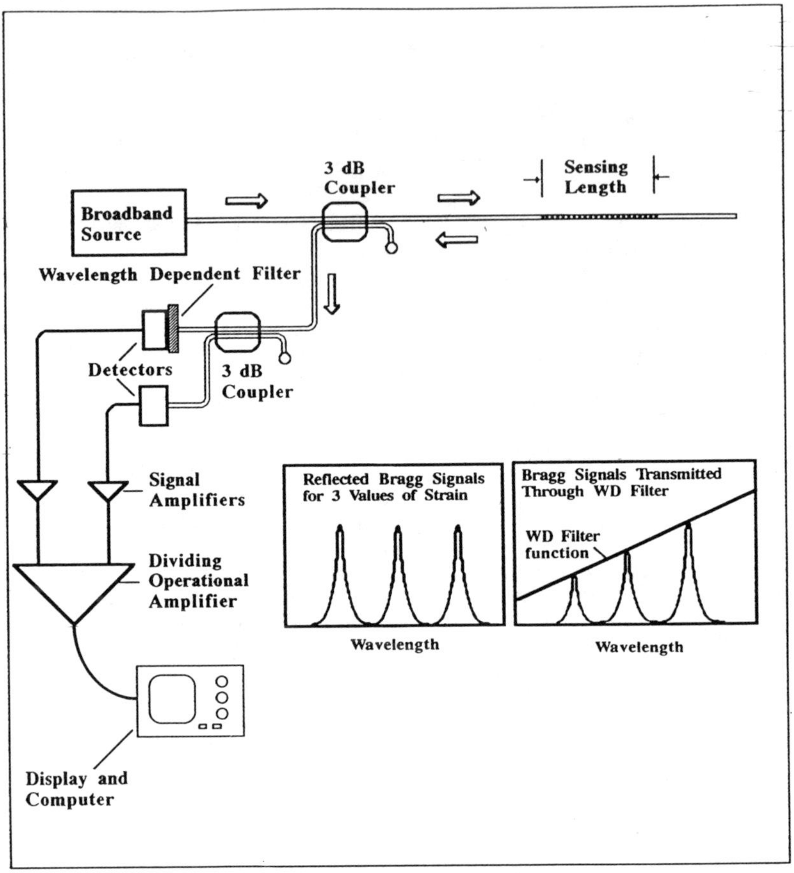

Figure 3. Intracore Bragg grating fiber optic sensor. Inset{bottom left} indicates Bragg reflected signals for three values of strain. Inset {bottom right} reveals these three signals after transmission through a wavelength dependent linear filter. Wavelength demodulation of the Bragg sensor is achieved from the ratio of signals from two detectors, one receiving its light after passage through the wavelength dependent filter, the other receiving its light directly and serving as a reference.

Another important virtue of the Bragg grating is that it can be used to control the wavelength of a laser when used as one of the laser cavity mirrors, **figure 4**. In this configuration the measurand, in effect, tunes the laser and when used in conjunction with our simple wavelength demodulation system could lead to the development of an "optoelectronic chip," **figure 5**, that would handle a large array of such sensors and solve the difficult structural interconnect problem [**Measures, 1992**].

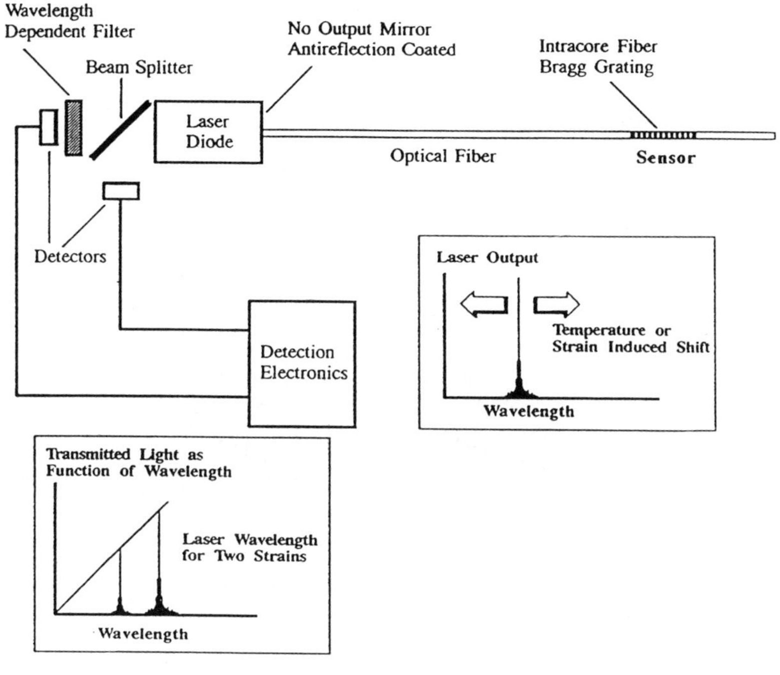

Figure 4. Bragg grating laser sensor in which an intracore fiber Bragg grating tunes the wavelength of a semiconductor laser in response to the measurand. Wavelength demodulation is achieved by means of a passive, spectral ratiometric technique.

A particularly attractive form of this solution is the use of a free space noncontact form of communication between the structural component with the resident sensing system and its support structure, **figure 6**. We have recently demonstrated the concept of a Bragg grating laser strain sensor, using an erbium-doped fiber laser, **figure 7**. The wavelength of the laser can be seen to be a linear function of the strain impressed on the Bragg grating which was adhered to the surface of a cantilever beam, **figure 8**.

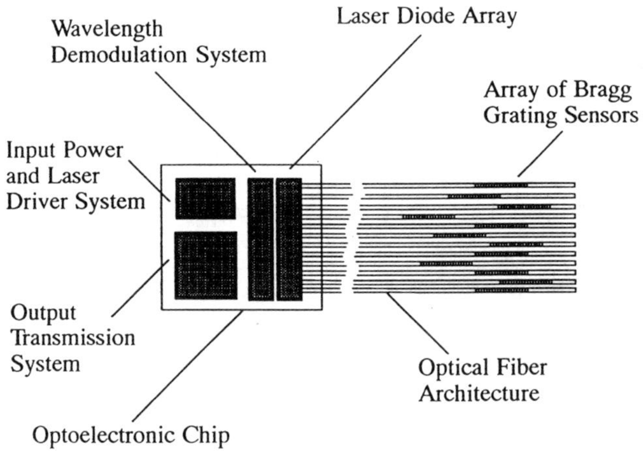

Figure 5. Schematic of a proposed optoelectronic chip that would be capable of: demodulating, processing, multiplexing and transmitting the sensing signals from an array of Bragg grating laser sensors, and would form the basis of an optoelectronic smart structure interface [OSSI].

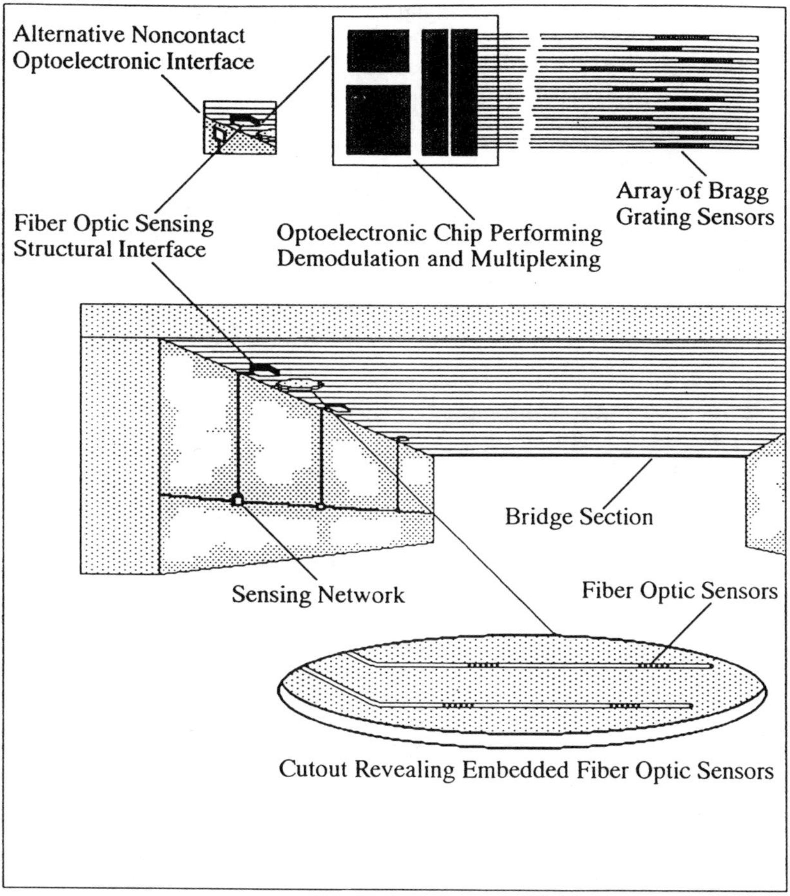

Figure 6. Cartoon illustrating a future bridge instrumented with a structurally integrated fiber optic cellular sensing architecture and optoelectronic smart structure interfaces.

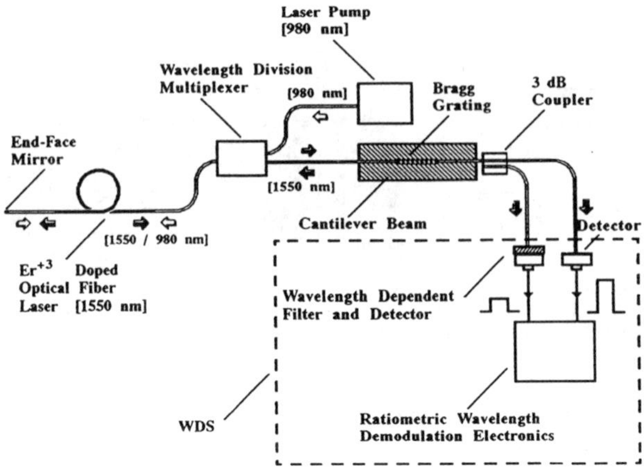

Figure 7. Schematic of an experimental arrangement used to demonstrate a Bragg grating fiber laser sensor.

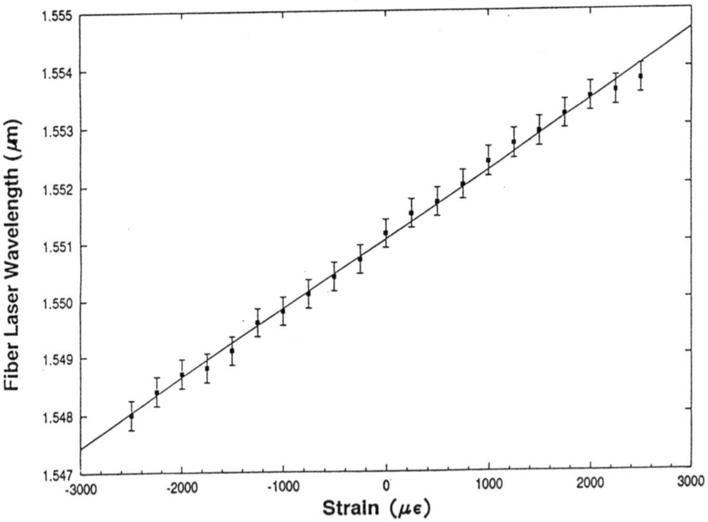

Figure 8. Variation in wavelength of Bragg grating fiber laser sensor with strain.

Structurally Integrated Fiber Optic Sensors and Measurements

Optical fibers can either be bonded to the surface of most structures or embedded within: advanced composite materials (ACM), concrete or low temperature melting metals. These sensors are capable of measuring: strain, vibration frequencies, structural deformation and temperature. They could possibly detect and locate internal damage generated by: impacts, manufacturing flaws, excessive loading or fatigue, and assess the growth in the extent of these damage zones. Structurally imbedded optical fibers sensors might also monitor the cure state of thermoset composites or concrete. This could lead to improvements in the quality control of structures made from these materials. In terms of ACM the combination of improved quality control and a resident sensing system could greatly enhance confidence in their use and expand their role as a primary building material. This might avoid structural overdesign and provide the opportunity of evaluating the true fatigue life based on the actual load history of each component. Monitoring the dynamic response of structural components during operation might also provide insight into the structural integrity of the system.

Although *Structurally Integrated Fiber Optic Sensor* (SIFORS) technology [**Measures, 1989**] is in its infancy, optical fiber sensors are fairly well developed. Fiber optic sensors that are well suited for Smart Structures should ideally be: *intrinsic* in nature for minimum perturbation and stability; *localized* for remote operation with insensitive leads; *all-fiber* for operational stability; able to provide a linear response; *single-ended* and involve a single optical fiber for ease of installation and connection; *nonperturbative* to the structure and *robust* for installation; *interrupt immune* and capable of *absolute measurement* . For cost minimization a single sensor could be developed for all of the measurements required by Smart Structures. Such a sensor might have to be sufficiently sensitive to detect acoustic emission, while exhibiting a measurement range adequate for monitoring the general loading of the structure. The two fiber optic sensors that appear best suited for this work are the intrinsic Fabry-Perot and the intracore Bragg grating sensor.

Critical Issues and Developments for Smart Structure Sensing

Implementation of Smart Structure technology will require a number of critical issues [**Measures et al., 1992**] to be addressed. These can be divided into micromechanic issues and system architecture issues.

• *Influence of Embedded Optical Fibers on the Material Properties*
If optical fibers are to be embedded within practical ACM or concrete structures they must not: compromise the tensile or compressive strength, increase the damage vulnerability or reduce the fatigue life of these materials. Although preliminary evidence suggests that degradation of the material properties is minimal [**Czarnek et al 1988, Blagojevic et al., 1989, Loken 1990, Roberts and Davidson, 1992**] providing that the diameter of the optical fiber is less than about 125 μm, fractographic studies reveal that in the case of ACM optical fibers embedded at an angle to the adjacent ply directions of the composite material create resin cavities (termed "resin-eyes" because of their shape). The formation of resin-eyes leads to high stress concentrations at the host/optical fiber interface which

may over a period of time and under occassional high loading conditions lead to debonding of the optical fiber from the host. Clearly, more definitive research will be needed before optical fibers can be imbedded with confidence within structures intended to have a 20 (plus) year working life.

• *Sensor/Host Interface, Coatings and Sensor Performance Life*
It is possible that this high stress concentration around the optical fiber may be diminished with coatings of appropriate size and stiffness, [**Sirkis and Dasgupta, 1990**]. Debonding between the optical fiber and the host represents a potentially serious concern in terms of sensor performance. Careful consideration will have to be given to the diameter of optical fibers and their type of coating if they are to be embedded within composite structures and function correctly with no performance degradation for the useful life of the structure. When the optical fibers are embedded collinear with the ply direction there is no appearance of a resin eye and minimal stress concentration is expected. However for this configuration a resin void is formed on the end of the optical fiber and this could lead to initiation of debonding from the host material. This suggests that it may not be prudent to locate the sensing region at the end of an embedded optical fiber intended for extensive use. More research is needed to ascertain the seriousness of this problem and we have initiated a program of study to address this issue. Initial result suggest that the solution may lie in the use of specially designed coatings, [**LeBlanc and Measures, 1993**].

• *Sensing System Damage Vulnerability and Degradation*
A sensing system within a practical Smart Structure will have to be fairly robust and degrade gracefully when the structure suffers modest damage. Special coatings, a judicious choice of location and orientation may help to reduce premature fracture of the optical fiber and use of a cellular sensing architecture minimize the loss of any particular set of optical fibers.

• *Sensing System Architecture*
The type of measurement to be undertaken will dictate whether the fiber optic sensors should be localized or distributed, while the nature of the structure will determine if they are multilayered or limited to form a single layer. Optical fiber orientation and placement, especially in an ACM layup, spatial resolution and constraints imposed by the finite bend radius are all important factors to be considered. It will also be important to identify any special structural features or regions of high stress concentration. The power budget and signal to noise factors will certainly play a key role in defining the sensing system architecture. Sensing system damage vulnerability and ease of fabrication represent other considerations to be taken into account.

• *Multiplexing Strategies*
Multiplexing is the merging of data from several channels into one channel, while demultiplexing is the inverse. The primary parameters used in optical multiplexing schemes are: wavelength, time, frequency, phase and space. Consequently, there are five multiplexing techniques. In the case of large civil structures, Time Division Multiplexing, can be practical and is also the least expensive. However, as discussed above there are other reasons for not placing all of the sensors on a single optical fiber and winding it throughout the structure. It may make good sense to

have a few sensors on each optical fiber and use a cellular sensing architecture.

• *Structural Interconnect/Interface*
The nature of the structural interconnect problem hinges on whether the output from the structure is optical or electrical. Current thinking is predicated on optical signals flowing into and out of the structure via the structural interface. In general this interface must have minimal structural perturbation, be easy to fabricate and introduced during fabrication of the structure. If multiplexing is not used each sensor would have its own input/output and a ribbon or bundle of optical fibers would have to egress from the structure. Nevertheless, the structural interconnect problem might be simplified if each major sensing cell were connected through a single electrical cable to the central computer facility charged with interpreting the data. The development of a multisensor signal processing optoelectronic chip would certainly simplify the interconnect issue and provide a number of advantages, **figure 5**, [**Measures et al., 1992**]. One of the most important, however, is the great reduction of unit cost when so much of the system is reduced to the form of an optoelectronic chip. This includes: multiplexing, multisensor signal processing and possible conversion to single electrical output.

Smart Structure Applications to Civil Engineering

Advances in Bridge Technology
The Institution of Structural Engineers (1989) reported that severe corrosion of bridge prestressing strands in structures built only tweny years ago had been discovered and related failures had occurred. This has prompted research into the use of alternative noncorrosive materials, especially ACM, for bridges. In order to permit the continued use of some bridges that have suffered serious deterioration, ACM plates have been bonded to the structure. Although, these materials cost more than steel, their lighter weight, better corrosion resistance (especially to sodium chloride) and lower cost of fabrication and installation have encouraged their use.

In the last few years bridge designers have looked to the use of ACM (carbon fiber/epoxy and fiberglass/epoxy) in the construction of new bridges due to their corrosion resistance, especially in high saline environments. In particular, the use of ACM tendons for prestressing of concrete girders and decks represents one of the most promising applications of this new material for bridges. Although, there has been no long term studies on ACM use in bridge construction, the prospect of instrumenting each ACM tendon with a built-in fiber optic sensing system makes their use worthy of serious consideration. Indeed, several bridges have already been built in Europe using ACM as a major constituent, [**Mufti et al., 1991**]:

-- The Ulenbergtrasse Bridge in Dusseldorf, is the first highway bridge to be constructed using glass reinforced prestressing strands (Polystar Tendons).
-- The Marienfelde Bridge in Berlin, is the first civil engineering structure to be built with external prestressing polystar tendons.
-- The Schiessbergstrasse Bridge in Leverkusen, uses 27 polystar prestressing tendons, several of which have fiber optic sensors in their core.
-- Finally, the A19 Tees-Viaduct in Northeast England, is a composte steel-concrete bridge with the world's first major ACM bridge enclosure.

ACM prestressing tendons are made of parallel filaments of glass, carbon, or aramid embedded within a resin matrix. The general advantages of ACM tendons as compared to steel are:
(a) High strength to mass ratio, which ranges between 10 to15 times that for steel.
(b) Excellent fatigue characteristics, three times that of steel for carbon and aramid fiber tendons.
(c) Superior corrosion resistance, especially for high saline environments.
(d) Low thermal expansion coefficient, especially for carbon fiber tendons.

It should be noted that the glass fiber reinforced material tendons may have a fatigue strength which is significantly lower than that of prestressing steel. Unfortunately, the high ratio of the axial to lateral strength of ACM tendons, means that conventional systems used for steel tendons cannot ensure a reliable anchorage for ACM systems, and can result in premature anchorage zone failure.

Types of Measurement and Sensing Strategy

Types of Measurement Relevant to Civil Engineering
A selection of some of the more important measurements that should be undertaken with the implementation of Smart Structure Sensing to large civil structures:

-- Vibration frequencies of: support columns, floors, windows, bridge decks and cables....
-- Spatial vibration modes of: walls, floors, bridge decks and cables....
-- Thermal strain and deformations caused by sunlight to one side of the structure.
-- Construction loads (excessive loading, pressures, impacts....).
-- Wind monitoring and wind pressure on: bridges, buildings.....
-- Long-term health monitoring including load history and excessive loads.
-- Shear forces on bridge bearings.
-- Internal strain distribution and hydropressure for dams.
-- Onset of internal crack formation in concrete structures.
-- Impact detection and localization.
-- Damage assessment (delamination in ACM).
-- Debonding of reinforcing bars and prestressing tendons in concrete structures.
-- Corrosion degradation.
-- Chemical sensor for: acid rain, smog, bird droppings, deicing salt solution.......
-- Real time truck/aircaft weight and load distribution.
-- Traffic flow patterns (number of vehicles, weight and velocity).
-- Ground creep or seismic movement.
-- Temperature distribution and anomalous hot spot detection.

Some of the characteristic problems of facing the designer of a structurally integrated sensing system for monitoring concrete civil structures are:

-- Small strains so need for high strain sensitivity.
-- Rough handling and a hard casting process.
-- Harsh environment (high levels of moisture, large temperature excursions).
-- Large dimensions and exposure to electrical interference.

Types of Sensor and Sensing Strategy

Since it will be necessary to make the unit cost of this sensing technology as low as possible for its implementation to be economic, one sensor should be capable of undertaking as many different measurements as possible for then it could be used in a diverse range of applications. Some of the most important measurements for civil engineering include: strain distribution, acceleration, vibration frequency, spatial vibration mode, pressure, temperature, elongation, cure for ACM or concrete, corrosion degradation, damage assessment. It may also be necessary, in certain situations, to be capable of undertaking chemical sensing.

It has been established that a Fabry-Perot (and possibly a Bragg grating) fiber optic sensor has sufficient sensitivity to undertake: strain distribution, acceleration, vibration frequency, spatial mode, pressure, temperature and elongation measurements, either directly or by suitable mounting, **figure 9.** First the Fabry-Perot fiber optic sensor represents an extremely sensitive strain gauge if bonding to, or embedded within, a structure. If such a Fabry-Perot optical strain gauge is mounted on the diaphragm of a suitably constructed cell it can be used as an effective pressure gauge or it can serve as an accelerometer if mounted so that it measures the force exerted by a small mass that is free to accelerate. For many practical situations, the temperature has to be measured to account for thermally induced apparent strain. This can be quite considerable (of the order of 1000 μstrain) for structures like bridges which may be subject to 60 ^{0}C temperature excursions. A Fabry-Perot fiber optic sensor can measure the temperature directly by preventing it from experiencing any strain coupling from the structure, **[Hogg et al., 1992]**. Cure monitoring of ACM with Fabry-Perot optical strain gauges could be feasible through the use of optoacoustic probing, **[Ohn et al., 1992]**. It is also possible that some form of optoacoustics may also enable cure monitoring of concrete to be undertaken with an embedded fiber optic sensing system, **[Fuhr et al., 1992]**. It may even be possible for a Fabry-Perot fiber optic sensor to undertake some degree of chemical sensing.

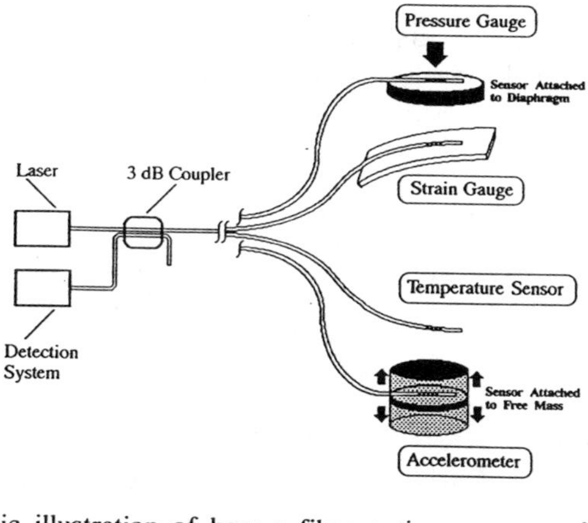

Figure 9. Schematic illustration of how a fiber optic sensor can be used as a: pressure gauge, strain sensor, temperature sensor, and an accelerometer.

The optimum sensing strategy for any given situation is very dependent on the measurements to be undertaken. For example, if it desirable to measure the wind loading on a tall building the best sensing configuration might be to use a distributed fiber optic sensor (or many time division multiplexed sensors along an optical fiber) that is bonded to an ouside wall of the building. On the other hand, a highly flexible sensing cell architecture may be more desirable in the case of a bridge where: various loads are to be recorded; structural vibrations monitored; the traffic flow is tracked; and corrosive degradation is assessed.

Current Integrated Sensing for Bridges and Other Civil Structures

Most bridges comprise three basic structural components: one or more deck spans; piers (or supports); and bearings. The bearings serve to: transmit load from the deck to the supports; provide a degree of freedom from physical constraints and damp dynamic effects. The Bearings are constructed of alternating layers of elastomer sheet and steel plates and are subject to vertical compression, horizontal shear and rotation. Special hard cladding, multimode microbend optical fiber sensors were integrated into the elastomer sheets [**Caussignac et al., 1992**]. Although the attenuation of light varied both with normal and shear forces applied to the Bearing, its nonlinear and inconsistent dependence limits its usefulness. Nevertheless, this initial work clearly indicates the value of pursuing this research with reliable, calibrated optical strain gauges having a linear response.

The Schiessbergstrasse triple span road bridge (53m long,10m across) in Leverkusen is one of the first of a new generation of civil structures which will use ACM tendons and built-in fiber optic sensors, **figure 10**. A combination of microbend sensors and elongation sensors are designed to evaluate the load on the tendons and monitor strain within the concrete. There are, in addition, built in chemical sensors and the bridge deck is protected by an epoxy coating and foil sealing achieved with polyurethane. [**Wolf and Miesseler, 1992**]. The bridge is designed with limited prestressing using 27 glass fibers prestressing tendons, three of which are instrumented with fiber optic microbend strain sensors to monitor excessive loading and deformation. These sensors check the integrity of the tendons or locate the damage. The tensile zone above the piers and the spans are monitored

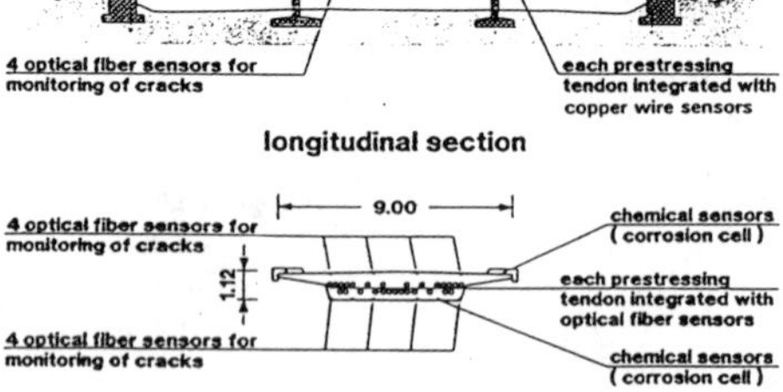

Figure10. Arrangement of optical fiber sensors and chemical sensors in the triple span Schiessbergstrasse road bridge, [Wolf and Miessler, 1992].

permanently with integrated optical fiber sensors with an accurancy of ± 0.15mm. There are also nine fiber optic sensors integrated into the concrete to track the chemical changes taking place. These corrosion cells monitor the progression of carbonation and eventual penetration of chlorides. Information extracted by an on site computer is transmitted to client along a phone line. This constant survelliance permits early detection of problems and reduced cost of maintenance and repair.

Detailed knowledge of the reinforcement strains in reinforced concrete structures is a prerequisite to a thorough understanding of how such structures actually behave. Strain distributions are profoundly affected by the formation of cracks and can change during the life of a structure due to effects such as creep, shrinkage and load history. To avoid affecting the steel/concrete interface and thereby weakening the structure, a set of electrical strain gauges were mounted within the rebars by milling two solid bars and bonding them together with the gauges and wiring included, [**Scott and Gill, 1992**]. The technique has been refined to the point where 3mm long strain gauges can be mounted about 12 mm apart (100 gauges to bar). These were able to pick up cracks in the concrete before they reached the surface. It is expected that strain gauge instrumented concrete structures could warm if any component was approaching its overload stress and in general monitor the state of internal strain. This might provide insight into weakness developing due to corrosion. Although this useful research has been undertaken to date using conventional foil strain gauges it demonstrates the kind of information that could be attained with much less work using fiber optic strain gauges.

Excessive displacements or deformations in large civil structures can lead to critical states of stability that can endanger public safety. The use of electrical based measurement systems in hydraulic plants is undesirable as they are susceptible to: corrosion of the metallic guides (due to the high atmosphic humidity), and electromagnetic interference (from lightening strikes, overvoltage). These problems are magnified by the vast distances involved (km) in major civil structures, like Dams. Two types of optical fiber system for monitoring span joints and structural stability in Dams is currently under consideration, [**Holst et al., 1992**].

Extinsic Fabry -Perot optical fiber sensors have been developed and used with short gauges lengths to measure strain and temperature within concrete specimens [**Kruschwitz et al., 1992**]. Although these sensors can be spatially multiplexed along a linear sensor data bus they are likely to be limited to short gauge lengths and represent somewhat of a perturbation if embeddedwithin composite materials. Nevertheless, strain direction ambiquity can be overcome in these sensors using two slightly different wavelengths or two adjacent sensors with different cavity lengths. Physically decoupling one end of the sensor head from the structure also permits the temperature to be determined. Two such sensors were embedded in 15cm x 20cm x 122cm reinforced cement concrete specimen during fabrication. Cylindrical metal washers ensured effective longitudinal strain coupling for sensors embedded directly in concrete, while the other sensors were attached to metal reinforcing rods prior to embedding. A 5% difference in the strain was observed between the rod mounted sensors and those directly embedded-suggesting some slip between the sensor, the rod, and the matrix, or nonparallel alignment of the sensor axes. Four of these sensors were attached, using two part epoxy adhesive, to the under surface of a simple span bridge made of reinforced concrete with a wooden bottom. A maximum reading of 12 μstrain was recorded when a 12,000 lb truck was driven slowly over the bridge and an asymmetry between the front and rear wheel loading was evident, **figure 11, [Kruschwitz et al., 1992]**.

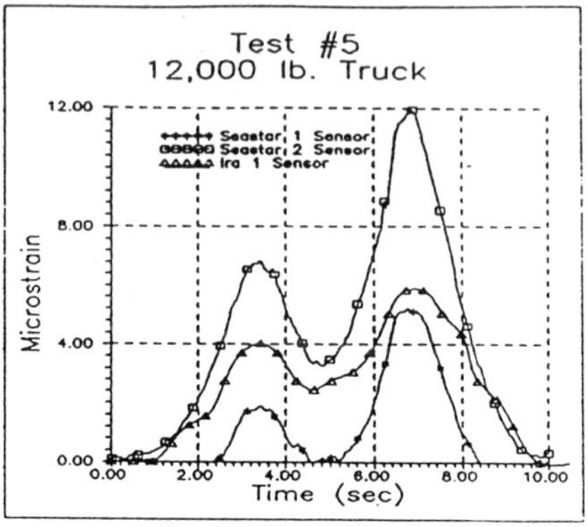

Figure 11. Microstrain signals versus time recorded by a pair of extrinsic Fabry-Perot fiber optic sensors adhered to a concrete bridge deck subjected to a 12,000 lb truck with an asymmetric load. This strain variation is compared with a conventional strain gauge, [Kruschwitz et al., 1992].

The University of Vermont is in the process of constructing a five-story, 65,000 square foot concrete (Stafford) Building, that has been equipped with fiber optic and conventional sensors embedded into the concrete superstructure [**Fuhr et al., 1991**]. Specificially, these sensors were embedded into the concrete itself - in the floor, the load bearing support columns, the walls and ceilings. This network of sensors will provide information about the buildings structural integrity, its response to both microvibrations (traffic) and macrovibrations (earthquakes), the strain levels of the reinforcement bars within the concrete, local wind pressures being applied to the outer surface of the building, and any thermal deformations of the building caused by sun induced heating of only a part of the building. Both conventional sensors (e.g. strain gauges and accelerometers) and fiber optic and optical sensors are being used. Although, not all of these sensors require components to be embedded into the concrete structural elements, a key aspect of developing a Smart Structure arises from the presence of the sensing components within the structure itself. This fact is particularly evident when attempting to monitor the curing of the concrete and stresses incurred during the construction phase, or determine if a section has developed internal cracks which aren't visible on the outside [**Fuhr et al., 1992**].

Concluding Remarks and Future Prospects

An enormous financial investment is associated with the established civil infrastructure: bridges, highways, railways, dams, pipelines, port facilities, airports and buildings. Aging of these structures represents a financial problem of immense proportions. Instrumentation of these structures with fiber optic sensors that could accurately assess their useful remaining life and schedule repair or replacement when required, could lead to substantial cost savings. Furthermore, we are witnessing a major change in the design and fabrication of large civil structures, like bridges, based on the replacement of steel with composite materials. It just happens that these new materials are quite conductive to the incorporation of resident fiber optic sensing systems which will allow the future custodians of these engineering investments to monitor the structure in a comprehensive way. This built-in inspectability will enhance confidence in using these new materials so that the two advances (use of composites and sensing) will reinforce each other and greatly accelerate the implementation of Smart Structure technology. In the last few years considerable progress has been made in the development of fiber optic sensors and their associated signal processing systems. Rapid and successful implementation of sensing systems into major new civil structures will best be achieved if there is the correct marriage of sensor developers with the civil engineers charged with the responsibly for these structures. There is a danger that progress could be impeded if there is not strong communication and cooperation across the disciplines.

In the 21st Century Smart Structure technology could lead to a civil infrastructure that is self monitoring and even self scheduling of its maintenance and repair. This level of self monitoring might make modular design more attractive, wherein a section is replaced as soon as it degrades beyond a certain point. This revolution in technology could usher in a "new age of civil engineering". This era would see the marriage of fiber optic technology and artificial intelligence with material science and structural engineering. Major structures constructed in this period would be constructed with built-in optical neurosystems and active actuation control that would make them more like living entities than the inanimate edifices we are familiar with today. To carry this biological paradigm one step further, it may even be possible to contemplate future structures that have the capacity for limited self repair. Indeed, if we think of the new frontiers for civil engineering as being in space or underwater, self diagnosis, self control and self healing may not be so much esoteric as vital.

Of course the future highway that senses and alerts a truck driver that one of his tires is dangerously underinflated, or that his load is incorrectly balanced, will prevent many accidents........it will most certainly also eliminate our freedom to speed!

Acknowledgements

This work was supported by the Ontario Laser and Lightwave Research Centre, the Natural Science and Engineering Research Council of Canada and the Ontario Centre for Materials Research. Some of the work reported in this paper was undertaken by: T. Alavie, D. Hogg, M. LeBlanc, K.Liu, S. Melle, M. Ohn and T. Valis.

References

Blagojevic, B., Tsaw,W., McEwen, K., and Measures, R.M., (1989), " The Influence of Embedded Optical Fibers on the Interlamina Fracture Toughness of Composite Materials", Review of Progress in Quantitative NDE, Brunswick, Maine, July 23-28.

Blake, J.N., Huang, S.Y. and Kim, B.Y., (1988), "Ellipitical: Core Two-Mode Fiber Strain Gauge", SPIE, 838-862.

Caussignac , J.M., Chabert, A., Morel, G., Rogez, P., and Seantier, J., (1992) "Bearing of a Bridge Fitted with Load Measuring Devices Based on Optical Fiber Technology", 207-210, 1st Europ. Conf. Smart Structures & Materials, Glasgow.

Czarnek, R., Guo, Y.F., Bennett, K.D., and Claus, R.O., (1988), "Interferemetric Measurements of Strain Concentrations Induced by an Optical Fiber Embedded in a Fiber Reinforced Composite", SPIE, Vol 986, Fiber Optic Smart Structures & Skins I, Boston, 43-54.

Dakin, J. P., "Distributed Optical Fiber Sensor Systems," Chap. 15, (1989), 575-598, Optical Fiber Sensors, Ed. B. Culshaw and J. Dakin, Artech House.

Fuhr, P.L., Huston, D.R., Kajenski, P.J., Ambrose, T.P., (1991) "Performance and Health Monitoring of the Stafford Medical Building Using Embedded Sensors", Private Communication.

Fuhr, P.L., Huston, D.R., Kajenski, P.J., Ambrose, T.P.,W.B. Spillman, (1992), " Installation and Preliminary Results from Fiber Optic Sensors Embedded in a Concrete Building", 409-412, 1st Europ. Conf. Smart Structures & Materials, Glasgow.

Hogg, D.,Valis, T., Mason, B. and Measures, R.M., (1992) "Temperature Compensated Strain Measurements Using Fiber Fabry-Perot Strain Gauges", Optical Fiber Sensor-Based Smart Materials and Structures Conference, Blacksburg, Virginia.

Holst, A., W. Habel and Lessing, R., (1992) "Fiber Optic Intensity Modulated Sensors for Continous Observations of Concrete and Rock-Fill Dams", 223-226, 1st Europ. Conf. Smart Structures Materials, Glasgow.

Housner, G.W., Masri, S.F., and Soong, T.T., (1992) "An Overview of Active Structural Control", 201-206, 1st Europ. Conf. Smart Structures & Materials, Glasgow.

Kruschwitz, B., Claus, R.O., Murphy, A., May, R.G., and Gunther, M.F., (1992) "Optical Fiber Sensors for the Quantitative Measurement of Strain Concrete Structures", 241-244, 1st Europ. Conf. Smart Structures & Materials, Glasgow.

LeBlanc, M., and Measures, R.M., (1992), "Impact Damage Assessment in Composite Materials with Embedded Fiber Optic Sensors," J. Composite Engineering, Vol 2, 573-596.

Liu, K., Ferguson, S., McEwen, K., Tapanes,E., and R.M. Measures, (1990) " Acoustic Emission Detection for Composite Using Embedded Ordinary Single-Mode Fiber Optic Interferometric Sensors," SPIE OE/Fibers' 90 - Conference: Fiber Optic Smart Structures and Skins III, San Jose, September 16-21.

Liu, K., and Measures, R.M., (1992) "Signal Processing Techniques for Localized Interferometric Fiber-Optic Strain Sensors", Intelligent Material Systems and Structures, Vol.3, 432-461.

Loken, D., (1990), "Effect of Fiber Straightness on Fatigue of Aligned Continuous CFRP Composites ", University of Toronto (Chemical Eng.) M.A.Sc Thesis.

Lu, Z.J., and Blaha, F.A., (1989), "A Fiber Optic Stgrain and Impact Sensor System for Composite Materials", SPIE, vol. 1170, 239-242.

Lu, Z.J., and Blaha, F.A., (1991), "Application Issues of Fiber Optics in Aircraft Structures", SPIE. vol. 1588, 276-281.

Mathews, C.T., and Sirkis, J.S., (1990), " The Interaction of Interferometric Optical Fiber Sensors Embedded in a Monolithic Structure, "SPIE, Vol. 1370, Fiber Optic Smart Structures and Skins III, San Jose, 142-153.

Measures, R.M.," Smart Structures with Nerves of Glass", (1989), Progress in Aerospace Sciences, Vol 26, 289-351.

Measures, R.M., Alavie, T., Liu, K., and Melle, S., (1992), " Optoelectronic Smart Structures Interface.....A Key Development for Practical Smart Structures" Paper 1798-?, SPIE International Symposium OE/Fibers'92, Boston, 8-11 September.

Measures, R.M., (1992), "Smart Composite Structures with Embedded Sensors", Composites Engineering, Vol. 2. 597-618.

Measures, R.M., Melle, S.M., and Liu, K., (September 5-6.1992), "Wavelength Demodulated Bragg Grating Fiber Optic Sensing Systems for Addressing Smart Structure Critical Issues", Smart Materials and Structures, Vol.1., 36-44.

Melle, S.M., Liu, K., and Measures, R.M., (1991), "Strain Sensing Using a Fiber Optic Bragg Grating," SPIE, Vol. 1588, OE/Fibers' 91 - Conference: Fiber Optic Smart Structures and Skins IV, Boston.

Melle, S. M., Liu, K., and Measures, R. M., (1992), "A Passive Wavelength Demodulation System for Guided-Wave Bragg Grating Sensors", IEEE Phot.Tech. Lett. vol.4, 516-518.

Morey, W.W., Meltz ,G., and Glenn, W.H., (1989), "Fiber Optic Bragg Grating Sensors", SPIE Vol. 1169, Fiber Optic & Laser Sensors VII, 98-107.

Mufti, A.A., Erki, M-A, Jaeger, L.G., Editors, Advanced Composite Materials with Application to Bridges, (1991), "Anchorages for Prestressing Tendons and Cables", 238-239, Pub. Canadian Society of Civil Engineers.

Murphy, K. A., Gunther, M. F., Vengarkar, A. M., and Claus R. O. , Optics Letters 16, 273.

Murphy, K.A., Miller, M.S, Vengsarkar, A.M., Claus, R.O., and Lewis, N.E., (1989), "Embedded Modal Domain Sensors Using Elliptical Core Optical Fibers", SPIE. vol. 1170, 566-573.

Ohn, M., Davis , A., Liu, K. , and Measures, R. M., (1992), "Embedded Fiber Optic Detection of Ultrasound and its Application to Cure Monitoring", Optical Fiber Sensor-Based Smart Materials and Structures Conference, Blacksburg, Virginia, April.

Roberts, S.S.J., and Davidson, R., (1992) "Short Term Fatigue Behaviour of Composite Materials Containing Embedded Fiber Optic Sensors and Actuators", 255-262,
1st Europ. Conf. Smart Structures & Materials, Glasgow.

Scott, R.H. and Gill, P.A.T., (1992) " Possibilites for the Use of Strain Gauged Reinforcement in Smart Structures", 211-214, 1st Europ. Conf. Smart Structures & Materials, Glasgow.

Sirkis, J.S., and Dasgupta, A., (1990), "Optimal Coatings for Intelligent Stucture Fiber Optic Sensors", SPIE Vol. 1370, Fiber Optic Smart Structures & Skins III, San José, 129-140.

Sirkis, J.S., and Haslach Jr., H.W., (1990), "Full Phase-Strain Relation for Structurally Embedded Interferometric Optical Fiber Sensors", SPIE, Vol. 1370, Fiber Optic Smart Structures and Skins III, San Jose, 248-259..

Turner, R.D., Valis, T., Hogg, W. D., and Measures, R. M., (1989), " Fiber Optic Strain Sensors for 'Smart ' Structures," J. of Intelligent Material Systems and Structures, Vol. 1, 26-49.

Valis, T., Hogg, D., and Measures, R. M, (1990), "Composite Material Embedded Fiber Optic Fabry-Perot Strain Rosette," SPIE, Vol. 1370, Fiber Optic Smart Structures and Skins III, San Jose, 154-161.

Wolf, R. & Miesseler, H.J., (1992) "Monitoring of Prestressed Concrete Structures with Optical Fiber Sensors", P.23-29, 1st Europ. Conf. Smart Structures & Materials, Glasgow.

The OSTIC programme - its achievements and their impact on instrumentation in civil engineering

Professor Brian Culshaw, Dr W C Michie[1]

Abstract

OSTIC is a recently completed European collaborative project which represents the first concerted effort in Europe on fibre optic sensing technologies for use in carbon and glass fibre reinforced plastic material. The scope of the project embraced both the characterisation of material interactions and the derivation of fibre optic sensing techniques for use within these materials. The project demonstrated successfully that optical fibres may be embedded within 1mm thick composite samples without any degradation in the mechanical performance of the composite panel. Fibre optic sensing in composite materials was also demonstrated showing distributed measurement of dynamic strain over a sample of the order of 1m in length with linear spatial resolution of the strain measurement of approximately 20cm. Additionally the successful simultaneous measurement of temperature and quasi static strain over similar gauge lengths were experimentally demonstrated.

These basic techniques for strain and temperature measurements may be applied to a wide range of structures, including the typically large scale structures found in civil engineering. Additionally the techniques may be modified to for example monitor corrosion or grout in fill processes.

1. Introduction

The OSTIC programme (Optical Sensing Technologies for Intelligent Composites) was the first programme in Europe to focus collaborative research into the techniques of fibre optic sensing for monitoring the physical condition of

---

[1]University of Strathclyde, Department of Electronic & Electrical Engineering, 204 George Street, Glasgow G1 1XW, Scotland

composite material structures. The broad objectives of the programme were to:

- demonstrate distributed fibre optic sensing techniques in composite materials with particular reference to measurement of temperature and strain and secondary reference to cure monitoring

- demonstrate embedding procedures which will enable optical fibres to be fabricated within composite but which will retain the mechanical integrity of the original material

The members of the consortium included Bertin and EDF of France, CISE and Alenia from Italy and AEA Technology Harwell and Strathclyde University from the United Kingdom. AEA Technology and Alenia were the major contributors to the materials engineering within the consortium whilst the remainder concentrated on optical system problems. Bertin was project manager and also has teams active in both materials and optics areas.

In this paper we review the technologies which were investigated under the OSTIC programme including quasi static strain and temperature measurement, dynamic strain measurement, elementary cure monitoring and embedding procedures compatible with CFRP and GFRP resin matrices. The principal achievements of the project included the demonstration of materials compatible embedding procedures, combined temperature and strain measurement in a single fibre with resolutions of approximately 1°C and 50 microns of displacement and distributed dynamic strain sensing with a sensitivity of tens of microstrain over a gauge length of approximately 20cm. some initial results were also obtained in cure monitoring but these require further characterisation.

The programme was formally completed in Autumn of 1991. Whilst the technologies which were demonstrated were successfully implemented the programme also pointed to significant requirements for future activity. Of these the most dominant is the need for field related experience using embedded sensors in simple structural systems. More complex systems will eventually require the availability of a panel to panel interconnection system which will enable the establishment of a continuous optical fibre link between composite sections. There is also considerable scope for an overall system approach to the sensing problem which will also require that intelligent networks capable of analysing sensor array data be implemented.

Whilst the applications in civil engineering differ in detail the generic need for temperature and strain measurement remains. In particular the techniques demonstrated within the OSTIC programme offer the potential of extremely accurate strain and length measurement within structures capable in principle of sub millimetre accuracy over lengths of the order to 100 metres. Additionally the cure monitoring techniques which are mentioned within the main text may also be reconfigured to permit the monitoring of setting procedures in grout and concrete. There are many examples of major concrete structures, especially bridges, in both Europe and the United States where the integrity of the structure cannot be certified with confidence. Intelligent and effective application of these instrumentation systems coupled with the development of new emerging concepts, particularly for corrosion monitoring, will enable much more confident evaluation of the condition and safety of concept structures.

2. Fibre Optic Sensors In Composites - Some Basic Considerations

2.1 Requirements in Sensor Instrumented Structures

Mechanical structures have been extensively instrumented for test purposes for many years. The traditional technique is to attach an array of strain sensors and/or thermocouples to the structure and record the outputs from these individual sensors whilst exercising the structure through its performance range. This approach is adequate for laboratory testing but, largely due to the complexity of the wiring harness, is totally incompatible with field use.

Consequently any sensor technology which is used for structural monitoring must be capable of making measurements with comparable spatial and thermal/strain resolution to the individual sensor array but should also minimise the wiring harness and be compatible with the mechanical properties of the material under tests.

These are powerful reasons why fibre optics has been the usual choice for the sensor technology. The principal ones include:

- the availability of a wide variety of fibre optic measurement techniques especially for monitoring strain and temperature

- the capacity of a single fibre to measure distributions of strain and temperature along its length thereby eliminating the wiring harness

- the compatibility of the mechanical properties of an optical fibre with those of modern fibre reinforced plastics. In particular an optical fibre will withstand the strain history which may be imposed upon a modern composite material. Most other strain sensing elements fracture before reaching the strain limit of fibre reinforced plastics.

Consequently there has been significant emphasis on optical fibre sensor technology as the means for implementing distributed sensing in composite materials. There are of course alternative techniques of which probably the most important are ultrasonic probing and visual NDT. The possibility of using the carbon fibre in CFRP as a resistive probe has also been discussed. These are likely to complement rather than compete with embedded optical fibre measurements. For example an ultrasonic signature analysis detected using embedded optical fibres as the response to a known input probe signal could form a very useful material analysis tool.

The general concept of the sensor based structure using fibre optics is shown in Fig. 1. The signals received by the sensor array may originate either within the structure itself (for example, vibration mode spectra) or via the measurement of responses to external stimuli such as calibrated thermal or mechanical impulse testing. The optical fibre sensor has tremendous capability for making the multipoint measurements of the variables of interest along its length and it is this capability which gives the optical fibre sensing approach unique advantages in the smart structure context.

2.2 Fibre Optic Sensor Technologies in Smart Structures

A considerable range of fibre optic sensor technologies have been assessed in smart structures. The diversity of fibre optic sensor approaches may be gleaned from Fig. 2 which illustrates the basic sensing techniques which can be utilised in smart structures. The details of these techniques may be gleaned from elsewhere [1].

This paper will concentrate on the particular sensing techniques utilised during the OSTIC programme. However, many other techniques are worthy of mention including the fibre optic strain rosette using Fabry-Perot interferometers [2], the use of index matched fibres for cure monitoring [3] and distributed strain monitoring using microbend effects in relatively large structures [4].

The technologies utilised in OSTIC included microbend signature analysis and polarimetric strain/temperature sensing for cure monitoring, polarimetry and mode:mode interference for combined temperature and strain measurement and white light polarimetry for distributed dynamic strain measurement. However before analysing the sensing phenomena in detail it is useful to briefly examine the embedding processes and their effects upon material properties.

2.3 Embedding Fibre Optics in CFRP and GFRP

An essential part of the OSTIC programme was the thorough evaluation of the effects of embedding optical fibres into composite materials. The basic potential problem is that the fibres which are used as reinforcing members are typically a few microns in diameter. The optical fibre is itself typically 80 to 125 microns in diameter. Consequently the optical fibre represents a potentially large inclusion into the composite structure.

The experimental procedures involved investigating a considerable range of optical fibre diameters and coatings, including in the latter carbon, polyimides and acrylates as provided by fibre manufacturers. Preliminary work was also undertaken with metal coated fibres using both aluminium and indium as the coating materials.

A very considerable accumulation of both experimental data and finite element modelling of the optical fibre/CFRP or GFRP structure arrived at the following basic results:

- optical fibres with thin coatings usually embed more satisfactorily than optical fibres with thick coatings

- of the coatings used in both CFRP and GFRP the polyimide appeared to produce the best results

- the mechanical transfer characteristics between thin coated fibre and the composite overall appeared to be much more reliable than those with thick coated fibre

- an optical fibre laid parallel to the reinforcing fibres of the composite caused minimal changes to the mechanical properties thereof. However a fibre laid perpendicular to the composite reinforcing members could cause significant local stress concentrations in the inevitable local resin rich region and consequently introduce structural vulnerability

- the optical fibre should have a diameter which is significantly less than (say at most 20%) the thickness of the layup in a cross ply composite.

The conclusions of this investigation are quite simply that good reliable embedding procedures are possible and that with reasonable care both the mechanical integrity of the structure overall and the transfer characteristics between the sensing fibre and the structure may be assured.

2.4 Embedding Technologies for Concrete

Optical fibres are remarkably strong and also remarkably flexible with grating strains of a few percent. Experimental trials involving embedding a simple polymer jacketed glass or silica fibre into concrete have been conducted in several laboratories with the perhaps surprising result that the integrity of the fibre has not been affected by the pouring the setting of the concrete around it. Of course in this simple form the fibres do not act as sensors either since in order for this to happen they must be coupled into the concrete and the cables are designed to minimise such coupling.

There are several potential mechanisms whereby the coupling process may be implemented. The simplest is shown is Fig. 2a which illustrates the microbend device used in civil engineering applications by organisations such as SICOM [4] and G2 Systems. This is essentially a strain and/or temperature sensor. The basic concept can be reconfigured to produce a fire monitoring (Ericsson of Sweden) in which the transduction process is effected via the expansion of a wax during its melting transition introducing localised microbend loss.

Alternatively the fibre may be loose tube mounted and anchored at the ends thereby enabling a direct length measurement between the anchoring points which in turn can allow for strain and/or settlement measurement within the structure.

The materials compatibility issues have largely been resolved. Despite the high pH experienced in typical concrete the cabling structures proved to be resistant to this environment and for some systems experience extending over several years has been accumulated.

3. Fibre Optic Sensor Technologies in OSTIC

Three principal techniques have been investigated. These are:

- microbend induced analysis of curing signatures
- combined polarimetric and mode:mode interference sensing for simultaneous temperature and strain measurement
- distributed white light polarimetric sensing for the measurement of distributed dynamic strain.

3.1 Cure Monitoring Using Microbend Signatures

It is well known that a periodic microbend applied to the external surface of an optical fibre can induce loss within the fibre transmission path. This is for example used as the basis for the strain measurement systems described in Ref. 4. Here a spiral wound member external to the fibre is coupled to the mechanical strain field. However if a bare optical fibre is embedded within a composite material then the loss through the fibre reflects the stresses applied by the composite to the fibre itself.

Since the setting of the resin matrix in a composite structure from the liquid phase will obviously cause changes in the local stress distribution then in principle this could be used as a means whereby the resin curing process can be monitored. A simple attenuation measurement through a fibre placed within the structure in the manner shown in Fig. 1 could then possibly provide an indication of the state of cure of the resins.

A comprehensive range of on-line measurements of signal transmission through an optical fibre with a variety of carbon fibre reinforced composite samples using the experimental apparatus shown in Fig. 3 showed a number of consistent features. An example is shown in Fig. 4a. In particular:

- for a given sample undertaking a given thermal cycle with the optical fibre in a given position within the sample, the time/transmittance curves fibre were repeatable

- depending on the orientation within the sample the transmittance curves could show small or large variations (see Fig. 4b)

- there are pronounced differences in behaviour between PEEK and lower temperature composites.

The conclusions of this investigation were that even though a repeatable signature could be obtained its form could not be unambiguously determined. Perhaps in the future there is scope for a neural network analysis of such signatures in order to provide insight into their meaning.

An alternative approach has also been investigated. This involves using the polarimetric techniques to monitor the combined temperature/strain field. This again produces a repeated signature (Fig. 4b) but detailed unambiguous interpretation requires further research.

3.2 Quasi Static Strain and Temperature Measuring

The results described very briefly in the previous section concern observations on optical attenuation using particular types of optical fibres aimed at giving microbend signatures. Other types of fibres can be designed to resist microbend effects.

Early in the OSTIC project the potential for phase/delay based fibre optic sensors was recognised. Ideally any sensor network should give a signal which is independent of intensity transmittance through it since this is known to be a relatively sensitive function of other external influences. Furthermore, the delay measurement is attenuation free, and therefore sensor multiplexing is considerably simplified compared to loss inducing techniques such as microbending. Delay and polarisation signals satisfy these basic requirements.

Phase delay and polarisation measurements usually necessitate the exploitation of single mode fibre as a transmission medium. Such fibres have the inherent advantage that the core is well displaced from the surface of the fibre so that intensity fluctuations due to microbending can be minimised especially with high numerical aperture core materials such as heavily germania doped silica. The fibres can also be made polarisation selective by inducing form or stress birefringents into the core. Observations on the intensity transmittance of such fibres during the curing process have already shown that very little change in transmittance occurs under such circumstances.

However, it is fundamental that both strain and temperature fields produce the same apparent effect in the delay (or differential delay) in a particular mode (or between modes) within the fibre. Consequently on a single measurement it is impossible to differentiate between the effects of change in temperature and the effects of changes in mechanical strain in the same bandwidth. A key feature is then to evolve a technique whereby strain and temperature may be independently measured on the same length of optical fibre.

In principle, this obviously requires two measurements to be made and in general these measurements m_1, m_2 can be expressed in the form:

$$\begin{bmatrix} m_1 \\ m_2 \end{bmatrix} = \begin{bmatrix} A & B \\ C & D \end{bmatrix} \begin{bmatrix} \Delta T \\ \Delta\varepsilon \end{bmatrix} \qquad (1)$$

where A, BC and D.are constants and ΔT and $\Delta\varepsilon$ are changes in temperature and mechanical strain respectively [5].

If we wish to extract ΔT and $\Delta\varepsilon$ then in principle all that is required is to invert the matrix is equation 1 and use the measurements m_1 and m_2. The snag is that for many practical situations, the matrix $ABCD$ is ill-conditioned so that the inversion process introduces significant errors.

The techniques used for these measurements was that m_1 should be a birefringence measurement on length of optical fibre and m_2 a differential phase measurement between the highest order mode and the fundamental mode in the same birefringent fibre, but at a shorter wavelength, than the birefringence measurement. Given the correct type of fibre then conditions may be found under which the matrix in equation 1 is reasonably well conditioned so that the inversion process can be assured.

The schematic diagram for combined temperature strain measurement is shown in Figure 5. The birefringent fibre used has a cut off at around 750nm so that it is dual moded at 633nm (one measurement wavelength) and single moded at 850nm (the other measurement wavelength). The mode:mode signal is derived at the shorter wavelength. A typical temperature:strain history for a sample under test is shown in Figure 6 and the apparent measurements derived using this technique after matrix inversion are superimposed on these figures.

As a general observation the temperature may be resolved usually to within ± 1°C but always within ± 3°C and the strain measurement can be implemented to resolve a distance change over the gauge length of approximately 50μm.

These measurements could be refined by improving the basic measurement technique and optimising the fibre design. However, they already demonstrate performance capability which is compatible with the requirement in composite material testing.

3.3 Dynamic Distributed Strain Measurement

The technique chosen for distributed dynamic strain measurement involves launching light into a birefringent fibre along one of its principal axis and arranging for the coupling of a small amount of light from this principal axis to the other axes at the beginning of each gauge length within the sensor structure (see Fig. 7). The gauge length is designed such that by the time the light has propagated along the two axes, the contributions in these axes are mutually incoherent. This obviously necessitates the use of a quasi-white light source in the interferometer structure. The receiver is a multi-channel Michelson interferometer with path lengths in each channel matched to the differential lengths between the first coupling point in the array and the coupling point of interest.

The application of a strain field to the sensor array causes small (usually less than 2π) changes in the differential phase delay between the two paths within the gauge lengths. These small changes are detected as changes in the polarisation state of the light in the corresponding channel in the white light receiver array. The outputs from each element in the multi-channel receiver array are then subtracted sequentially to give the strain in each section.

For the reasons outlined in the previous section, this approach is inherently unsuitable for monitoring quasi static strain since only one measurement is available and temperature and strain fields produce the same effect. However, it is eminently suitable for monitoring dynamic strains and since, in many situations, dynamic strain analysis is exactly what is required, this approach is well matched to the application in composite material monitoring.

One particular feature of this array is the requirement to introduce controlled coupling between the light launched in one particular principal axis and additional light propagating in the orthogonal axis. The total amount coupled should be of the order of 1% or less and the coupling factor should be stable and controllable. A number of approaches have been evaluated within the OSTIC programme. The initial one used compressive metal rings [6] but these were found to have unpredictable properties when embedded. Finally for experimental purposes, a simple "twist and fuse" was adopted. Indeed the twist was found to be unnecessary and gentle heat treatment of a limited length of the fibre was found to produce the necessary coupling. In the longer term[7] a more controlled approach based upon photo-refractively produced diffraction gratings within the fibre core may prove to be a good workable solution.

Again extensive tests have been conducted on a prototype system (see, for example Fig. 8). This system comprised six sensor elements and used a super luminescent diode as the optical source operating at 850nm. The system proved to have a sensitivity of tens of microstrain in a gauge length approximately 25cm. In its present form the receiver is a bulk optic system with inherent sensitivity to environmental perturbations. However, there is no reasons why a micro-optic and/or guided wave approach to the receiver design should not be eventually adopted thereby providing a rugged and practical solution to the quasi distributed sensor problem.

4. <u>Discussion - Relevance to Civil Engineering</u>

The results obtained under this programme have demonstrated both the potential offered by the basic measurement techniques involving microbend induced loss modulation and optical delay modulation monitored either through direct interferometry or through differential interferometry. The former may be used to measurement strains and temperature or, more interestingly, to monitor curing processes. Whilst the latter has yet to be seriously evaluated in civil engineering materials, especially concrete, the changes in dimension involved during the setting process are almost certain to produce measurable microbend signatures. Indeed laboratory tests conducted by ourselves and others have confirmed that this is the case. Traditionally moisture sensitive coatings may be used to assess the drying process within the material. Given the slightly surprising observation that the setting procedures in concrete are purely empirical, the possibility for such transduction systems is immense.

As for strain monitoring, the precision with which this may be executed is only limited by the precision of time standards. These can be readily obtained to parts in 10^6 so distance measurement accuracy to a few microstrain is possible with optical ranging techniques. Indeed laser range finders are already in common use in surveying and the overall principle involving using fibres is similar with the obvious enhanced flexibility enabling the fibre to be incorporated within the structure between predetermined points.

There has been significant activity in the civil engineering area already using fibre optic sensors to monitor icing, cracking processes, dynamic and static strains. Most of the experimental work to date has relied upon microbend related transducers. The results of the OSTIC programme demonstrate the availability of much more accurate transduction processes and thereby coulc significantly enhance the flexibility and utility of instrumentations for the next generation of civil engineering structures.

5. Conclusions

The OSTIC programme started in Summer 1988 was completed in Autumn of 1991. During that time, the partners in the programme successfully demonstrated that fibre optics can be reliably embedded into composite material structures and that the basic temperature and strain sensing functions may be realised with the necessary resolution accuracy and dynamic range.

The investigations into cure monitoring which were outwith the original specification for the project showed only preliminary results but did indicate that curing signatures could be derived from experimental data.

The programme took the first steps in the realisation of sensory structures within Europe. The programme also identified a broad range of requirements for future activity in the area. Of these, the principal one is probably the need for the acquisition of field data on evaluation sensor systems. This is, of course, already underway in a preliminary form with instrumented pre-stressed concrete bridge structures using microbend based strain sensors both in Europe and the US. Expanding this work into polymer matrix composites on the one hand and structural metals on the other would provide significant useful additional data.

There are also a range of technological factors which require addressing. Of these the most important at the optical level is that of connecting fibres between panels. To date there have been a few mentions in the literature of potential techniques whereby panels may be interconnected for multimode fibres but nothing to date for single mode fibre or for birefringent fibre. Such interconnect is essential if the power of fibre optic distributed sensing is to be fully exploited.

The other principal technological need for fibre optic sensors in composites is that of reliable signal processing of the data which is returned from the sensor array. This data is, in its very nature, very complex and it seems that only a neural network approach to analysing this data will begin to produce usable and reliable results. The presence of systematic spurious signals, which arise because of the nature of the embedding process, makes this processing approach even more important. To date much has been said but little has been published on the usefulness of neural network processing in this context. The need for processing applies not only to temperature and strain measurements but also to cure signature measurements if these are ever to prove to be useful. The performance specification could also be refined. At the beginning of the OSTIC programme the objective was set to measure temperature and strain with spatial resolutions of centimetres. We almost achieved that but there is a need to more closely approach this ideal. Similarly the temporal bandwidths of such measurements could - depending upon the structural size - extend into the few kHz region and whilst this is in principle not a problem, some elements of the mechanical scanning in the Michelson receiver array for the distributed network are limited in their operational bandwidth. The onset of structural damage is another parameter of great importance. This was not addressed during our programme though a number of approaches could be identified of which, perhaps the most promising, is ultrasonic signature analysis using an embedded monitoring array.

In conclusion, the principles have been demonstrated but there is still a great deal to be done before practical systems may be realised.

Acknowledgements

This work was partially supported by the BRITE/OSTIC programme and includes contributions from Pierre Sansonetti, Scott Roberts, Roger Davidson, Bruno Fonari, Mario Martinelli, Valeria Gusmeroli, Priscilla Escobar, Pierre Ferdinand, Joseanne Planté and others. Without the active collaboration of all these partners, the programme could not have succeeded. The Strathclyde team would also like to acknowledge the NATO travel grant which enabled constructive and useful interactions with their counterpart Virginia Polytechnic.

References

1. Dakin J, Culshaw B, (Editors), 1988. "Optical Fibre Sensors: Principles and Components", vols. 1 & 2.

2. Valis, Thomas, 1990. "Localised and Distributed Fibre-Optic Strain Sensors Embedded in Composite Materials", PhD thesis, University of Toronto.

3. Aframowitz M A, Lam K Y, 1988. "Fibre optic cure monitor for thermoset composites", Proc. Fibre Optic Smart Structures and Skins, SPIE vol. 986, pp. 135-139.

4. Sales Literature SICOM, Gremberge-Strasse 151a, D5000, Cologne 91.

5. Vengsarkar A, Michie W, Jankovic L, Culshaw B, Claus R, 1990. "Fibre Optic Sensor to Simultaneous Measurement of Strain and Temperature", SPIE vol 1367, Fibre Optic and Laser Sensors VIII, pp. 249-260.

6. Gusmeroli V, Vavassori P, Martinelli M, 1989. "A Coherence Multiplexed Quasi-Distributed Polarimetric Sensor suitable for STructural Monitoring", OFS-6, pp. 513-518.

7. Guerin J J, Maillard J M, Sansonetti P, 1992. "Polarisation Mode Couplers made by Photo induced Grating for White Light Quasi distributed Polarimetric Sensors", Optical Fibre Sensors, Monterey.

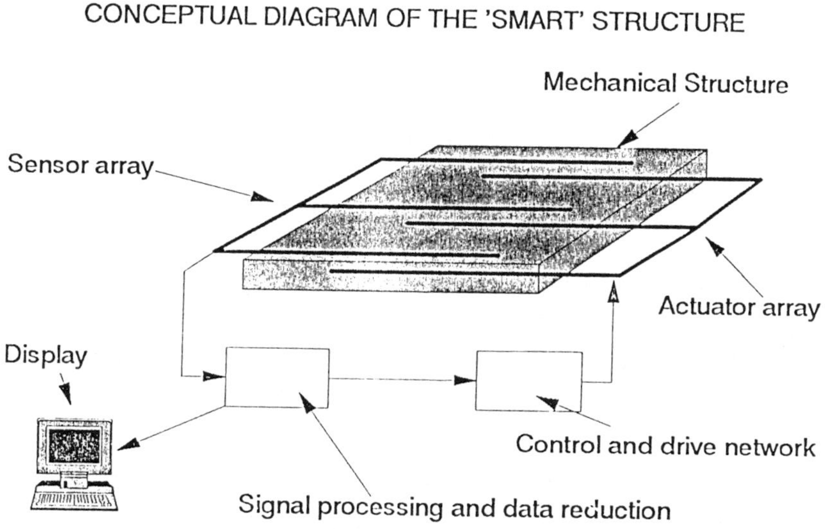

Figure 1

GENERAL PRINCIPLE OF MICROBEND CABLE SENSORS

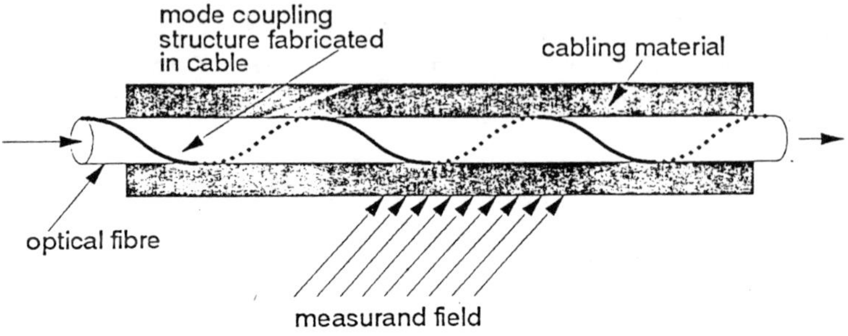

Figure 2a Deformations to the cabling material are transferred to microbending induced losses via the coupling structure

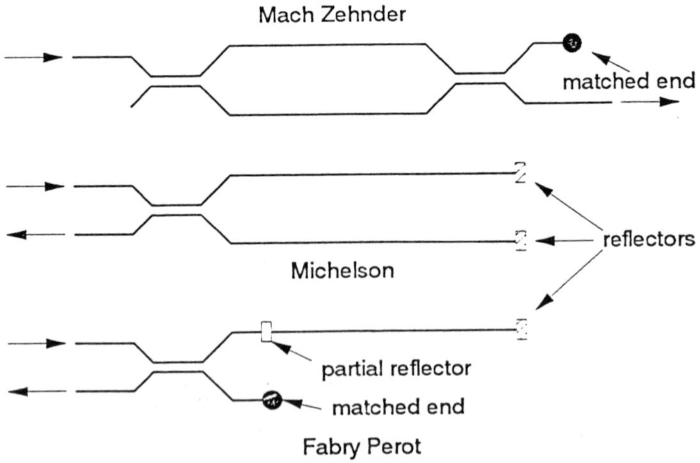

Figure 2b OPTICAL FIBRE INTERFEROMETERS

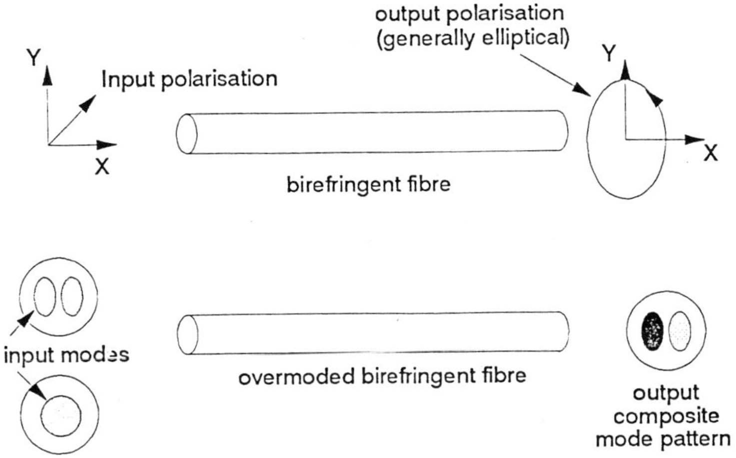

Figure 2c

EVANESCENT APPROACHES TO CHEMICAL SENSING

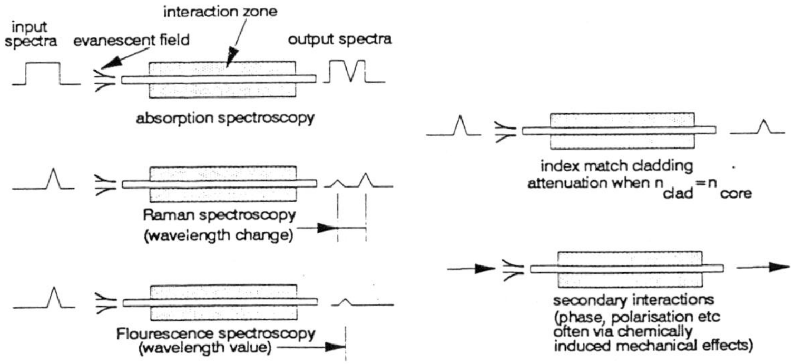

Figure 2d

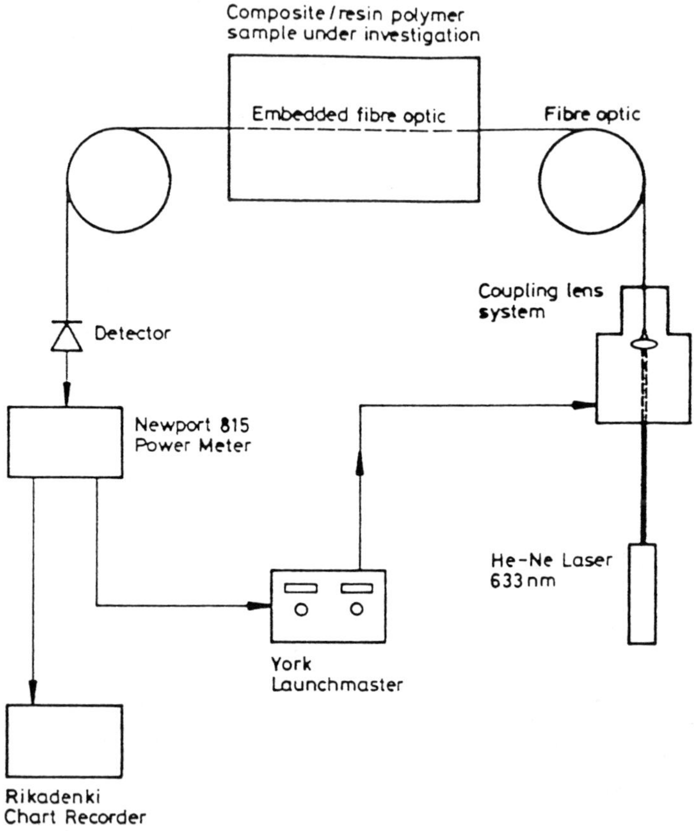

Figure 3 Experimental set up for 'on-line' attenuation measurements

4a

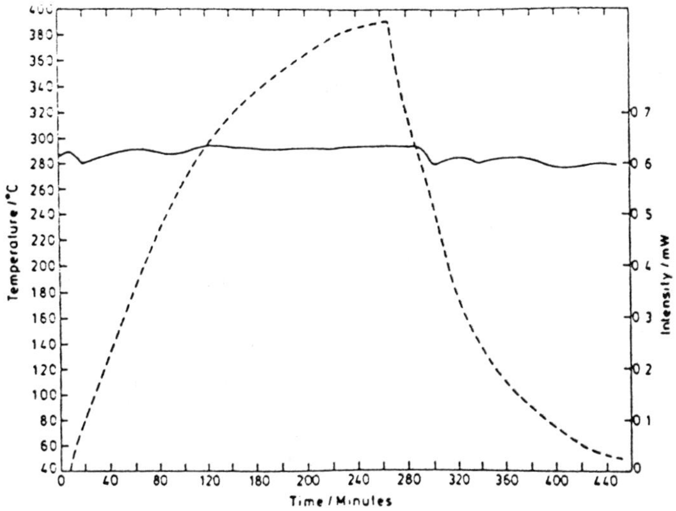

4b

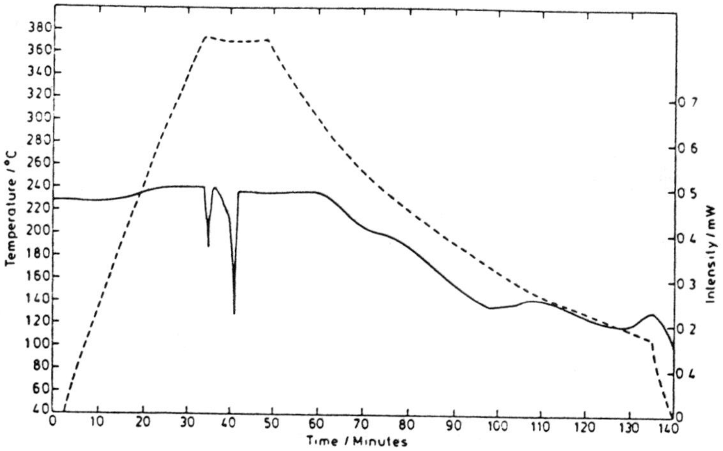

Figure 4a: Signal modulation through 80/93 um polyimide coated fibre - parallel to reinforcements

Figure 4b: Signal modulation for 80/93 um plyimide coated fibre perpindicular to reinforcements

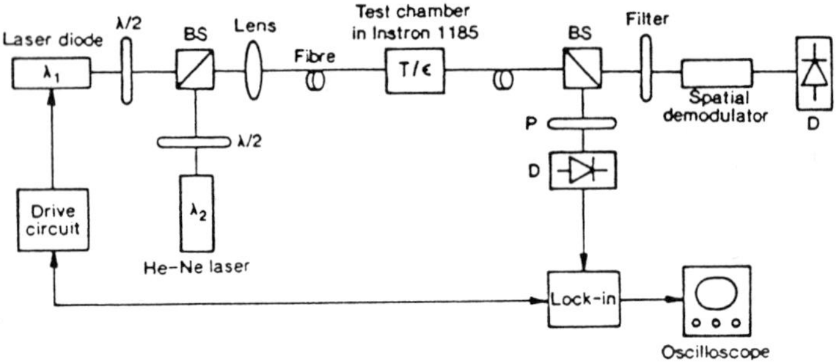

Figure 5: Experimental arrangement for dual mode/polarimetric sensor

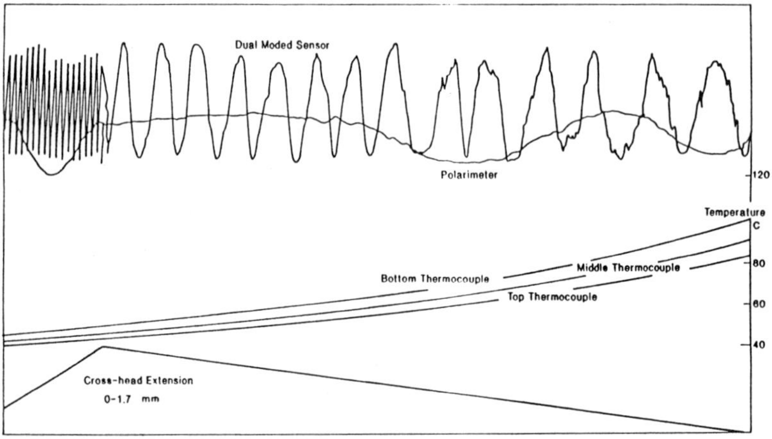

Figure 6a Temperature/Strain Measurement on Andrews E-Core Fibre
Fibre Batch Number : 890731A

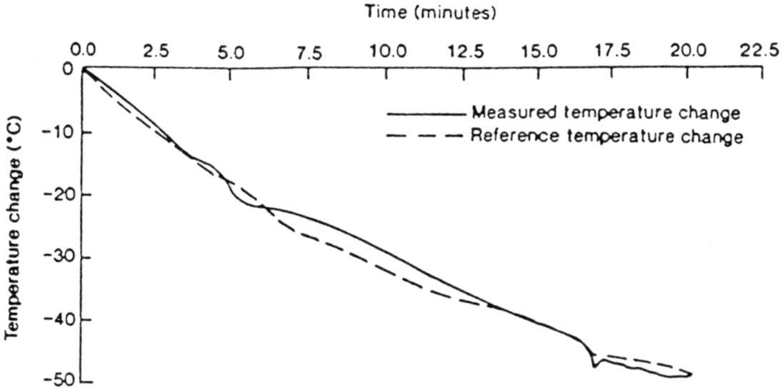

Simultaneous temperature / strain measurement (temperature recovery)

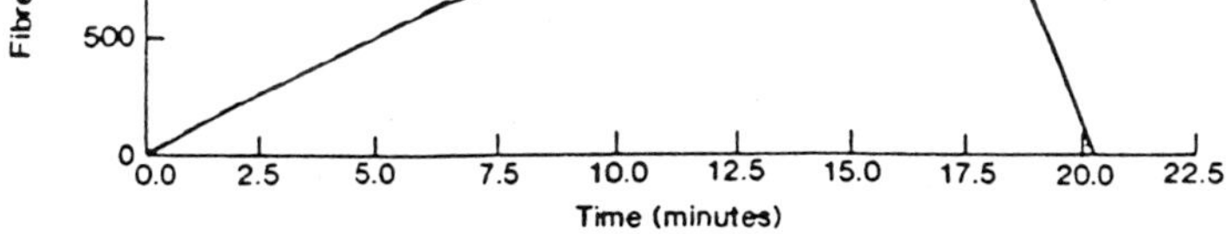

Simultaneous temperature / strain measurement : fibre elongation (μm)

<u>Figure 6b</u> <u>Recovered temperature and strain data from experiment depicted in Figure 6a</u>

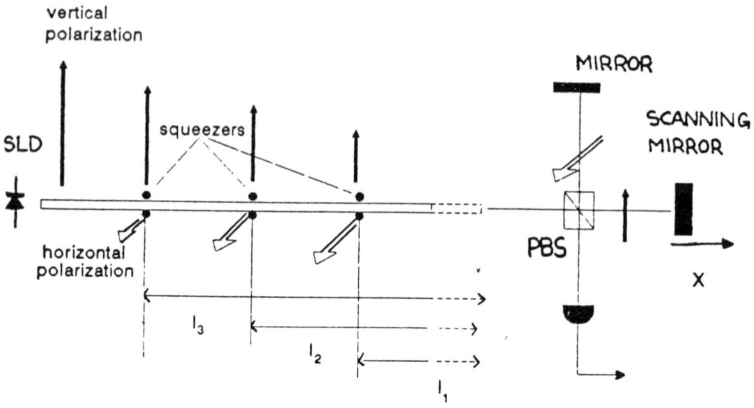

Figure 7: Schematic of Quasi-distributed polarimetric measurement

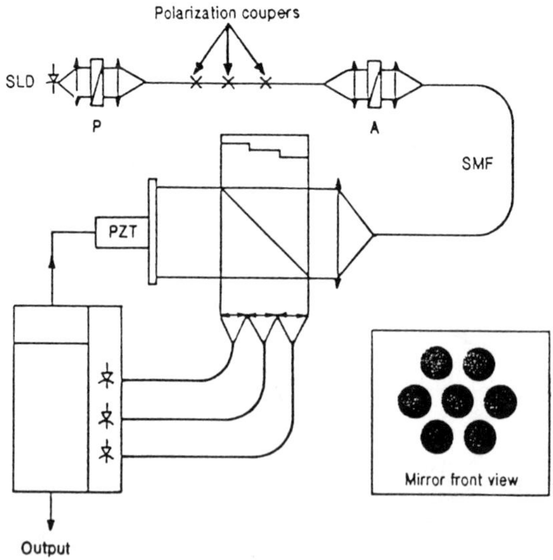

Figure 8: SCHEMATIC OF QUASI DISTRIBUTED POLARIMETRIC INTERFEROMETER

II - MEASUREMENT OF STRAIN

Extrinsic Fabry-Perot Sensor for Structural Evaluation

R. O. Claus, M. F. Gunther, A. B. Wang, K. A. Murphy and D. Sun[1]

Abstract

We summarize the design and applications of a fiber optic extrinsic Fabry-Perot interferometric sensor element. As a strain sensor, the device has been demonstrated at temperatures ranging from -200 to 900°C. It is shown that a Fabry-Perot strain gage with a one centimeter gage length can be operated in differential mode with 0.01 microstrain resolution in real-time, and in an absolute mode with 0.5 microstrain resolution with a three second scan time.

Introduction

Numerous types of optical fiber sensors for materials evaluation have been proposed during the past fifteen years. Of these, fiber optic Fabry-Perot (FP) sensors allow local, highly sensitive, short gage length measurements of temperature, mechanical vibration, acoustic waves, and magnetic fields [Yoshino]. Techniques to create the FP cavity have varied from the fabrication of Bragg gratings in or on the fiber [Belsley] to the use of air-glass interfaces at the fiber ends as the reflectors [Kersey]. A relatively new technique described by Lee and Taylor involves fabricating pairs of semireflective splices at locations in an otherwise continuous length of fiber [Lee]; since the propagating electromagnetic field does not leave this fiber in this situation, the etalon sensor has been described as an intrinsic Fabry-Perot interferometer.

An extrinsic Fabry-Perot interferometer may be designed instead to allow the input light field to be back reflected at surfaces on either side of an air gap along the length of the fiber. The main advantages of this extrinsic method over the intrinsic method are avoidance of polarization problems, and the detection of only axial strain components. A brief description of the operation of this sensor is given in Section 2, several demonstrations are considered in Section 3, and methods for absolute strain sensing described in Section 4. A discussion of implications is given in Section 5.

Extrinsic Fabry-Perot Interferometric Sensor Operation

A drawing detailing the construction of the extrinsic FP interferometer is given in Figure 1. A single mode fiber , used as the input/output fiber, and a

[1] Fiber & Electro-Optics Research Center, Bradley Department of Electrical Engineering, Virginia Tech, Blacksburg, VA 24061-0111 USA

multimode fiber, used purely as a reflector, form an air gap that acts as a low-finesse FP cavity. The far end of the multimode fiber is shattered so the

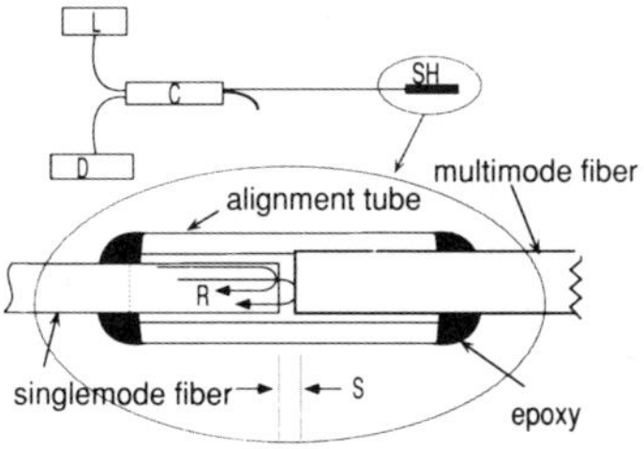

Figure 1. Schematic of extrinsic FP interferometer.

reflections there do not add to the detector noise. The Fresnel reflection from the glass/air interface at the front of the air gap (reference reflection) and the reflection from the air/glass interface at the far end of the air gap (sensing reflection) interfere in the input/output fiber. Although multiple reflections occur within the air gap, the effect of reflections subsequent to the ones mentioned above are negligible. The two fibers are allowed to move in the silica tube and changes in the air gap length cause changes in the phase difference between the reference reflection and the sensing reflection. This change in phase difference modulates the intensity of the light monitored at the output arm of a fused biconical tapered coupler.

The interference of the two-wave interferometer can be evaluated in terms of a plane-wave approximation. A coherent, approximately-plane wave detected at the output of the sensor can be represented in terms of its complex amplitude U_i (x, z, t), given by

$$U_i(x, z, t) = A_i \exp(j\phi_i), \quad i = 1, 2,$$

where the variable A_i can be a function of the transverse coordinate x and the distance traveled, z, and the subscripts i =1, 2 stand for the reference and the sensing reflections, respectively. Assuming that the reference reflection coefficient $A_1 = A$, the sensing reflection coefficient A_2 can be approximated by the simplified relation [Keiser]

$$A_2 = A\left(\frac{ta}{a + 2s\tan(\sin^{-1}(NA))}\right),$$

where a is the fiber core radius, t is the transmission coefficient of the air-glass interface (≈ 0.98), s is the end separation and NA is the numerical aperture of the single-mode fiber, given by NA = $(n_1^2 - n_2^2)^{1/2}$. The terms n_1 and n_2 are the refractive indices of the core and the cladding, respectively. The observed intensity at the detector is a superposition of the two amplitudes and is given by

$$\begin{aligned} I_{det} &= |U_1 + U_2|^2 \\ &= A_1^2 + A_2^2 + 2\,A_1A_2 \cos\left(\phi_1 - \phi_2\right)' \end{aligned}$$

which can be rewritten as

$$I_{det} = A^2\left[\frac{2ta}{a + 2s\tan\left[\sin^{-1}(NA)\right]}\cos\left(\frac{4\pi s}{\lambda}\right) \times \left(\frac{ta}{a + 2s\tan\left[\sin^{-1}(NA)\right]}\right)\right]$$

where we have assumed that $\phi 1 = 0$ and $\phi 2 = 2s\,(2\pi/\lambda)$, and λ is the wavelength of operation in free space. The simplified loss relation for the misalignment of two fibers is sufficient for understanding the decrease in the output intensity of the sensor as a function of the gap displacement. For a strain sensor, it is useful to plot the detected intensity versus gap-separation s, as shown in Figure 2. The fringe contrast decreases as the displacement increases; this is to be expected because the relative intensity of the sensing reflection begins to decrease with respect to the intensity of the reference reflection.

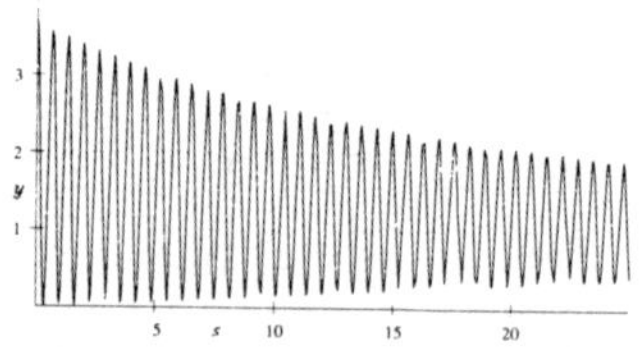

Figure 2. Calculated FP output as function of gap separation.

The extrinsic FP interferometer was tested as a displacement sensor by attaching one fiber to a stationary block and the second fiber to a micropositioner which was adjusted to produce a known displacement between the fiber ends. For comparison with the theoretical results, Figure 3 shows an oscilloscope trace of the continuously monitored output intensity of the sensor for s = 0 to s = 203 μm. The experimentally counted number of fringes for the displacement was 310.5 which corresponds to a displacement of 202 micrometers [Claus].

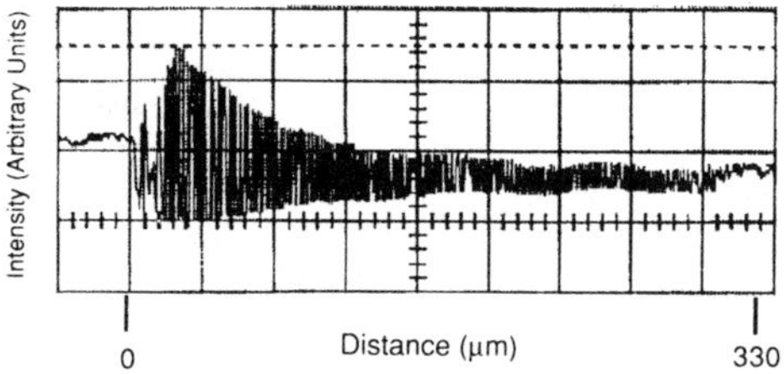

Figure 3. Experimental result of the sensor output as a function of gap separation.

Most FP sensors described in the past have been useful in the measurement of quasi-static strain alone. That is, when the induced strain changes its polarity, conventional FP interferometers are not be able to detect this change if the switch in direction takes place at a maximum or minimum of the transfer function curve. Recently, methods for obtaining directional strain information using thin-film, or resistive gages have been demonstrated [Putz]. It can be shown that two strain signals shifted 90° in phase will resolve this issue [Murphy] and can be used with EFPI strain sensors to measure the relative direction of dynamically varying strain. A theoretical description of the sensor operation is presented and two different configurations to obtain a quadrature phase-shifted operation are described.

Figure 4 shows a typical sinusoidal variation of the FP sensor optical output intensity with respect to changes in the phase difference between the reference and the sensing reflections. If the phase difference ϕ varies sinusoidally with time and the peak-to-peak variation is large enough to push the sensor out of its linear range, fringes are observed at the output of the detector as shown in the figure. The basic principle of operation of the detection scheme for a dynamic strain measurement system can now be described by considering two sinusoidal transfer functions out of phase by 90°. Assuming that the transfer function for the first detector (D1) leads that of the second (D2), we find that the output waveform for D1 leads that of D2 until time t_1. At time t_1, the phase ϕ changes direction because of a change in direction of the strain, and the output waveforms switch their lead-lag properties. Notice the output from D2 now leads that of D1 until time t_2 when the strain changes direction again. Keeping track of the lead-lag phenomenon between the two detectors gives unambiguous information about the relative direction of the strain.

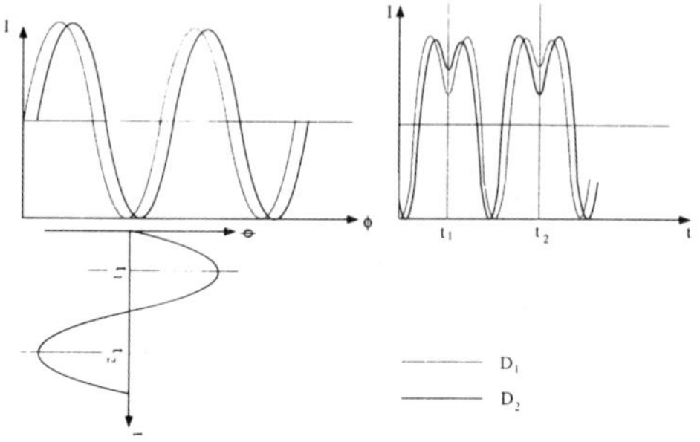

Figure 4. Sensor outputs for different operating wavelengths.

If only one detector were used, the switch in direction would not be detectable if the strain changed direction at a peak of the transfer function curve. With two signals out of phase by 90°, if the direction change occurs at one peak (of either D1 or D2), the other transfer function curve will provide information about the direction change.

A practical method of obtaining two signals 90° out of phase with respect to one another is shown in Figure 5. We adjust the quadrature phase-shift of the two separate sensors actively by moving the fibers in and out of the tube

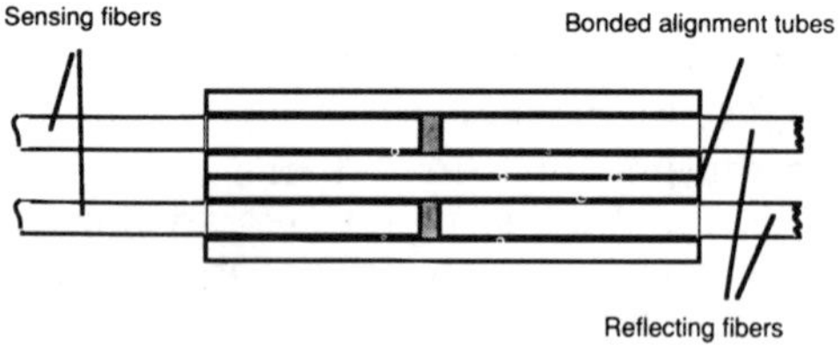

Figure 5. Side-by-side QPS sensor geometry.

until an approximately 90° phase shift is achieved at the output. Since the two tubes have external diameters on the order of only a few hundred micrometers, the two sensors monitor almost the same environmental perturbations.

To test the validity of the quadrature phase-shifted (QPS) sensors, the sensor was attached, using the scheme shown in Figure 5, to a cantilevered titanium beam with an epoxy [Claus]. The fiber sensor was attached along the length of the beam and beam vibrations were monitored. A typical oscilloscope trace is shown in Figure 6. The figure clearly shows the shift in the lead/lag properties of the two signals as the relative direction of the strain at the location of the sensor on the beam changes from increasing to decreasing.

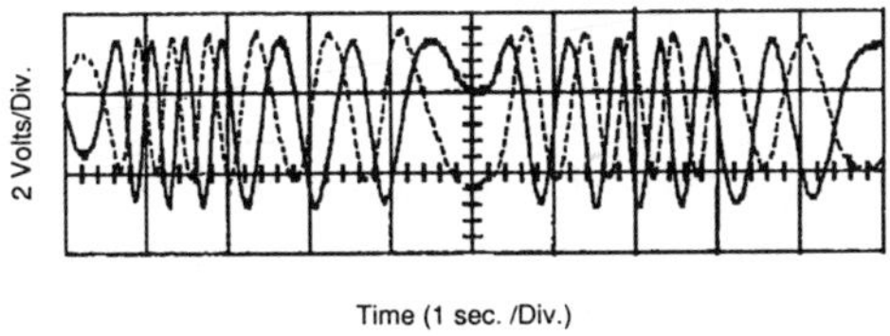

Figure 6. Oscilloscope trave of QPS signal outputs.

Experimental Results for Strain Not Parallel to Sensor Axis

An additional advantage of the extrinsic Fabry-Perot fiber sensor over all other types of currently investigated fiber sensors [Sirkis] is its sensitivity to axial strain alone. By analytically modeling the construction of extrinsic Fabry-Perot, intrinsic Fabry-Perot, Mach-Zehnder, polarimetric and grating-based sensors, Sirkis [Sirkis] has found that in general all sensors other than extrinsic FP sensors effectively produce output signals proportional to linear combinations of the strain component parallel to the long axis of the fiber and the two additional strain components perpendicular to that axis. Unambiguously determining single axis strain component information from such sensors thus is cumbersome due to the required need for multiple redundant sensor elements and complicated signal processing.

Extrinsic Fabry-Perot sensor elements produce an output phase change signal proportional to the superimposed contributions of mechanically-induced axial strain and thermally-induced axial elongation of the element.

Separation of the these effects of mechanical stress and temperature may be achieved using two optical wavelengths to address a single FP etalon, or two closely space sensor elements with one sensor mechanically isolated from the mechanical stress field inside a small diameter hollow tube.

A simple experiment to demonstrate the off-axis strain sensitivity of a single non-QPS extrinsic FP element was performed by attaching a sensor to an aluminum plate and mechanically loading the plate in tension in directions gradually varying with respect to the long axis of the sensor [Sun]. Here the temperature of the sensor element was maintained constant enough during the experiment that the output signal should not contain large temperature-induced components.

The results of this experiment are shown in Figure 7. For axial strain on the sensor element, the sensor indicates the axial strain. For strain not parallel to the sensor axis, the sensor gives an output signal that is a combination of the strain vector projected on that axis, and the projection of the strain vector caused by the perpendicular contraction of the material due to Poisson effects. Here, since QPS sensing was not used, the sensor cannot absolutely resolve the difference between tensile and compressive loads, so the output is differentially measured as proportional to $|\varepsilon_{z,\,load}| + |\varepsilon_{z,\,Poisson}|$. When the direction of the applied mechanical load is perpendicular to the axis of

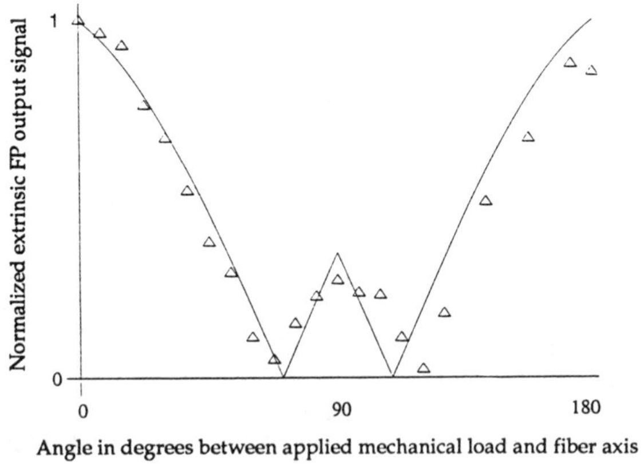

Figure 7. Angular strain sensitivity data for FP sensor.

the fiber, the output signal is effectively a measure of the Poisson effect alone. For this experiment, an aluminum support plate was used, and the output data indicates a Poisson ration for aluminum of approximately 0.21.

Absolute Strain Sensing Techniques

The difficulty encountered above in determining absolute rather than relative displacements may be solved using signal processing methods. This section describes a method for absolute measurement of displacement using an EFPI sensor. The sensors described previously are differential in nature and cannot give an absolute measure of displacement. To produce an absolute sensor, a wideband optical source must be used rather than a monochromatic laser diode source. By scanning the various wavelength components of the sensor output, an absolute measurement of the gap separation is obtained.

The output of the EFPI is described by

$$I = I_o \sin\left(4\pi Ln/\lambda_s + \phi_o\right)$$

where L is the distance between the two endfaces of the fibers, n is the refractive index of the material filling the gap (for air, n = 1), λ_S is the wavelength in vacuum of the ideal monochromatic source, and ϕ_O is some initial phase.

Let λ_S take on two different values, λ_1 and λ_2, where $\lambda_1 < \lambda_2$ and redefine the argument of the sinusoid to be

$$\phi_{1,2} = 4\pi L/\lambda_{1,2}$$

then the phase difference becomes

$$\Delta\phi = \phi_1 - \phi_2 = 4\pi L\lambda_1\lambda_2/\left(\lambda_2\text{-}\lambda_1\right)$$

ϕ_O is assumed to be a constant resulting from the initial gap separation. Rewriting Equation (6) gives

$$L = \Delta\phi\lambda_1\lambda_2/4\pi\Delta\lambda$$

where $\Delta\lambda$ is the difference of the two wavelengths to be scanned. By scanning the phase difference accumulated between λ_1 and λ_2, this equation can be used to compute the corresponding gap length.

A white light source was used to provide the wideband excitation necessary to operate the sensor in an absolute configuration [Claus]. An EFPI was attached to a translation stage to provide accurate displacements and the sensor output was monitored on an Ando AQ 6103C Optical Spectrum Analyzer. The EFPI leads were constructed from 1300 nm circular core fiber with a cutoff wavelength of approximately 1100 nm for single mode operation. A representative output scan, as well as the theoretical prediction, is provided in Figure 8.

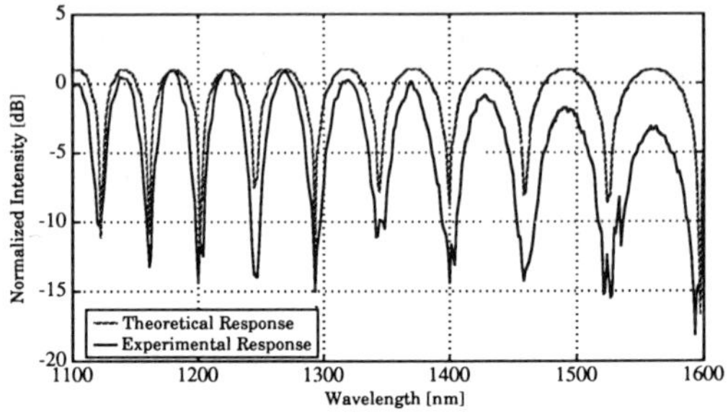

Figure 8. Sensor output as a function of wavelength.

There are three parameters that must be measured to compute the gap L, these parameters are the two wavelengths of interest and the associated phase between them. Two methods were used to compute the gap separation, 1) the constant phase method - λ_1 and λ_2 were found for $\Delta\phi = 4\pi$, and 2) the constant wavelength method - $\Delta\phi$ was found between $\lambda_1 = 1100$ nm and $\lambda_2 = 1300$ nm. The results can be seen in Figure 9, the dotted

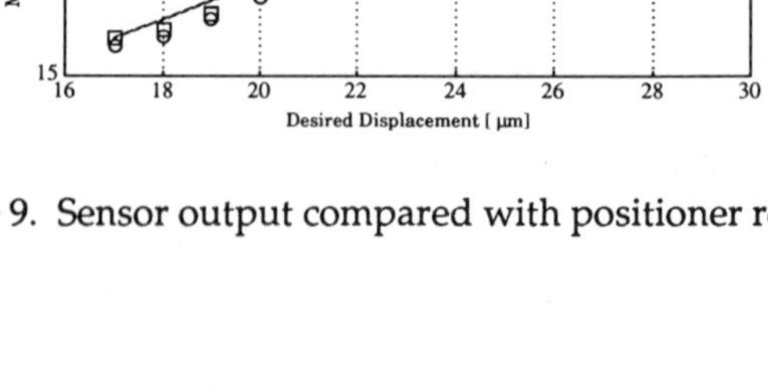

Figure 9. Sensor output compared with positioner reading.

line indicates the measured gap displacements produced using the translation stage. The accumulated error is due to inaccuracies in reading actual displacements from the micrometer.

This method is limited to static measurements because of the three second scan speed of the optical spectrum analyzer. Finally, the Ando AQ 6103C Optical Spectrum Analyzer has a resolution of 0.1 nanometers for wavelength measurements, which corresponds to an absolute gap measurement resolution of 5 nm.

Conclusions

We have discussed the operation of extrinsic Fabry-Perot sensors for the measurement of strain. Such sensors have been applied to the characterization of aerospace, hydrospace and civil structures at temperatures ranging from -200 to 900°C. Absolute displacement measurement methods allow unambiguous determination of strain amplitude and direction.

Acknowledgments

This work was supported in part by the Virginia Center for Innovative Technology and Fiber and Sensor Technologies, Inc.

References

K. L. Belsley, J. B. Carroll, L. A. Hess, D. R. Huber, and D. Schmadel, "Optically multiplexed interferometric fiber optic sensor system," Proc. - SPIE Int. Soc. Opt. Eng., vol. 566, pp. 257-264, 1985.

R. O. Claus, M. F. Gunther, A. Wang and K. A. Murphy, "Extrinsic Fabry-Perot sensor for strain and crack opening displacement measurements from -200 to 900°C," Smart Materials and Structures, vol. 1, pp. 237-242, 1992.

G. Keiser, Optical Fiber Communications, McGraw-Hill Inc., New York, 1983.

A. D. Kersey, D. A. Jackson, and M. Corke, "A simple fibre Fabry-Perot sensor," Opt. Comm., vol. 45, pp. 71-74, 1983.

C. E. Lee and H. F. Taylor, "Interferometric optical fibre sensors using internal mirrors," Electron. Lett., vol. 24, pp. 193-194, 1988.

K. A. Murphy, M. F. Gunther, A. M. Vengsarkar, and R. O. Claus, "Quadrature phase shifted extrinsic Fabry-Perot fiber optic sensors," Opt. Lett., vol. 16, pp. 273-275, 1991.

J. Putz, J. Putz, A. Wicks, and T. Diller, "Thin-film shear stress gage", Proceedings of ASME Winter Annual Meeting, Symposium on Micro Structures, Sensors and Actuators, Dallas, November 26, 1990.

J. Sirkis, "Phase-strain-temperature model for structurally embedded interferometric optical fiber strain sensors with applications," Proc. SPIE, vol. 1588, 1991.

D. Sun, S. Christopher, P. Schultz and R. Claus, "Angular strain sensitivity of extrinsic Fabry-Perot sensors," Proc. Optical Fiber Sensor-Based Smart Materials and Structures Workshop (Blacksburg, VA), April 1993.

T. Yoshino, K. Kurosawa, K. Itoh, and T. Ose, "Fiber-optic Fabry-Perot interferometer and its sensor applications," IEEE J. Quantum Electron., vol. QE-18, pp. 1624-1632, 1982.

Etched fibers for strain and vibration measurements

Masoud Vaziri, Ralph Coan and Chin-Lin Chen[1]

Abstract

We report a new class of optical fiber intensity sensors that do not require the use of external structures to impose perturbations on fibers. Instead, the topological structures are etched directly onto the fiber cladding. The responses of the optical fiber strain gauges are linear, repeatable, and have no measurable hysteresis. These sensors are very sensitive with a gauge factor as high as 170 and a flat frequency response to at least 2.7 kHz. They can be used to monitor a strain as large as 0.4%. The gauge factor depends on the asymmetry of the etched structures, the number of etched sections, and the amount of cladding material removed.

Presented in this work are the basic principles underlying the operation of etched fiber sensing elements, the experimental confirmation, comparison with resistive strain gauges, and possible applications of these sensors.

Introduction

The development of new materials, designs, and construction techniques in engineering are putting more stringent demands on sensor technology to monitor structural safety and integrity. The economical and commercial viability of existing and future structures can only be assessed with a complete knowledge of the forces affecting the structure. Thus, the measurement of strain and vibration is becoming increasingly important to the various applications. Examples include the short-term and long-term monitoring of structural deformations, the spectral measurement of tensile strains in flexible pavements, pile driving analysis, and various other static and dynamic structural parameter measurements (ICE 1981; IMC 1987).

Recently, optical fiber sensors have been shown to be useful in the measurement of strain. These sensors offer many advantages over existing strain gauges. Optical fibers are rugged, flexible, and lightweight so that they would not reinforce or otherwise affect the material to be measured. They are immune to environmental conditions such as temperature variations, rain, snow, or lightning strikes, all of which can seriously affect the performance of electric and piezoelectric sensors. Other advantages include increased sensitivity, immunity to electromagnetic interference, small size, and the fiber's inherent ability to serve as both the sensing element and the signal transmission medium, allowing the sensitive electronics to be remotely located from the measurement site.

Various types of optical fiber strain sensors have been proposed (Giallorenzi et al. 1982; Culshaw 1984). Many of these sensors use single mode fibers and

----

[1]School of Electrical Engineering, Purdue University, West Lafayette, IN 47907-1285

interferometric techniques to measure strain. These techniques offer the highest sensitivity at the expense of increased complexity and cost. The simplest types of optical fiber sensors use multimode fibers with LEDs and simple PIN detectors to monitor intensity variations (Harmer 1979). These intensity sensors use the concept of microbending, whereby small bends are induced in the fiber by the force to be measured, and the decreased light intensity at the fiber output gives a measure of the applied force (Asawa et al. 1982). However, the external microbending structures required to induce bending in the fiber also cause drift, hysteresis, and other problems. In addition, the microbending structures surrounding the sensitive part of the fiber increase the size and weight of the sensor and may therefore restrict its use in certain applications. If corrugation structures can be built onto and become integral parts of fibers, no external structure would be needed. The aforementioned problems can then be greatly reduced, if not eliminated (Weiss 1989; Vaziri et al. 1992). The strain sensors described in this work have topological structures etched directly onto the fiber cladding, thus reducing the sensor's complexity and bulk, as well as eliminating hysteresis and drift problems. These etched fiber sensing elements may be used as a basis to build simple and reasonably sensitive sensors.

Optical fibers have also been shown to be useful as accelerometers (Tveten et al. 1980; Kersey et al. 1982; Bucholtz et al. 1986; Gerges et al. 1989; Freal et al. 1987). Many of these studies have demonstrated accelerometers with sensitivities of less than 1 μg (1 g=9.8m/s^2). Interferometric techniques are usually employed to reach this level of sensitivity. These techniques require costly electronics and optics, and may also introduce problems with increased susceptibility to unwanted environmental disturbances. Fiber optic accelerometers based on microbending have also been shown to be highly sensitive (Freal et al. 1987). However, the bulky structure required to induce microbending may be troublesome in certain applications. An accelerometer with etched fiber elements overcomes these problems, yet still provides sensitivity to a few mg.

In the following sections, the basic principles underlying the operation of etched fiber sensing elements as a strain gauge or as an accelerometer are discussed. This is followed by a detailed description of the experimental investigations. In the conclusion section, comparison with existing strain gauges and possible applications are discussed.

Etched Fiber Strain Sensing Element

A schematic diagram of a sensor with an etched fiber sensing element is shown in Figure 1. A LED is used as the light source and a PIN diode as the light detector. A section of multimode fiber is subjected to tensile stress.

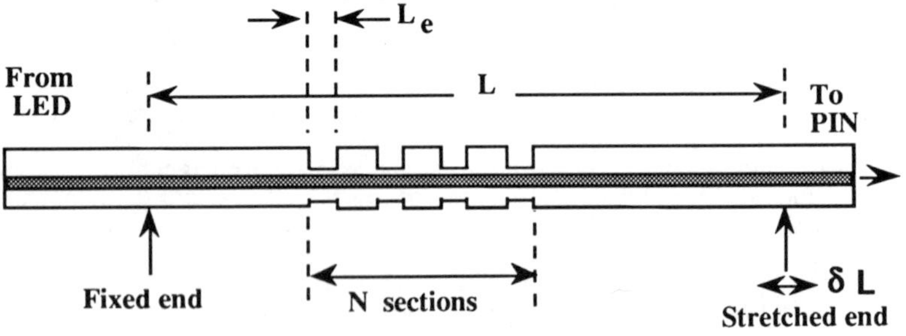

Figure 1. Optical fiber strain gauge with an etched fiber sensing element.

When the sensing element is stretched, transmitted power is reduced. By monitoring the change of power transmitted through the fiber, the strain on the sensing section is inferred. The attenuation of optical signal is caused by the change of mode volume and lateral bending when the etched fiber sections are stretched. The mode volume change is due to the changes in the core index and core radius caused by stretching. The lateral bending is also induced by axial stretching if the etched structures are asymmetric. In the presence of lateral bending, the sensitivity is greatly enhanced.

Operation of Fibers with Symmetric Structures

Qualitatively, the operation of fiber sensing elements with symmetric structures may be understood as follows. When a fiber is stressed longitudinally, the increase in length is accompanied by a reduction in the cross sectional area. The refractive index also changes as a result of the photoelastic effect. The changes in the core radius and index lead to a change in the mode volume. Since the etched sections have a smaller cross section, the strain in the etched sections is larger. The radius and index changes in etched sections are larger than those in unetched sections. As a result, the change in the mode volume in etched sections is larger than that of unetched sections. There is a power loss in joining fibers with different mode volumes. This is the principle underlying the operation of etched fiber sensing elements with symmetrically etched structures.

Quantitative considerations proceed in three steps. First we relate the mode volume change to changes of the core index and radius which are due to the strain applied to the fiber. Next we consider the power loss in joining two fibers with different mode volume. Lastly we estimate the difference of the tensile strain components due to the difference in the cladding radii. Throughout our discussions, graded index fibers with a core radius a and a cladding radius b are considered. The core and cladding indices are n_{co} and n_{cl} respectively. We label the core radii of the etched and unetched sections as a_e and a_u, and the cladding radii as b_e and b_u. When a longitudinal force F is applied to the fiber, the tensile strain component in the etched and unetched sections are $F/(\pi b_e^2 Y)$ and $F/(\pi b_u^2 Y)$ where Y is the Young's modulus. Consider a sensing element with a total length L (Figure 1). There are N etched segments and each has a length L_e. The total length of etched and unetched fiber sections are NL_e and $L-NL_e$. In terms of the parameters introduced, the total power loss due to N fiber junctions is (Vaziri et al. 1992)

$$\frac{\delta P}{P} = - NQ\,[n_{co}^2 (p_{11} - 2\,\sigma\, p_{12}) + 1\,]\,\frac{\delta L}{L} \tag{1}$$

where p_{11} and p_{12} are the photoelastic constants , σ the Poisson's ratio of the fiber and Q is a geometry enhancement factor

$$Q = \frac{L\,(b_u^2 - b_e^2)}{Lb_e^2 + NL_e\,(b_u^2 - b_e^2)}\ . \tag{2}$$

Following the customary practice, we use the gauge factor as a measure of the sensor sensitivity. Suppose the gauge output V is changed by δV when a strain $\delta L/L$ is applied, then the gauge factor is defined as

$$G = \frac{\delta V}{V} \Big/ \frac{\delta L}{L}\ . \tag{3}$$

For optical fiber strain gauges, the output of the photodetector is proportional to the optical power P. Thus the gauge factor of the optical sensing element with N etched segments is

$$G = \frac{\delta P}{P}\frac{L}{\delta L} = -NQ\,[n_{co}^2 (p_{11} - 2\,\sigma\, p_{12}) + 1\,]\ . \tag{4}$$

For silica fibers, $n_{co}^2 (p_{11} - 2\sigma p_{12}) + 1$ is about 1.06. So the gauge factor is essentially the product of the number of etched sections and the geometry enhancement factor. As a numerical example, we consider etched fiber sensing elements with L = 6.4 cm and $b_e / b_u = 0.5$. Figure 2 depicts the gauge factor as a function of number of etched segments for four values of L_e. Initially, G increases linearly with N. Then, it levels to a value of $1.06(L / L_e)(b_u^2 - b_e^2)/(b_u^2 + b_u^2)$ as N approaches $L/(2L_e)$. Also superimposed on the graph are dots corresponding to the measured gauge factors for L_e=1.25 mm. Calculations also reveal that as b_e/b_u increases, the sensitivity decreases drastically. For a fixed number of etched segments N, the gauge factor decreases as L_e increases.

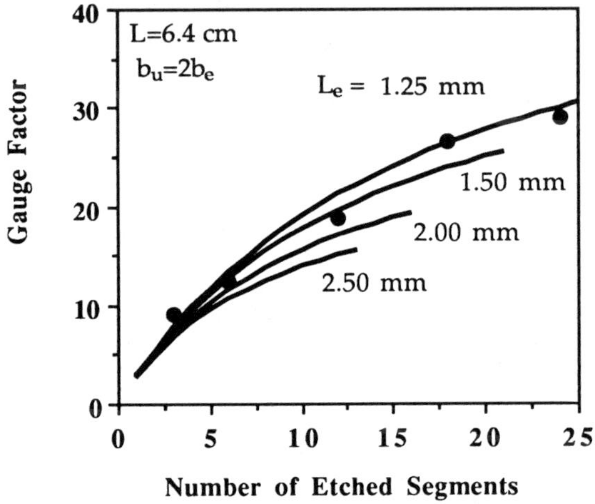

Figure 2. Gauge factor as a function of the number of etched sections. Experimental results are shown as dots.

Operation of Fibers with Asymmetric Structures

Strain sensing capability of asymmetrically etched multimode optical fibers is primarily the result of two physical mechanisms: lateral bending and mode volume mismatch. Structural analysis, to be presented in the next section, demonstrates that when fibers with asymmetric structures are stretched longitudinally, small lateral bends are induced in the stretched section. This is depicted schematically in Figure 3. Both the mode volume mismatch and the lateral bending lead to the reduction of the transmitted power. For asymmetrically etched fibers, effects due to lateral bending are much stronger than the effects of the mode volume mismatch. In the following sections, we will concentrate mainly on the effects of fiber bending on the sensitivity of asymmetrically etched optical fiber strain sensors.

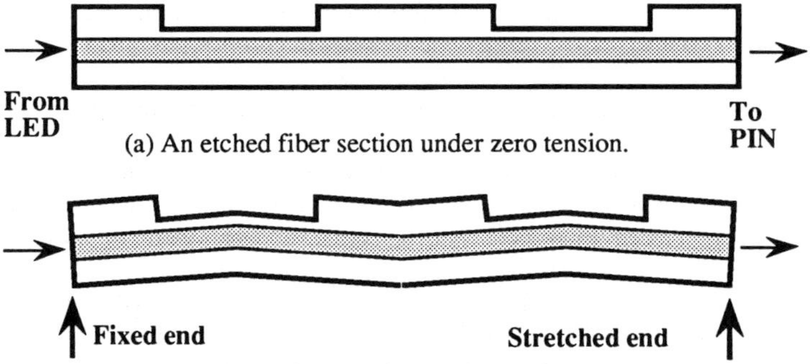

(a) An etched fiber section under zero tension.

(b) An etched fiber section under tension.

Figure 3. Lateral bending of an etched fiber under tension.

Microbending Effect in multimode fibers

Since fiber bending is the main factor affecting the power loss in etched fiber sensing elements with asymmetric structures, the physical mechanisms involved in these sensors are similar to those involved in microbend sensors (Harmer 1979; Asawa et al. 1982). Microbending refers to a large number of infinitesimal random deformations of fibers. According to Marcus' coupled mode analysis, supported by many experimental studies (Marcus 1973; Lagakos et al. 1987), the power loss increases sharply when the distribution of random distortions correlates with a characteristic distance of the fiber. The characteristic distance is commonly referred to as the critical periodicity of a fiber. For step index fibers, the critical periodicity is $\Lambda_c=\sqrt{2}\,\pi a\, n_{co}/\,NA$ and for graded index fibers $\Lambda_c=2\pi a\, n_{co}/\,NA$ where NA is the numerical aperture of the fiber (Lagakos et al. 1987).

We use a ray tracing technique to relate the effects of the spatial periodicity Λ, number of bends N, and the bending amplitude on the sensitivity of fiber sensing elements. We begin with a large number of rays with equal power emanating from the entrance to a waveguide section and trace the rays through the distorted waveguide section until they exit from the waveguide. These rays are reflected and refracted at the waveguide boundaries. Power loss due to each reflection is calculated. The total power arriving at the output plane is calculated. Figure 4 shows the loss versus Λ for a waveguide with 10 sinusoidal bends and a bend amplitude of 0.5 µm.

For a step-index waveguide with a core index of 1.4884, a numerical aperture of 0.2, and a waveguide width of 50 µm, the calculated peak attenuation is 30% and it occurs when Λ is near 800 µm. For the particular waveguide studied, the critical periodicity Λ_C is 827 µm. In other words,the peak attenuation occurs at $\Lambda/\Lambda_C = 0.96$. To accentuate the role played by the critical periodicity, we plot the attenuation as a function of Λ/Λ_C. We note that the attenuation is greater than 15%, i.e., half of the peak attenuation, for Λ/Λ_C in the range of 0.75 and 1.2. This range may be viewed as the "width" of the peak attenuation.We have also studied the effects of the number of bends and the bend amplitude on the attenuation with $\Lambda=\Lambda_C$. The attenuation increases monotonically with the number of bends and the bend amplitude. Further studies showed that by increasing the number of bends, the "width" of the attenuation peak narrows. These results are in agreement with the coupled mode analysis (Marcus 1973).

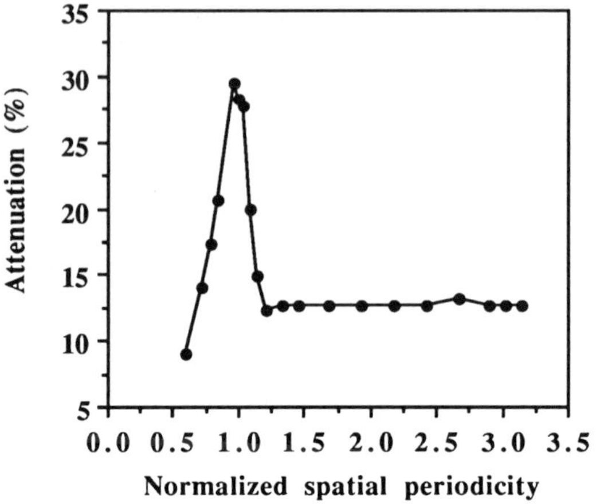

Figure 4. Bending loss dependence on the spatial periodicity.

Structural Analysis of Asymmetrically Etched Fibers

We have used two software programs to investigate the structural deformations of waveguides with asymmetric structures under tensile strain. The programs (SSTAN 1987; ANSYS 1990) are written for the structural analysis of elastic beams and they are based on finite-element methods. The results of these two programs were almost identical. We model the etched optical fibers as elastic beams with square cross-sections and assume sharp transitions from etched to unetched sections. The equivalent depth of etching is chosen such that the amount of the material removed from a fiber is the same for a square beam.

First, we studied the effects of the location of etched sections on the waveguide deformation and found that the bend amplitude is the largest when the etched section is located at the center of the stretched section. Next, we investigated the profiles of the induced deformation as a function of the spatial periodicity. Figure 5 is a plot of the induced deformations of a waveguide with five etched sections and a fixed length (L=16.2 mm). To display the structural deformations graphically, the lower boundary of the waveguide is plotted. It should be noted that the scale of the vertical direction is in μm while that in the horizontal direction is in mm. It is interesting to note that all bends have the same bend amplitude when $L \sim 5\Lambda$. The maximum lateral displacement occurs at the center of each etched section. Similar results were obtained for waveguides with 2, 3, ... 9 etched sections. In all the cases studied, uniform bends are obtained when $L=N\Lambda$. These lateral deformations may be compared to the vibration of elastic strings under harmonic excitation.

In general, the bend amplitude is a function of N, Λ, etching depth, width W of the etched sections and a relative etching width defined as $R=W/\Lambda$. However, we found that when the etched sections are evenly spaced in the sensing section, the bend amplitude is independent of N. We have also studied how the bend amplitude changes as the applied strain varies. It is satisfying to note that the bend amplitude increases linearly with the strain.

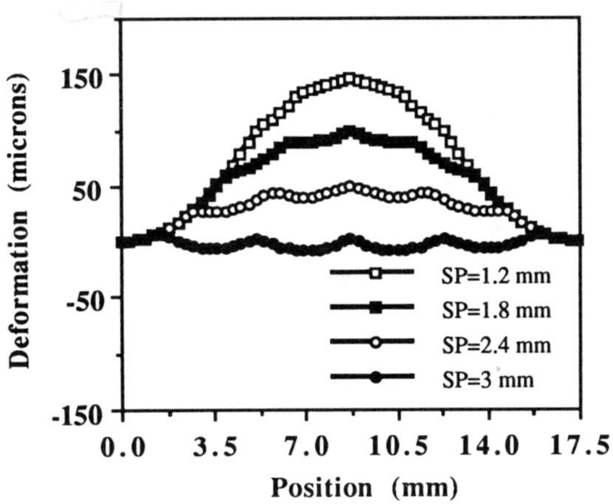

Figure 5. Induced deformations for four spatial periodicities (SP) with N=5.

Experimental Investigations

Standard integrated circuit techniques were applied to form topological features directly onto the fiber cladding (Vaziri et al. 1992). Throughout our experiments, multimode fibers with core and cladding diameters of 52 and 125 µm, respectively, are used. A typical etched fiber section with symmetric etched structures are shown in Figure 6.

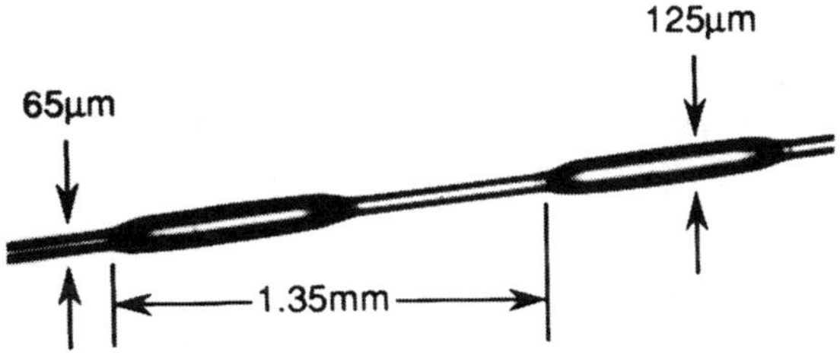

Figure 6. Photograph of a symmetrically etched fiber section.

To test a sensing element, the portion of fiber containing the etched sections was identified. One end of the fiber section was glued on a stationary stage and the other end was fixed to a movable stage equipped with a micrometer. A LED and a PIN diode were used as the light source and photodetector, respectively. As the micrometer was turned, the fiber was stretched and the output of the photodetector was monitored.

Strain Sensing with Symmetric Etched Optical Fibers

A typical response of a strain sensing elements with ten etched sections is shown in Figure 7. For the tested sensor, the spatial periodicity of the etched structure was 1.25 mm and the diameter of the etched and unetched sections were 65 μm and 110 μm, respectively.

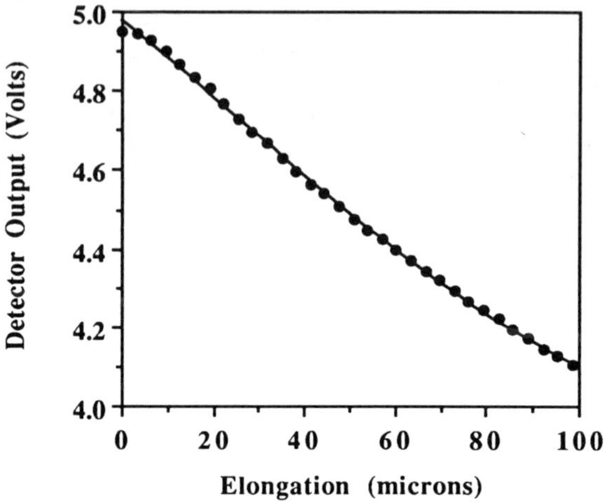

Figure 7. Typical response of a symmetrically etched optical fiber strain sensor.

The effect of the cladding diameter on the sensitivity of the sensing elements was studied to find the optimum cladding diameter of the etched sections. Our measurements showed that sensing elements with etched cladding diameter of less than 58 μm and larger than 75 μm were not very sensitive. On the other hand, samples with etched cladding diameters about 63 μm showed the highest sensitivities. We also investigated the dependence of the gauge factor on the number of the etched segments and found that the gauge factor increases with increasing number of etched sections. More importantly, the responses of the sensing elements tested were repeatable and had no measurable hysteresis. These sensors can be used to detect strains as large as 0.4%. An optical fiber strain gauge was tested for its cyclic endurance. After straining the sensing element more than 13000 times, no measurable degradation was observed.

Strain Sensing with Asymmetrically Etched Optical Fibers

Static response

The transmission of optical power through asymmetrically etched sensing elements decreases linearly as the strain increases. To investigate the effect of the spatial periodicity on the gauge factor, several samples with the same number of etched sections (N=10), identical etching depths, R=0.4, and the same tensile strain were tested. However, the spatial periodicities Λ for these sensing elements were different. As shown in Figure 8, a maximum gauge factor is observed when the spatial periodicity of the etched structure is about 1.25 mm. This spatial periodicity is very close to the theoretical critical periodicity of the fiber being tested (Λ_c=1.22 mm). These results are in agreement with the numerical predictions presented in previous sections.

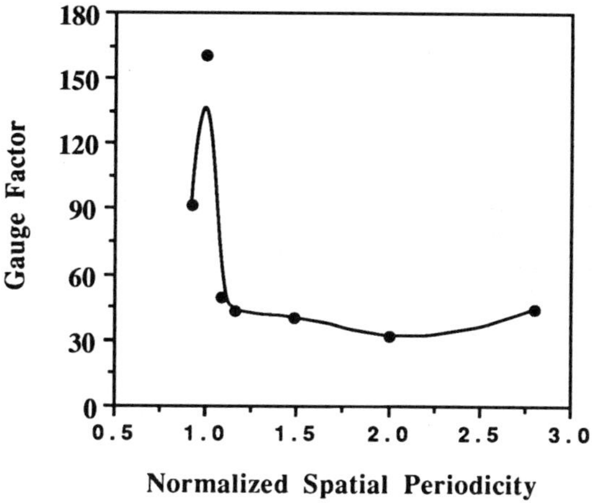

Figure 8. Gauge factor vs. spatial periodicity of the asymmetrically etched structure.

The effect of etching depth on the gauge factor was also studied experimentally. The results show that the gauge factor increases linearly with increasing etching depth. The maximum etching depth is limited by the cladding diameter of the original fiber and the minimum cladding diameter required to confine the optical fields in the core. For the multimode graded-index fiber used in our experiments, the maximum etching depth is about 30 μm. For such an etching depth, a gauge factor of 170 was obtained.

Dynamic response

To measure the dynamic response of asymmetrically etched optical fiber strain sensors, an actuator was used to stretch a sensing element periodically. To monitor the displacement of the actuator, a separate non-contact optical fiber displacement sensor was built and used. The basic components of the displacement sensor are a 3-dB optical fiber directional coupler and a small mirror attached to the actuator. The same fiber used to illuminate the mirror is also used to receive the reflected light. The output of the etched fiber strain sensor was monitored while the response of the displacement sensor was held constant. The frequency response of a sensing element with a gauge factor of 100 is shown in Figure 9. Also shown in the figure is the response of the displacement sensor. Both responses have been normalized for the purpose of comparison. As noticed in the figure, the frequency response of the etched fiber remains flat up to 2.7 kHz. This upper limit is set by the lowest mechanical resonance frequency of the setup.

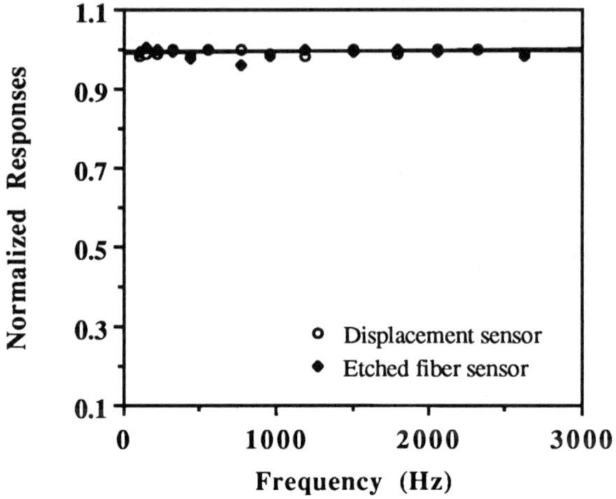

Figure 9. Frequency response of an asymmetrically etched optical fiber strain sensor.

Etched Fiber Accelerometer

A simplified experimental structure was used to demonstrate the principle of operation of the accelerometer. The experimental setup is shown in Figure 10. The fiber optic structure was modeled after that given in (Tveten et al. 1980), but in the present case we use asymmetrically etched multimode fiber rather than (unetched) single mode fiber to take advantage of the simplicity demonstrated in our etched fiber strain sensors. When the device is accelerated in the axial direction, the motion of the mass m induces a strain in the fiber, thus creating lateral bends and causing optical loss in the fiber. The change in detector voltage gives a measure of the acceleration a of the structure:

$$a = \frac{Y \pi b^2}{mG} \frac{\delta V}{V} \qquad (5)$$

where πb^2 is the cross-sectional area, m is the mass, G is the gauge factor, and $\delta V/V$ is the relative change in detector voltage. For our fiber accelerometer, $Y=7\times10^{10}$ N/m^2, m=2.7 grams and G is about 60. The complexity of the electronics can determine the sensitivity of the accelerometer. A very simple setup with just a LED source and PIN detector with no signal processing electronics can give sensitivities of about 5 mg.

The fiber is given an initial strain of about 0.001 while attaching it to the structure. This allows operation well within the range of the strain sensor, and gives a transverse fundamental frequency of about 1.5kHz. The frequency range is limited by the longitudinal resonance frequency given by

$$f_L = \frac{1}{2}\sqrt{\frac{Yb^2}{\pi mL}} \qquad (6)$$

where b is the cladding radius and L is the length of the structure. In our case, L=5 cm and f_L=385 Hz. Useful operation is restricted to below this resonance.

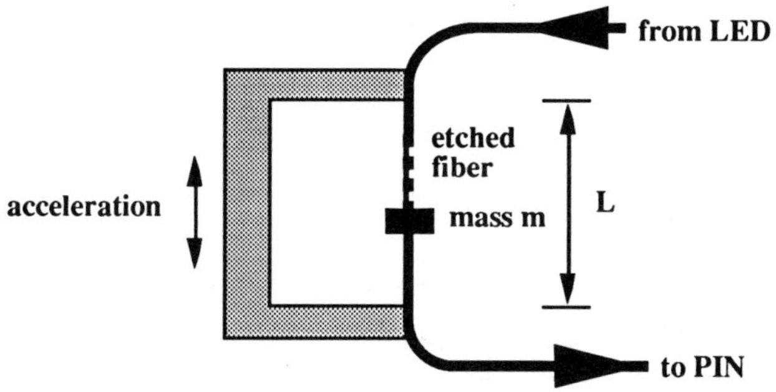

Figure 10. Setup for acceleration measurements.

The acceleration was produced by driving an actuator with a sinusoidal voltage. The actuator was calibrated using the same non-contact displacement sensor used previously in the dynamic response measurements. Figure 11 shows the response of the accelerometer at frequencies of 100, 200, and 300 Hz. The detector voltage is shown to be linear with applied acceleration with slopes of 6.1, 7.1, and 7.7 mV/g respectively. The precision of the actuator and the simple electronics limited our measurements of the minimum sensitivity of the accelerometer. Nevertheless, the operation principle and the linearity have been demonstrated.

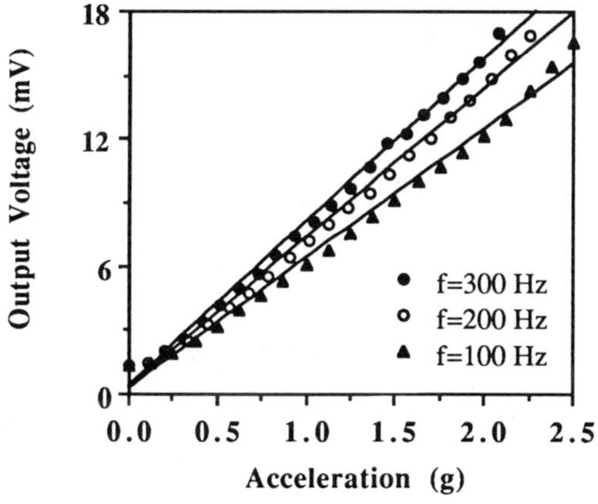

Figure 11. The accelerometer response at frequencies below resonance.

Comparison with Resistive Strain Gauges

Electrical resistive strain gauges are the most widely used strain sensors (Kobayashi 1987; Dally 1978). Before comparing the performance of the optical fiber strain sensing elements with that of resistive strain gauges, it is appropriate to review the performance of this well-accepted strain gauge technology. An ideal strain gauge should have many of the following characteristics:

- small size and light weight;
- high strain sensitivity;
- negligible reaction of gauge on substrate, small stiffness;

- stability of calibration with time and dynamic loading;
- immunity from temperature, humidity and other environmental fluctuations;
- insensitive to acceleration;
- high temporal resolution and sensitive to high frequency dynamic strains;
- ease of attachment, calibration and remote readout;
- low cost.

All electrical resistors are strain sensitive. When a resistor is stretched, its length is increased and the cross-section reduced. Thus, when a resistor is stretched, its electrical resistance is increased. The variation in the electrical resistance is a measure of the applied strain. There are two types of resistive strain gauges: metallic and semiconductor strain gauges.

Metallic strain gauges are of the form of small wires or thin foils. Wire strain gauges are used for simple uniaxial strain measurements and foil strain gauges for biaxial, shear, or more complicated strain determination. Although some metallic strain gauges may be strained to 10%, their useful and reliable range is typically 0.2% (Webster 1988). The minimum gauge length of a metallic strain gauge may be as small as 0.2 mm. However, it has been reported that gauges shorter than 3 mm have a lower maximum elongation, reduced fatigue life, less stable behavior, and greater installation difficulty (MG 1988). Heat dissipation is another problem of short resistive strain gauges. Although the operation and the construction of metallic strain gauges are simple, strain measurements, in practice, are not routine. The most important factor limiting the accuracy of metallic strain gauges is their low gauge factor, typically about two. The change in the resistance of a typical metallic strain gauge with a resistance of 200 Ω is about 0.2 mΩ for an applied strain of 10^{-6}. To measure such a small resistance change, a Wheatstone bridge followed by an amplifier is typically used. Another limiting factor of metallic strain gauges is their nonlinear response to the temperature change. This not only limits their useful temperature range, but also sets an upper limit on the voltage and current applied to the Wheatstone bridge and the detecting circuit. If the applied signal is too strong, heat dissipation in the strain gauge could cause a change in the gauge resistance even in the absence of applied strain. Drift and hysteresis could also cause problems for metallic strain gauges. In strain gauge literature, drift is referred to as the zero shift. A zero shift could be induced by a cyclic strain, static strain over time, or temperature change. Thus, for an accurate determination of strain, it is necessary to reset the bridge circuit prior to each measurement.

There have been many efforts to minimize or correct the problems affecting the performance of the resistive strain gauges (MG 1988; Dean 1962). Attempts in solving these problems have resulted in the design of many alloy strain gauges. The most frequently used alloys are constantan, annealed constantan, isoelastics, and nichel-chromium. Each alloy is designed for specific strain measurements and has problems of its own. To date, no all-purpose strain gauge has been realized.

Semiconductor strain gauges are very sensitive and have gauge factors ranging from 50 to 200. However, their response is highly nonlinear and temperature dependent. These problems seriously limit the application of semiconductor strain gauges to measurements where metallic strain gauges cannot be used due to their minimum length limitation.

The optical fiber strain gauges reported in the present work have many advantages over the resistive or semiconductor strain gauges. The gauge factor of the sensing elements is much larger than those of wire and foil strain gauges. In addition, the fiber optic sensing elements do not need an inert atmosphere to realize the optimum sensitivity and their response is not affected by moisture in air (MG 1988). Other notable features include the linearity, repeatability, and lack of hysteresis. Since optical fiber components are inherently immune to electromagnetic interference or noise, no sophisticated electronic equipment is needed for signal detection. Clearly, these sensing features make the etched fiber sensing elements ideal in many applications.

Summary

Optical fiber strain sensors with etched structures have been fabricated and tested. These sensors are based on commercially available multimode fibers. The main mechanisms responsible for the observed sensitivity are the formation of lateral bends in the sensing sections and the mode volume mismatch when the fiber is stretched. Structural analysis shows that when the etched sections are distributed uniformly in the sensing region, uniform bends are formed when the fiber is stretched. The optimum ratio of the etched width to the spatial periodicity is about 0.4. A ray tracing study shows that etched fibers with spatial periodicities close to the critical periodicity of the fiber are most sensitive to longitudinal strains. Deeper etching depth also leads to an improved sensitivity. The results are in qualitative agreement with the experimental studies.

The highest gauge factor measured is 170. This should be compared with a gauge factor of 1 to 3 for typical metallic strain gauges. Our experimental investigations show that the dynamic response of these sensors is flat to at least 2.7 kHz. Since the etched fiber sensing elements have no backing materials, the frequency response of etched fiber sensors is limited only by the elastic resonant frequency of the bare fiber itself. Based on the elastic properties of silica fibers, we estimate that the etched optical fiber sensors are useful up to 20 kHz. Because of their wide frequency range and the high gauge factor, the etched fiber sensing elements are ideal for monitoring vibrations. They can also be embedded in and become parts of intelligent structures or smart skins.

Acknowledgement

This project has been supported by NSF under Grant CDR 8803107 to ERC for Intelligent Manufacturing Systems.

References

ANSYS, (1990), revision 4.4, Swanson Analysis Systems, Inc., PO Box 65, Johnson Road, Houston, PA 15342-0065.

Asawa, C. K., Yao, S. K., Stearns, R. C., Mota, N. L., Downs, J. W., (1982), "High sensitivity fiber optic strain sensors for measuring structure distortion." *Electron. Lett.*, 18(9), 362-365.

Bucholtz, F., Kersey, A. D., and Dandridge, A., (1986),"DC Fibre-Optic Accelerometer with Sub-micro g Sensitivity." *Electron. Lett.*, 22(9), 451-453.

ICE, (1981), "Sensors in Highway and Civil Engineering." Conf. Proc., The Institution of Civil Engineers, London.

IMC, (1987), "Measurement of Static and Dynamic Parameters of Structures and Materials," Conf. Proc., International Measurement Confederation (IMEKO), Pizen, Czechoslovakia.

Culshaw, B., (1984), *Optical fibre sensing and signal processing,* Peter Peregrinus Ltd., London, UK.

Dally, J., (1978), *Experimental Stress Analysis*, McGraw-Hill.

Dean, M., (1962), *Semiconductor and Conventional Strain Gauges*, Academic Press.

Freal, J. B., Zarobila, C. J., and Davis, C. M., (1987), "A Microbend Horizontal Accelerometer for Borehole Deployment." *IEEE J. Lightwave Tech.*, 5(7), 993-996.

Gerges, A. S., Newson, T. P., Jones, J. D. C., and Jackson, D. A., (1989), "High-sensitivity fiber-optic accelerometer." *Optics Lett.*, 14(4), 251-253.

Giallorenzi, T.G., Bucaro, J.A., Dandridge, A., Sigel Jr., G.H., Cole, J.H., Rashleigh, S.C., Priest, R.G., (1982), "Optical fiber sensor technology." *IEEE J. Quantum Electronics*, 18, 626-665.

Harmer, A. L., (Aug. 7, 1979), "Optical strain gauge." US Patent, 4 163 397.
Kersey, A. D., Jackson, D. A., and Corke, M., (1982), "High-sensitivity Fibre- Optic Accelerometer." *Electron. Lett.*, 18(13), 559-561.
Kobayashi, A., (1987), *Handbook on Experimental Mechanics*, Prentice-Hall.
Lagakos, N., Cole, J. H. and Bucaro, J. A., (1987), "Microbend fiber-optic sensor." *Appl. Opt.*, 26(11), 2171-2180.
Marcus, D., (1973), "Losses and impulse response of a parabolic index fiber with random bends", *B.S.T.J.*, 52(8), 1423-1437.
MG, (1988), Measurement Group, Micro-Measurement Division, Technical Data, Part B.
SSTAN, (1987), Marc Hoit, Department of Civil Engineering, University of Florida, Gainesville, FL 32611.
Tveten, A. B., Dandridge, A., Davis, C. M., and Giallorenzi, T. G., (1980), "Fiber Optic Accelerometer." *Electron. Lett.*, 16(22), 854-856.
Webster, J.G., (1988), *Tactile Sensors for Robotics and medicine*, Wiley.
Weiss, J. D., (1989), "Fiber optics strain gauge", *IEEE J. Lightwave Tech.,* 7, 1308-1318.
Vaziri, M. and Chen, C.-L., (1992), "Etched Fibers as Strain Gauges," *IEEE J. Lightwave Tech.*, 10, 836-841.

Interpretation of Embedded Optical Fiber Sensor Signals

James S. Sirkis[1]

1 Abstract

This paper attempts to provide a fundamental basis for interpreting the optical signal recovered from structurally embedded optical fiber sensors. The series of transformations that allow one to infer data indicative of structural performance from the scalar optical data provided by embedded sensors is discussed. The most challenging technological barriers that must be overcome in order to achieve these transformations are emphasized during the ensuing discussion. Specific topics that are emphasized include forward and inverse phase-strain-temperature relations, the difference between fiber and structure measurands, the uniqueness and stability of inferred structural data, and the concepts of transverse strain sensitivity and thermal apparent strain.

2 Introduction

Optical fiber sensors have been embedded in structural elements since the early 1980's in hopes of inferring information regarding the internal stress, strain, or temperature states of the structure [Claus and Cantrell, 1980]. Much of the research effort in the years that followed concentrated on the development of sensors, signal recovery techniques, specialty fibers and coatings, and embedding procedures. All of this technology is intended to produce sensor systems that will measure "stress," "strain," "temperature," or any other measurands that can provide some indication about the overall state of the host structure. The discussion in this paper concentrates on monitoring the temperature and strain state of the host material, but the concepts are generally applicable to other scalar and tensor quantities. By strain and temperature *state* we mean the full thermomechanical strain tensor (thermal plus mechanical strains) and the temperature at a given instant in time. The discussion is further limited to only interferometric and Bragg grating optical fiber sensors since these sensors dominate "smart structures" applications.

Interpretation of signals produced by structurally embedded optical fiber sensors experiencing general thermomechanical loading is complicated because the fiber is subjected to a strain state, which is a tensor quantity, in addition to the scalar temperature field. Therefore, the scalar signal produced by an embedded optical fiber sensor has contributions from a minimum of four quantities

[1]University of Maryland, Department of Mechanical Engineering, College Park, Maryland 20742.

(three principal strains and the temperature). Interferometric optical fiber sensors; therefore, do not measure "strain" or "temperature," rather they measure optical phase shifts from which the strain and temperature states are inferred. This brings to light the first hurdle facing structurally embedded optical fiber sensors. A single scalar output is linearly proportional to a minimum of four independent measurands. The inversion of this relationship is in general non-unique, and it is this non-uniqueness that leads to classic instances of transverse strain and thermal apparent strain sensitivity.

Let us put aside for the moment the uniqueness issue of inverting the so-called phase-strain-temperature relationships. For the sake of argument, assume that a unique principal strain and temperature state can be inferred from the optical fiber signal. The next question that must be addressed is "to which spatial location does this inferred data correspond?" The answer, of course, is in the optical fiber core (the fiber core diameter is generally small enough that it can rightfully be considered a single spatial location). We will simply refer to this location as *in the fiber* . The stated goal of embedded sensors is not to measure strain and temperature states in the fiber, but rather, to monitor the same quantities in the host structure. Therefore, the strain and temperature state in the fiber must somehow be related to the strain and temperature state in the host material. The temperature relationship presents little difficulty, but the strain state relationship is quite complex. The optical fiber is an elastic inclusion, therefore the strain in the host immediately adjacent to the optical fiber is different from that in the fiber; and even worse, the host strain state is a function of position. Now one is forced to ask, "where in the structure is strain state information required?" Again the question of uniqueness arises. Finally, since the fiber is an elastic inclusion, the local strain state in the host is inevitably altered by the very presence of the optical fiber. We simply accept this perturbation as a natural price of developing *fiber optic smart structures* .

This paper addresses the *forward* relationship between the recorded optical signal (phase changes) and the strain and temperature *state* in the fiber. Methods of creating an *inverse* relationship that will provide a unique state of strain and temperature in the fiber are then described, and the ramifications of using non-unique inverse relations are discussed in detail. This presentation is followed by a discussion outlining the requisite techniques required to infer host strain and temperature state from that of the fiber sensor, and the associated uniqueness issues.

3 Forward Phase-Strain-Temperature Relation

The interferometric optical fiber sensors that are most prominent in the *smart structures* arena include intrinsic and extrinsic Fabry-Perot, Dual-mode, Michelson, Mach-Zehnder, polarimetric, and Bragg grating sensors [Turner et al, 1989; Morey et al, 1989]. With the exception of Bragg grating sensors, all of these sensors record changes in optical phase. The Bragg grating sensor records changes in the Bragg wavelength. For the sake of brevity, optical phase changes will be specifically discussed, but the results directly translate to Bragg wavelength changes [Sirkis, 1993a]. The general form of the forward relationship between the normalized phase change (or Bragg wavelength change) and the strain and temperature state for a straight fiber sensor with a small gage length is [Sirkis, 1993a; 1993b; Sirkis and Haslach, 1990; Sirkis and Mathews, 1993]

$$\Delta\phi(S_j,T)=\beta_o L\{S_1^f-\frac{1}{2}n_o^2(P_{11}+P_{12})(S_2^f+S_3^f)]+\xi T^f\} \quad (1)$$

where β_o, n_o, P_{11}, and P_{12} are the propagation constant, refractive index and Pockels' constants; L is the gage length; S_1^f is the normal thermomechanical strain component tangent to the fiber axis, and S_2^f and S_3^f are orthogonal normal thermomechanical strain components in the plane defined by the 1-direction, T is the temperature, the superscript "f" refers to *in the fiber*, and $\xi = (1/n)(\partial n/\partial T)$ is the thermal-optic constant. This equation was found by neglecting the $S_1^f S_j^f$ [Sirkis and Haslach, 1990] and $\xi S_1^f T^f$ [Farahi et al, 1990] terms since they are both much less than S_1^f . It must be emphasized that S_j^f (j = 1,2,3) is the *sum* of the thermal and mechanical strain states. The exact form taken by the forward phase-strain-temperature relationship depends on the sensor type and state of polarization of the light propagating in the sensor, but all maintain the strain and temperature state dependence. The interested reader should see Sirkis [1993a, 1993b] for a detailed discussion of the many forms in which Eq. (1) can be found. Notice that the "form" of Eq. (1) is defined by the optical system and is invariant to the method in which the structural strain and temperature state is transferred to the fiber. The structural strains and temperature state influence the optical phase change in Eq. (1) by determining the fiber strain and temperature states via thermoelastic formulations for cylindrical elastic inclusions [Sirkis and Mathews, 1993].

It is common practice in the *smart structures* community to simplify Eq. (1) by setting the transverse strains equal to Poisson contractions ($S_2^f = S_3^f = -\nu S_1^f$) producing the so-called Butter and Hocker relation [Butter and Hocker, 1978]. However, this is only valid for surface mounted [Sirkis and Haslach, 1990] or embedded fibers experiencing axis-symmetric, generalized plain strain loading scenarios [Sirkis, 1993b; Sirkis and Haslach, 1990]. One can quickly disprove Butter and Hocker's relation for general use by considering an optical fiber sensor in plane strain diametral compression. According to the Butter and Hocker model the phase shift produced by such a sensor should be zero. Fig. 1 shows the results of just such an experiment where a standard low-birefringence

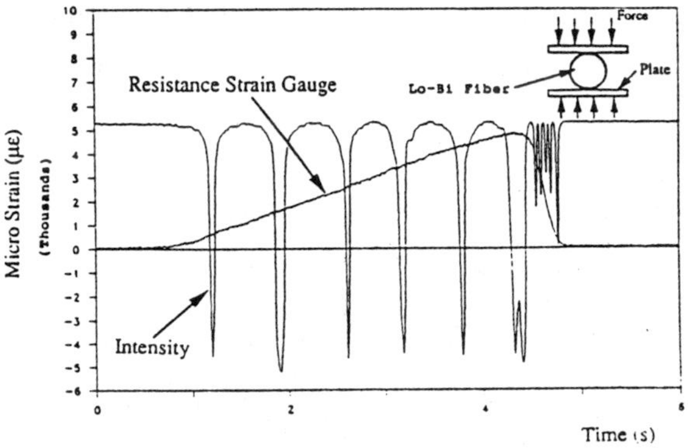

Figure 1. Intensity Trace Produced by a Polarimetric Sensor Subjected to Plane Strain Diametral Compression.

optical fiber is used as polarimetric sensor. The intensity trace clearly shows fringes as the load is applied and released; thereby showing that optical phase change is indeed occuring. In this scenario, the entire response of the embedded sensor is attributable to transverse strain sensitivity. The degree to which this transverse strain sensitivity causes errors in the sensor measurements depends solely on the interpretation of the sensor signal. If, for example, the sensor is considered to contain axial strain information only (Butter and Hocker), then the entire sensor response would be erroneous.

In a similar example, consider the embedded optical fiber sensor depicted in Fig. 2 in which the entire system is subject to uniform thermal loading. For convenience the coefficient of thermal expansion of the host is taken to be larger than that of the optical fiber (this is generally true since the α_f is .5e-6). The scenario in Figure 2 illustrates the thermomechanical nature of the sensor

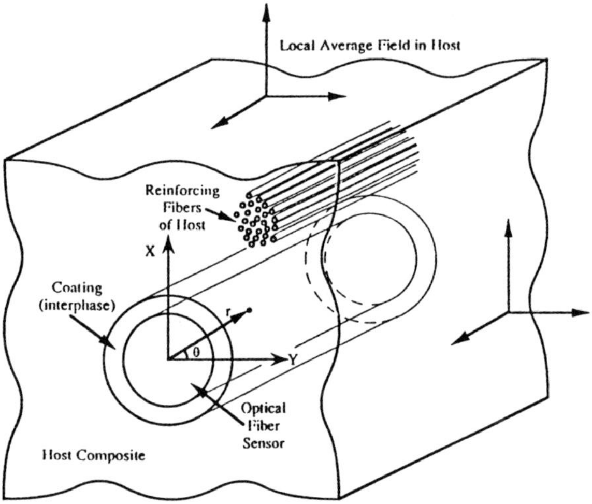

Figure 2. Optical Fiber Embedded in a Fiber Reinforced Composite Material.

response. It is clear that the applied temperature field will induce an optical phase change since a strain state is developed in the fiber and the refractive index changes since it is an independent function of temperature and strain. Whether or not this thermally induced phase change introduces error into the sensor measurements again depends on the interpretation of the sensor signal. If the sensor is believed to be dependent on only mechanical strain, then the entire sensor signal would be erroneous since it would indicate mechanical strain when only thermal loading is present. This affect is known as thermal apparent strain since the fiber sensor apparently responds to a mechanical strain field that is, in actuality, a thermally produced strain and thermo-optic response. Regardless of the sensor interpretation, it is important to understand that the optical fiber strain state is not simply free thermal expansion. Under the temperature field, the optical fiber tries to expand in the radial and axial directions, only to be inhibited by the surrounding host medium. The strains in the fiber therefore contain contributions from the mechanical restraining forces applied by the host. This point is made to emphasize that any attempt to design "self-temperature compensated" (STC) embedded optical fiber sensors must account for the thermomechanical loading of the host on the optical fiber. This means STC sensors will have to be designed with specific host material systems in mind.

4 Interpretation Errors in Embedded Optical Fiber Sensors

The advantage offered by interpreting optical fiber sensor signals by following the Butter and Hocker approach is that a one-to-one relationship between optical phase change and the axial strain in the fiber is provided. This relation can then be directly inverted so that the axial strain can be obtained directly from the measured phase. However since the Butter and Hocker approach to interpreting optical fiber sensor signals is only approximate, it is appropriate to assess the associated error penalty. The error is manifested in transverse strain sensitivity and thermal apparent strain sensitivity, much in the same way that they are manifested in surface mounted resistance strain gages. Unfortunately, a complete discussion of thermal and mechanical apparent strains is beyond the scope of this section due to the many popular optical fiber sensor configurations. Therefore, a more generic presentation will be provided that imparts the important concepts.

4.1 Transverse Sensitivity in Resistance Strain Gages

A convenient introduction to transverse sensitivity and thermal apparent strain error in embedded optical fiber sensors is provided by quickly reviewing transverse strain sensitivity in resistance strain gages. Consider the resistance strain gage subjected to independent longitudinal and transverse in-plane strains shown

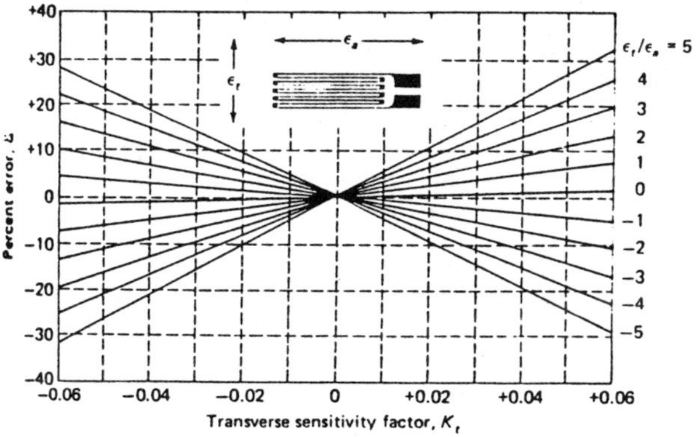

Figure 3. Transverse Strain Sensitivity in Resistance Strain Gages [Dally and Riley, 1991].

in Fig. 3. The change in resistance in this gage is given by

$$\frac{\Delta R}{R} = \gamma_a \varepsilon_a + \gamma_t \varepsilon_t \tag{2}$$

where γ_a and γ_t are the axial and transverse sensitivities of the gage. If one assumes that the gage is sensitive to axial strain (ε_t) only, then standard error analysis techniques can be used to derive the error associated with this gage [Dally and Riley; 1991]. The right side of Figure 3 shows the fruits of this analysis, where the transverse sensitivity factor is defined as γ_t/γ_a. It is clear from this result that even surface mounted resistance strain gages are susceptable to significant transverse sensitivity errors if the biaxality ration ($\varepsilon_t/\varepsilon_a$) is large. Similar arguments hold for thermal apparent strain [Dally and Riley, 1991]. Interestingly, surface mounted optical fiber sensors are immune to the transverse strain sensitivity error described by Eq. (2) [Sirkis and Haslach, 1990].

4.2 Thermal Apparent Strain and Transverse Sensitivity Of Embedded Optical Fiber Sensors

It is possible to follow resistance strain gage concepts to devise meaningful quantifications of errors due to transverse strain sensitivity and thermal apparent strain sensitivity. As with resistance strain gages, this approach requires certain preconceived notions regarding how an embedded optical fiber sensor should respond to general thermomechanical loading. The response of embedded phase-based optical fiber sensors can be generalized from Eq. (1) to

$$\delta = \frac{\Delta\phi}{\phi_o} = F_a S_1 + F_{t2} S_2 + F_{t3} S_3 + F_T T \tag{3}$$

where

δ = normalized phase change of the sensor
ϕ_o = absolute phase retardation in a load free state
F_a = sensitivity of the sensor to axial strain
F_{t2} = sensitivity of the sensor to transverse strain in the 2-direction
F_{t3} = sensitivity of the sensor to transverse strain in the 3-direction
F_T = sensitivity of the sensor to temperature

Expressions for these sensitivities in terms of optical properties are provided in [Sirkis, 1993b] for the seven popular phase based sensor types. For convenience, all sensitivities are normalized by F_a so that the sensor response becomes

$$\delta = F_a(S_1 + K_{t2} S_2 + K_{t3} S_3 + \gamma T) \tag{4}$$

where K_{t2}, K_{t3}, and K_T are respectively the two transverse strain sensitivity factors and the thermal apparent strain sensitivity factor. Again, Eq. (4) emphasizes that fact that the actual sensor response is dependent on more that just the axial strain. If one follows the precepts used by the resistance strain gage community, then embedded optical fibers are assumed (or more precisely, desired) to respond to axial strain only (Butter and Hocker approach). Using this approach, it is possible to define the normalized phase of the *ideal* optical fiber sensor as

$$\delta' = FS_1 \tag{5}$$

where the superscript prime signifies the ideal sensor and F is the ideal axial sensitivity. The value of F will depend on the sensor type and calibration procedure (Sirkis, 1993b). Errors due to transverse strains or thermal loading can then be quantified by simple finding the relative error between the actual and ideal sensor response

$$E = 100 \times \frac{\delta - \delta'}{\delta'} = 100 \times ((F_a - F) + K_{t2}(S_2/S_1) + K_{t3}(S_3/S_1) + \gamma(T/S_1)) \quad (6)$$

This expression reveals that the sensor error is a function of 1) the difference between the actual and the calibrated axial sensitivities, 2) the transverse strain sensitivity factors, 3) the thermal apparent strain sensitivity factor, and 4) the loading ratios S_2/S_1, S_3/S_1, and T/S_1. The form of Eq. (6) is similar to the error equations developed for resistance strain gages and quantifies the amount of error that is attributable to phase changes caused by any loading condition other than axial strain. Eq. (6) would predict, for example, an infinite error for the plane strain loading in Fig. 1 since S_1 is zero by definition. This simply means that even though the entire sensor signal is assume to be directly related to the axial strain (Eq.(5)), no portion of the phase change is actually caused by axial strain.

It is clear from the definition in Eq. (6) that error can be quantified only after 1) the sensor type is specified so that F_a, K_{t2}, K_{t3}, and γ can be calculated, 2) the calibration procedure is specified so that F can be calculated, and 3) the load regime of operation is specified so that limits can be placed on the loading ratios $S_2/S_1, S_3/S_1$, and T/S_1. Further, the strains in Eq. (6) have contributions from both the thermal and mechanical loads so that individual definitions for thermal apparent strain error and transverse strain sensitivity error can be derived [Sirkis, 1993b]. For the sake of example, consider an intrinsic Fabry-Perot optical fiber embedded parallel to the reinforcing fibers in a transversely isotropic laminated composite host material (Fig. 2). Let the calibration consist of isothermal uniaxial loading in the fiber direction [Sirkis, 1993b]. Fig. 4 shows

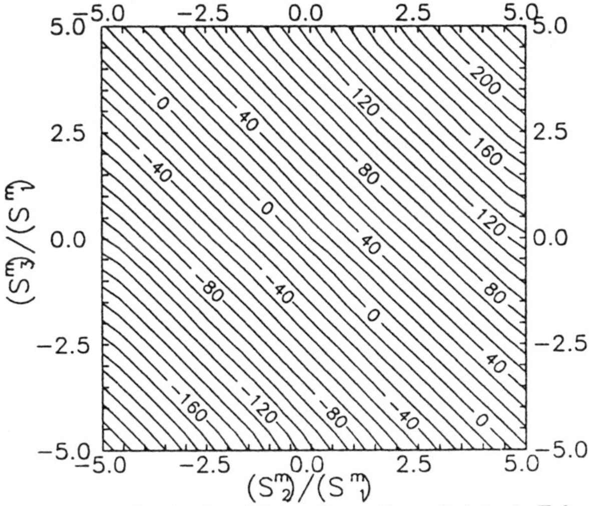

Figure 4. Transverse Strain Sensitivity Error In an Intrinsic Fabry-Perot Optical Fiber Sensor Embedded In a Transversely Isotropic Laminated Composite [Sirkis, 1993b].

the transverse strain sensitivity error computed for this sensor and calibration for arbitrary limits on the mechanical loading ratios. This graph shows that an embedded optical fiber sensor of this type can operate in a specific band on loading ratios without incurring significant transverse strain sensitivity. The zero error contour corresponds the the strain ratios produced by the calibration loads. This graph also shows that transverse strain sensitivity error becomes dominant for off-diagonal combinations of mechanical loading ratios. While the Mach-Zehnder, Michelson, Bragg grating, and dual mode sensors have a similar transverse strain sensitivity error as shown in Fig. 4, the polarimetric sensor possesses three orders of magnitude greater sensitivity of transverse strains, and the extrinsic Fabry-Perot sensor possesses no transverse strain sensitivity at all! The polarimetric transverse strain sensitivity error is enormous because the stress induced birefringence that governs this sensors response is dominated by the transverse strains. The non-existent transverse strain sensitivity in extrinsic Fabry-Perot sensors is attributable to its sensing region being composed of air whose refractive index is independent of strain [Sirkis, 1993a].

Thermal apparent strain sensitivity error is provided in Fig. 5 for Mach-Zehnder, Polarimetric and extrinsic Fabry-Perot sensors assuming the same host structure and calibration used to produce Fig. 4. For ease of presentation, the polarimetric and extrinsic Fabry-Perot errors are respectively divided and multiplied by ten. Fig. 5 shows that not only is the extrinsic Fabry-Perot sensor insensitive to transverse strains, it is also least sensitive to thermal apparent strains. However, a certain level of caution is required when viewing Figs. 4 and 5. They have been developed under very specific assumptions, namely that the sensor ideally is sensative to axial strain only, and that the sensor is embedded in a certain laminated composite system. These conditions may change depending on the sensor design. For example, it may be beneficial to design a polarimetric sensor that is ideally sensitive to only transverse strains as has been done by Bonniau et al [1991] and Bocquet et al [1991]. Axial strain sensitive error would then have to be addressed.

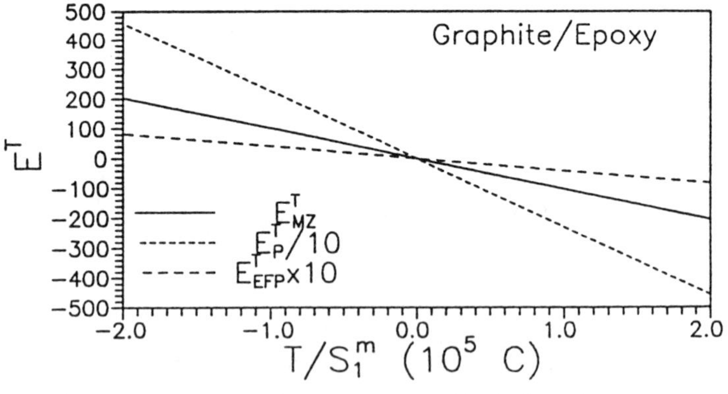

Figure 5. Thermal Apparent Strain Error For Mach-Zehnder, Polarimetric, and Extrinsic Fabry-Perot Optical Fiber Sensors [Sirkis, 1993b].

5 Exact Inverse Phase-Strain-Temperature Relations

It quickly becomes apparent that in order to extract a unique and accurate fiber strain and temperature state from the forward phase-strain-temperature relation (Eq. (1)) for all loading scenarios, one must simultaneously measure four *independent* phase quantities. These measurements must result in linearly independent equations, and can result from multi-fiber rosettes or multiparameter fiber sensors. Rosettes are one potential solution, but sensor-sensor mechanical interaction makes determining the structural thermomechanical strain state from those of the fibers in the rosette difficult. The individual sensors must be located far enough away from each other so as not to influence the measurements of neighboring fiber sensors. This, in turn, leads to an averaging of the strain state over a large region. Further, full strain state information is obtainable only with six axis volumetric rosettes. It is worth noting, however, that Valis et al [1992], Carman et al [1992], and Lesko et al [1992] have used embedded three element optical fiber rosettes with limited degrees of success.

The alternative to rosette configurations is a single optical fiber sensor capable of measuring strain and temperature state. There have been several examples in the literature where simultaneous axial strain and temperature measurements have been successfully performed (see for example [Meltz et al, 1987; Farahi et al, 1990; Vengsarkar et al, 1990; Wang et al, 1992]). Michie et al [1991] present one such example where the optical fiber is embedded in a unidirectional composite laminate. These two-parameter sensors generally use combined interferometric, polarimetric or dual-mode signals, or two wavelength interferometry to obtain the two independent phase measurements. In concept, one should be able to combine both of these approaches to simultaneously record four measurands. In practice, however, developing four *independent* equations of the form presented in Eq. (1) is an error prone proposition [Vengsarkar et al, 1990; Wang et al, 1992]. This is particularly true for short gage length sensors.

Potential sensor systems for achieving four simultaneous measurands include combined dual-wavelength polarimetric-interferometric sensors or dual-wavelength Fabry-Perot sensors in HiBi fiber [Valis, 1991]. The former is a direct combination of the ideas discussed in the preceding paragraph, while the latter creates simultaneous Fabry-Perot cavities along the two eigen axes in a HiBi fiber [Valis and Measures, 1992]. These two cavities can be interrogated with two wavelengths to achieve the four phase measurements. In either case, the use of dual wavelength interferometry requires that the two wavelengths be above cutoff, and be sufficiently separated so as to produce the independent equations. Either of the two above mentioned candidate systems can be used, in theory, to uniquely invert Eq. (1), and are mentioned as a starting point of discussion. The development of sensor systems that will ultimately lead to four independent measures of phase should be one of the primary goals of current and future *smart structures* research programs; for clearly, one can not otherwise expect to infer unique and universally accurate internal strain and temperature states.

Assuming for the present that four simultaneous and independent phase measurements can be made successfully, the inverse phase-strain-temperature relations can then be represented in the following matrix form

$$\{S^f\} = [A]^{-1}\{\Delta\phi\} \tag{7}$$

where $\{S^f\} = \{S_1^f, S_2^f, S_3^f, T^f\}^T$, $\{\Delta\phi\}$ is the 1x4 vector containing the phase change measurements, and [A] is the 4x4 *characteristic* matrix as termed by Michie et al [1991]. The rank of $[A]^{-1}$ and the exact mathematical form of its individual components depend on the number of simultaneously measured phase terms and on the type of sensor systems used to make the measurements.

The two-parameter optical fiber sensor systems that have been presented in the literature use a series of calibration tests to directly determine four of the sixteen components of $[A]^{-1}$ for given thermomechanical load situations [Meltz et al, 1987; Farahi et al, 1990; Vengsarkar et al, 1990; Wang et al, 1992; Michie et al, 1991, O'Keefe et al, 1992]. The components of $[A]^{-1}$ thus determined are valid only under conditions producing nearly identical strain and temperature states as those used during calibration because they necessarily omit contributions from the transverse strain components. As an example, unembedded sensors calibrated under uniaxial tension and isotropic thermal loading can only be legitimately used in surface mounted applications since surface mounted sensors are everywhere in a state of uniaxial tension [Sirkis and Haslach, 1990]. Also, calibrations of this type are limited to a temperature range near the calibration temperature [Vengsarkar et al, 1990]. This calibration approach is not acceptable for structurally embedded optical fiber sensors that are expected to perform in general thermomechanical loading environments. For example, aircraft structures in service, will not simply experience uniaxial tension, but rather time-varying, multidimensional thermomechanical loading depending on flight maneuvers, environmental conditions, altitude, and so on. Even so, the above mentioned two-parameter sensors represent a significant first step towards higher order parameter sensors.

To accommodate unknown service loading, one must determine the fully populated characteristic matrix (or its inverse). Measuring the propagation, Pockels', and thermo-optic constants for the given sensor material as a function of temperature and wavelength is one method of directly determining [A]. This approach is possible because the optical "constants" and the form of the forward phase-strain-temperature relation (Eq. (1)) are independent of the means in which strain and temperature is transferred to the optical fiber sensor. As a practical matter, determining β_o, P_{11}, P_{12} and ξ, as a function of temperature and wavelength is not a debilitating obstacle. The latter two have already been measured over wide temperature and spectral ranges for several silica (and other) glasses [Gunter and Closs, 1975; Shibata et al, 1981; Wray and Heu; 1968]. Bertholds and Dandliker's [1988] method of measuring P_{11} and P_{12} can easily be extended to other wavelengths and temperatures.

Given the assertion that a fully populated characteristic matrix must be determined in order for embedded optical fiber sensors to provide meaningful results, it is prudent to put in perspective the results found by Michie et al [1990]. They calibrated a dual-mode/polarimetric two-parameter sensor system under uniaxial tension and isotropic thermal loading prior to embedding it parallel to the reinforcing fibers in a unidirectional composite laminate. Their measured results correlated well with the applied strain (parallel to the reinforcing fibers) and temperature. These loading conditions are closely approximated by axi-symmetric, generalized plane strain, thermoelastic situations; therefore, the strain and temperature state in the embedded optical fiber will be very similar to that of an unembedded fiber [Sirkis, 1992; Sirkis and Haslach, 1991; Sirkis and Mathews, 1993]. As a result, the calibrated components of the partially populated characteristic matrix, [A], provided acceptable results. However, if the same composite laminate were subjected to transverse compression, for example, then the experimental results would be much less encouraging.

The final topic regarding the inverse phase-strain-temperature relations that merits discussion deals with the numerical stability of the inversion process. The need for four *independent* equations for $\Delta\phi$ has been repeatedly emphasized. Obtaining these equations is a challenging task, and depends directly on the sensor systems. The quality of the potential sensor systems can be gaged by the condition number of the characteristic matrix [Vengsarkar et al, 1990]

$$cond([A]) = \| [A] \| \| [A]^{-1} \| \tag{8}$$

where $\| [A] \|$ is the norm of [A]. This number is very large when the systems of equations is ill-conditioned, and small when the system is well conditioned [Atkinson, 1985]. Eq. (8) is the measure against which all multiparameter systems can be judged.

6 Relation to Structural Measurands

The strain and temperature state that is discussed in the preceding paragraphs has been that existing in the optical fiber. In fact, the state of the optical fiber is of little interest to design and reliability engineers. Instead, it is the state of strain and temperature in the host material that is the desired information. As a result, one must find a means of relating the optical fiber strain and temperature state to the corresponding strain and temperature state in the host material. Consider the finite element analysis of an optical fiber sensor embedded in an orthotropic laminated composite (loaded by in-plane tension) shown in Fig. 5. The contour plot of the von Mises' stress clearly shows a complex distribution, thus forcing the question "where in the host should the strain state be measured." The most reasonable answer is to measure the far-field (relative to the optical fiber) homogeneous average strain.

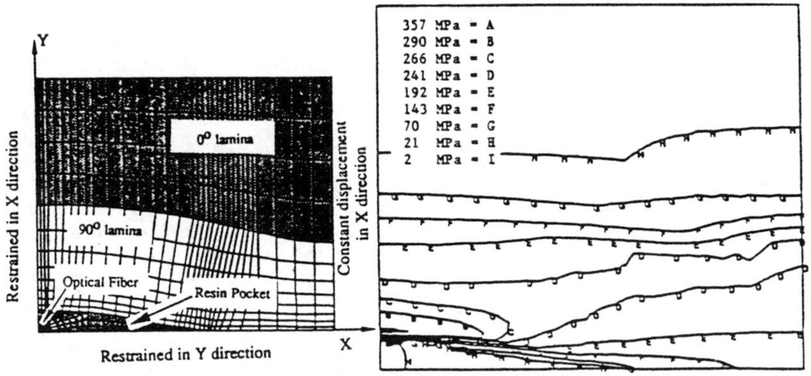

Figure 6. Finite Element Mesh of an Optical Fiber Embedded in a $[0_2/90_2/OF/90_2/0_2]$ Laminate and the von Mises' Stress Resulting From In-Plane Tension [Dasgupta et al, 1992].

This can be done by finding the fiber strain state in terms of the average far-field strains, and then inverting this relationship. In attempting this inversion, it is important to realize that the "local" average strain in the host (as seen on a structural length-scale), becomes the "far-field" average strain for the sensor (as seen on the microscale of the sensor's diameter). The obtrusiveness of the sensory inclusion causes a local perturbation of the host's average strain field being measured. This is not a problem, in and of itself, provided the perturbation field does not interact with any other finite boundary effects. In other words, if the sensor is far away from an external boundary or interfaces/heterogeneities in the host, and far away from any other perturbation fields due to similar sensors (in rosettes for example), then a closed-form solution can be obtained for the perturbations. In transversely isotropic laminate composites with the optical fiber parallel to the reinforcing fibers, these closed form solutions exist and are provided by [Pagano, 1972], Benviniste [1978] or Sirkis [1993b]. The advantage of a closed-form relation is that it not only solves the problem of computing the sensor strains in terms of the host strains, but more importantly, it solves the inverse problem of obtaining the host strains in terms of measured strains in the sensor. If there is any interaction with other perturbation fields, then the interaction between the sensor and host strain fields can still be obtained numerically, but the inverse problem to obtain the host strains from the sensor measurement cannot be solved easily.

7 Conclusions

This paper has discussed the interpretation of the optical signals produced by embedded optical fiber sensors. The discussion has been presented from both practical and idealistic perspectives. The forward relation between the *optical fiber* strain and temperature states and the resulting optical phase change has been presented. The problem arises when one tries to invert this relation in order to infer the strain and temperature state from the measured phase change. On a practical level, the inversion can be accomplished by considering the embedded sensor as being sensitive to axial strains only. While this approach is easy, it comes with significant penalties in transverse strain and thermal apparent strain error for most loading scenarios. The ideal sensor would provide four *independent* measurements of optical phase so that the characteristic matrix of the multi-parameter sensor could be directly inverted and the strain and temperature state can be determined exactly (at least in theory). This paper has outlined potential avenues for attaining this goal. However, even if multiparameter sensors systems that can produce invertible phase-strain-temperature relations, they still only provide strain and temperature state information *in the fiber* . Relating this data to the structural strain and temperature state requires that the thermoelastic solutions for cylindrical elastic inclusions be solved, and the issue of where in the structure do we desire the information be confronted. Provided closed form solutions exist for the inclusion problem and there are no heterogeneities within the zone of influence of the sensor, the host average strain field can be obtained from the multiparameter sensor data. If closed form solutions do not exists or if interactions occur, numerical solutions can be used to propose pseudo-closed-form solutions by curve-fitting techniques, to solve the inverse inversion problem.

8 References

Atkinson, K., 1985, ELEMENTARY NUMERICAL ANALYSIS, John Wiley and Sons, New York.

Benveniste, Y., Dvorak, G. J., Chin, T, 1989, "Stress Fields in Composites with Coated Inclusions," Mechanics of Materials, 7, pp. 305-317.

Bertholds, A., and Dandliker, R., 1988, "Determination of the Individual Strain-Optic Coefficients in Single-Mode Optical Fibers," Journal of Lightwave Technology, 6 (1), pp. 17-20.

Bocquet, J. C., Lecoy, P., and Baptiste, D., 1991, "Optical Sensors Embedded in Composite Materials," SPIE Vol. 1588, pp. 210-219.

Bonniau, P., Chazelas, J., Lecuellet, J., Gendra, F., Turpen, M., Le Pesant, J-P, and Brevingnon, 1991, " Damage Detection in Woven-Composite Materials Using Fiber Optic Sensors," SPIE Vol. 1588, pp. 52-63.

Butter, C. D. and Hocker, G. B., 1978, "Fiber Optics Strain Gage," Applied Optics, pp. 2867-2869, 17 (18).

Carman, G. P., Lesko, J. L., Case, S. W., Fogg, B., and Claus, R. O., 1992, "Development of an Embedded Fabry-Perot Fiber Optic Strain Rosette Sensor (FP-FOSRS)," Proc. Fiber Optic-Based Smart Materials and Structures Workshop, Blacksburg, pp. 45-50.

Claus, R. O., and Cantrell, J. H., 1980, "Detection of Ultrasonic Waves in Solids by An Optical Fiber Interferometer," Proc. IEEE Ultrasonics Symposium, 2, pp. 719.

Dally, J. W., and Riley, W. F., 1991, EXPERIMENTAL STRESS ANALYSIS, 3rd Ed., McGraw Hill, New York.

Dasgupta, A., Wan, Y. and Sirkis, J., 1992, "Prediction of Resin Pocket Geometry for Stress Analysis of Optical Fibers Embedded in Laminated Composites," To appear in Smart Materials and Structures, 1, pp. 101-107.

Farahi, F., Webb, D. J., Jones, D. C., and Jackson, D. A., 1990, "Simultaneous Measurement of Temperature and Strain: Cross-Sensitivity Considerations," Journal of Lightwave Technology, 8, pp. 138-142.

Gunter, R. C., and Closs, J. V., 1975, "Refractive Index of Nd:CaF and Some Nd Doped Glasses as a Function of Wavelength, Applied Optics, 14 (1), pp. 174-76.

Lesko, J. J., Case, S. W., Fogg, B. R., and Carman, G. P., 1992, "Embedded Fabry-Perot Fiber Optic Strain Rosette Sensor For Internal Stress State Assessment", Proc. 7th Annual ASC Conf., pp. 909-918

Meltz, G., Dunphy, J. R., Glenn, W. H., Farina, J. D., and Leonberger, F. J., 1987, "Fiber Optic Temperature and Strain Sensors," SPIE Vol. 798, pp. 104.

Michie, W. C., Culshaw, B., Roberts, R. S. J., and Davidson, R., 1991, "Fibre Optic Technique for Simultaneous Measurement of Strain and Temperature Variations in Composite Materials," SPIE Vol. 1588, pp. 342-355.

Morey, W. W., Meltz, G., and Glen, W. H., 1989,"Fiber Optic Bragg Grating Sensors," SPIE Vol. 1169, pp. 98-106.

O'keefe, C. V., Djordjevic, B. B., and Ranganathan, B. N., 1992, "In-Situ Acoustical and Optical Waveguide Sensors," Proc. of the First European Conference on Smart Structures and Materials, (B. Culshaw, P. T. Gardiner, and A. McDonach, Eds.), Glasgow, pp. 391-394.

Pagano, N. J., 1972, "The Stress Field in a Cylindrically Anisotropic Body Under Two-Dimensional Surface Tractions," J Applied Mechanics, pp. 791-796, September.

Shibata, N., Shibata, S., and Edahiro, T., 1981, "Refractive Index Dispersion of Lightguide Glasses at High Temperature," Electronics Letters, 17 (8), pp. 310-311.

Sirkis, J. S., and Haslach, H. W., 1990, "Complete Phase-Strain Relation for Structurally Integrated Interferometric Optical Fiber Strain Sensors," Journal of Intelligent Material Systems and Structures, 2 (1), pp. 3-25.

Sirkis, J. S., 1993a, "A Unified Approach to Phase-Strain-Temperature Models for Smart Structure Interferometric Optical Fiber Sensors: Part I - Development," to appear in Optical Engineering, April.

Sirkis, J. S., 1993b, "A Unified Approach to Phase-Strain-Temperature Models for Smart Structure Interferometric Optical Fiber Sensors: Part II - Applications," to appear in Optical Engineering, April.

Sirkis, J. S., and Haslach, H. W., 1990, "Interferometric Strain Measurement by Arbitrarily Configured, Surface Mounted, Optical Fibers," Journal of Lightwave Technology, 8, pp. 1497-1503.

Sirkis, J. S., and Mathews, C. T., 1993, "Experimental Investigation of Phase-Strain-Temperature Models for Structurally Embedded Interferometric Fiber Optic Sensors," to appear in Experimental Mechanics.

R. D. Turner, T. Valis, W. Hogg, and R. Measures, 1990, "Fiber-Optic Sensors of Smart Structures." Journal of Intelligent Material Systems and Structures, 1 (1), pp. 26-49.

Valis, T., 1991, Personal Communication.

Valis, T. and Measures, R. M., 1992, "Far-Field In-Plane Stress Measurement With an Embedded Eigenaxis-Flipped Fiber Fabry-Perot Strain Gage," Proc. Fiber Optic-Based Smart Materials and Structures Workshop, (R. O. Claus, Ed.), Blacksburg, pp. 23-29.

Valis, T., Hogg, D., and Measures, R. M., 1992, "Fiber-Optic Fabry-Perot Strain Rosettes," Journal of Smart Materials and Structures, 1, pp. 227-232.

Vengsarkar, A. M., Michie, W. C., Jankovic, L., Culshaw, B., and Claus, R. O., 1990, "Fiber Optic Sensor For Simultaneous Measurement of Temperature and Strain," SPIE Vol. 1367, pp. 249-260.

Wang, G. Z., Wang, A. B., Murphy, K. A., and Vengsarkar, A. M., 1991, "Two-Mode Fabry-Perot Optical Fibre Sensors for Strain and Temperature Measurements," Electronics Letters.

Wray, J. H., and Neu, J. T., 1969, "Refractive Index of Several glasses as a Function of Wavelength and Temperature," Journal of the Optical society of America, 59 (6), pp. 774-776.

Serpentine Optical Fiber Strain Gauge Evaluation

Steve Hiles[1]
Chuck Wilkerson[2]
J. Richard Houghton
Dale A. Wilson

Abstract

In the smart structures area of strain sensing, the equipment used is as important as the quality of the sensor itself. Strain sensing using fiber optics is a relatively new field of engineering, but a wide variety of optic equipment is available. The majority of the equipment, however, is still unable to provide the accuracy and resolution required to produce usable strain sensing results.

In this paper, various types of equipment and sensor configurations are presented. Limitations and advantages of these are also given for comparison as well as problems arising from sensor design and procedures for applying sensors in the smart structures area. Problems of optical fibers in general are discussed as well as the particular problems of Whisper Modes and their relative importance in transducer design.

Introduction

Optical fibers have been used in the communications industry for over fifteen years, but optical fibers used as strain sensors has only appeared in recent years. As communication tools, the optical fibers purpose is to transmit light from a source to a receiver, but this is not the case with smart structures. As sensors, optical fibers can be used to detect various properties such as temperature, surface quality, thermal expansion, and strain, to name a few. The fibers ability to detect strain is the property of most interest for structural designers.

---

[1]Research Assistant, Center for Manufacturing Research and Technology Utilization, Tennessee Technological University, Cookeville, TN 38505

[2]Project Engineer at NASA MSFC

As discovered by the communications industry, optical fibers are sensitive to large bends relative to the fiber diameter. These bends, known as macrobends act to reduce the amount of light which can travel through the optical fiber, thus attenuating the light. The attenuation is caused by factors such as Rayleigh backscattering and loss of light due to increasing the angle at which the light rays reflect from the internal surfaces of the fiber. The Rayleigh backscatter is due to micro discontinuities and foreign particles in the fiber's glass which reflect the light.

The singlemode optical fibers are best suited for strain sensing caused by bending since this form of light transmission only travels in one mode. This factor reduces the complexity of relating the optical attenuation to mechanical strain since mode shifts and loss and gain of higher order modes of light are not present.

Due to the optical fiber's sensitivity to bending, it provides a means to detect structural bending when attached to or embedded into a structure. Although this feature is the basis for optical strain sensing, it provides little information unless the fiber's orientation is correct. In order for the optical fiber to detect strain it must be bent to a degree much larger than any load bearing structure could withstand without failing. For this reason, serpentine shaped sensors are chosen to provide strain sensing instead of straight fiber sensors. Using a fiber with a pre-bend or serpentine allows the change in attenuation due to bending to start when the load is first applied since the fiber is already to the bend radius at which it is bend sensitive. As an example, a straight fiber in a structure subjected to a bend change from 2 inches to .5 inches would see the same radius change of 2 inches to .5 inches, but a serpentine fiber with a pre-bend of .21875 inches loaded such that the bend radius of curvature of the beam changes from 2 inches to .5 inches would have a bend change of .21745 inches to .20041 inches for the fiber[1]. Since the serpentine fiber is already at the point of high attenuation due to preset bending, the structure radius change would show an attenuation change whereas the straight fiber would never reach the bend radius required to attenuate the light. Due to this, serpentine optical fiber transducers are the preferred choice for strain sensing related to bending.

The two methods generally used to obtain attenuation measurements are by using Continuous Wave Through Transmission(CWTT) and Optical Time Domain Reflectometry(OTDR). An advanced OTDR method called a Photon Counting Reflectometer(PCR) is another category of equipment which can detect attenuation. The differences between CWTT testing and using the OTDR types of equipment are discussed in the following section.

Continuous Wave Through Transmission(CWTT)

The CWTT type of testing requires the most basic of the test setups. A laser source, usually a laser diode is connected to the optical fiber in question which is in turn connected to an Optical to Electrical(O/E) converter. The converter is then connected to a display device such as a sampling oscilloscope and the optical power output is shown. This type of setup requires access to both ends of the optical fiber to obtain attenuation measurements. Using the output power for the fiber when it is not deformed(P_{in}) and the output power with the load or bend applied(P_{out}), the equation used for attenuation measurements is:

$$\alpha(dB) = 10\log_{10}(P_{in}/P_{out}) \tag{1}$$

The value of α for attenuation is expressed in dBs. This method of detection produces the fastest and most accurate results since the signal-to-noise ratio is high and averaging is minimized. The greatest disadvantage of this form of testing is that the attenuation levels displayed are a net sum of all of the attenuating factors encountered by bending along the fiber. It is impossible for this method of detection to distinguish the location of the bends or the number of bends causing the attenuation. It can only produce a sum of the attenuating factors.

Optical Time Domain Reflectometry(OTDR)

The OTDR uses the same types of sources and O/E converters as the CWTT method except that the laser source is pulsed instead of being continuous. However, the setup of the components which are used in the OTDR is different than the CWTT in that the source and the receiver are both connected to the same end of the test optical fiber. This is accomplished by using an optical coupler to split the optical fiber into two pigtails, one for the source and one for the receiver. The OTDR detects the light that is scattered back through the fiber when points of attenuation are encountered. This scattered light falls under one of two types, either Rayleigh backscatter or Fresnel reflections. Rayleigh backscatter occurs when the rays of light traveling through the optical fiber hit impurities or micro-discontinuities within the fiber and are reflected backwards relative to the source. The Fresnel reflections occur when the light encounters macro-discontinuities such as the mirror like surfaces of splices or breaks in the fiber. When an optical fiber is bent, the stresses in the fiber cause micro-discontinuities which reflect the light. This reflected light is detected by the O/E converter and is displayed as levels of attenuation at that point. The laser source must be pulsed to allow the minute amount of reflected signal back to the detector. Since the detected signal is very small, compared to the source power, the signal to noise ratio is low and considerable averaging is required to determine the attenuation levels. Due to this, the processing speed is reduced as well as the accuracy of the attenuation readings.

Despite the disadvantages of the OTDR, several advantages make this type of equipment important in the study of optical attenuation and smart structures. Since the light is sent and received from the same end of the optical fiber, the positions of multiple bends causing the attenuation can be found. Also attenuation levels for individual bends can be found since the equipment displays the attenuation points separately. The only limitations of attenuation position detection are in the resolution obtainable by the OTDR and the "dead zones" caused by the Fresnel reflections. The resolution of the OTDR is based on the width of the optical pulse transmitted through the fiber. The smallest detectable spacing between points of attenuation is one-half of the width of the pulse. Typically this resolution is on the order of centimeters to meters which means that the spacing of the attenuation points is centimeters to meters. The second limitation of the OTDR's is the dead zone which occurs when the large Fresnel reflections saturate the detector with light and blind the circuitry. In this length of fiber immediately following macro-discontinuities, attenuation measurements can not be obtained. Typical dead zones for OTDRs range from several meters to several hundreds of meters. This dead zone hampers the design of optical fiber transducers since large amounts of optical fiber would be required after any splice. These two limitations show the need for equipment which has a higher resolution and better circuitry.

Photon Counting Reflectometry (PCR)

An advanced type of OTDR uses microwave technology to produce shorter pulse widths and highly accurate and stable trigger delays to detect the attenuation in optical fibers. This type of equipment known generally as a PCR is setup the same as a typical OTDR except that the circuitry is extremely advanced. The equipment uses a time delay trigger to turn the detector on and off exactly when the pulses from the source laser are returning. This, in effect, reduces the length of the dead zone since the detector is turned off during the Fresnel reflection and does not get saturated. Also the resolution of the PCR is higher because it can sample photons of light entering the detector at a specific time delay instead of averaging groups of data like the OTDR. The resolution of the PCR is based on the resolution of the display device so each pixel on the screen shows a level of attenuation at a point on the optical fiber. Using this method of detection allows for millimeter resolution spacing between points of attenuation and minimal undetectable fiber length loss due to the dead zone effects. The length of time required for averaging is still equal to, if not greater than, the typical OTDR's but the resolution is greatly enhanced. For these reasons the PCR is currently the best type of equipment to be used for smart sensing applications.

A "smart" structure is defined for this paper as one which has the built in ability to detect flaws in manufacturing, discrepancies in actual load and stress fields compared to theory, the ability to sense situations which could cause failure before it occurs, and display damaged regions of a structure. Essentially the structure has a nervous system which can determine what is happening to itself. Almost any structure or part which experiences stress or strain can have a strain sensing device added and thus become a candidate for a "smart" structure application.

Although sensor orientations are endless, some factors limit the ways in which optical fiber strain sensors can be used. The most demanding limit of optical strain sensors is that the transducer must not receive any inadvertent microbends in the fiber during manufacturing or physical application to a structure. If microbends are produced, the transducer would be rendered useless since they attenuate the light and false readings would occur. Microbends occur when the fiber is subjected to abrupt direction changes, such as when it is embedded between or attached to rough surfaces. Examples of microbends would be if the optical fiber exits a graphite fiber structure by crossing laminate layers, placed in a concrete column unprotected, or epoxied to an uneven surface and forced to conform to its shape. These are essentially the only limits of application for optical fiber sensors.

Research conducted by Tennessee Technological University and NASA-MSFC include optical fibers that have been embedded in graphite sheets, on fiberglass and plastic beams, in graphite pultrusion parts, in graphite filament wound bottles, and in mandrel wrapped tubes. The study reported here involves embedding optical fibers in mandrel wrapped graphite tubes such as the one shown in Figure 1 below:

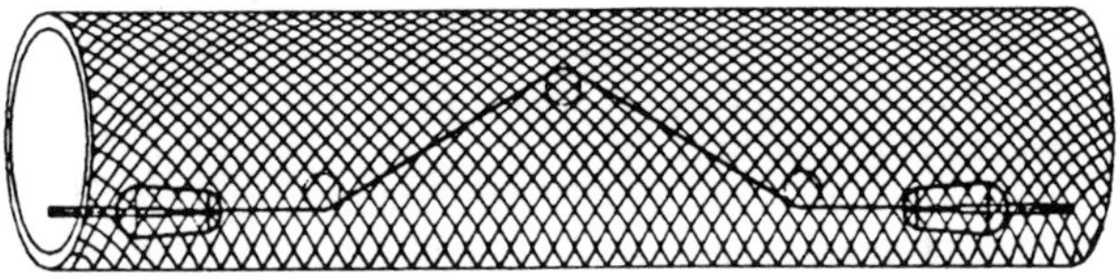

Figure 1. Schematic View of Graphite Mandrel Wrapped Tube

As stated above, the limits of microbends for this type of application require that the optical fiber exit the structure on the same plane with which the serpentine sensor is embedded. The tube is composed of five individually wound layers, each alternating from a hoop orientation to a helical one. The optical transducer is placed on a graphite fiber tape and secured to achieve uniform serpentine bends. The serpentine configuration is composed of three bends, which allow the center bend to fall at the middle of the tube length. The two outside bends act to orient the fiber for the center bend which is the strain detector. Each bend has a 7/16" diameter. The fiber lay-up for the tube is as follows: (1)the inside layer is of helical pattern, followed by (2)a hoop, (3)a helical, (4)the transducer patch, (5)a hoop, and finally (6)a helical layer. The transducer is attached to the graphite tape but is not cured. To reduce voids and delaminations, the transducers graphite patch is cured with the graphite filaments, which make up the tube, in the same process. After curing, the part has become our sample smart structure.

As with any smart structure the optical fiber must be accessible in order to connect equipment and perform tests. The easiest method for connecting to the optical fiber is to leave a length of fiber exposed and connect this to the equipment using a glass splicing technique such as epoxy or fusion fiber splicing. This method, known as pigtailing, is the most cost efficient but also the most hazardous type of fiber exit and termination. The fiber is exposed to all of the hazards of handling and environment since it is bare upon exit, so the likelihood of fiber breakage is extreme. Typically the fiber will break at the point of highest stress which is usually at the entrance of the structure. Obviously breaking the fiber makes the sensor useless. Sometimes, as with the graphite tube, the plane that the transducer is on can not be used as an exit since the ends of the part are machined to allow connection to larger structures.

These problems inspired us to engineer a fiber exit/terminator which can withstand the hazards of handling and environment. A fiber terminator has been developed which serves these purposes as well as providing a low loss connection. A schematic view of this connector is shown in Figure 2.

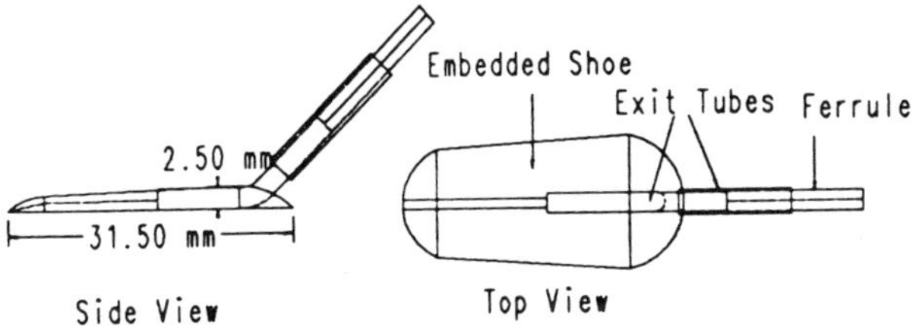

Figure 2. Schematic View of The Optical Exit/Terminator

This design protects the delicate fiber and is terminated with a ceramic ferrule compatible with standard connectors on the market today. Using this type of terminator virtually eliminates the need for delicate handling around the fiber exit and provides a simple and efficient method for fiber connection.

Using this type of connector along with the serpentine optical fiber embedded in the graphite tube, tests are conducted to validate the sensors abilities to detect strain in a simulated structure. Tests of other fiber orientations and different types of equipment such as the CWTT, the OTDR, and the PCR are used to optimize the fiber types and provide the basic requirements for equipment capabilities. Issues such as the Whisper Mode, its causes and effects, and ways to utilize its characteristics will also be discussed because of its significance on the transducers design.

Experimental Evaluation of Test Equipment

It was previously argued that the OTDR type equipment was best suited for smart sensing applications. This decision was validated using experiments which provide details of equipment limits through actual data acquisition procedures.

In order to obtain an accurate comparison of equipment, several brands of OTDRs were tested. The brands tested were: the Hewlett Packard OTDR Model 8146A, the Ando Optical Fiber Analyzer Model AQ-7130, and the Opto-Electronics Photon Counting Reflectometer. The equipment was used to find correlations between attenuation and bend radius. These were performed to compare the types of equipment and also show their ability to take the data required in the design of optical strain sensors.

Hewlett-Packard OTDR

The Hewlett-Packard OTDR Model 8146A was equipped with a 1300/1550 nm SM plug-in module. The fiber tested was a 1300/1550 nm SM communications fiber. Figure 3 shows the screen display of the equipment.

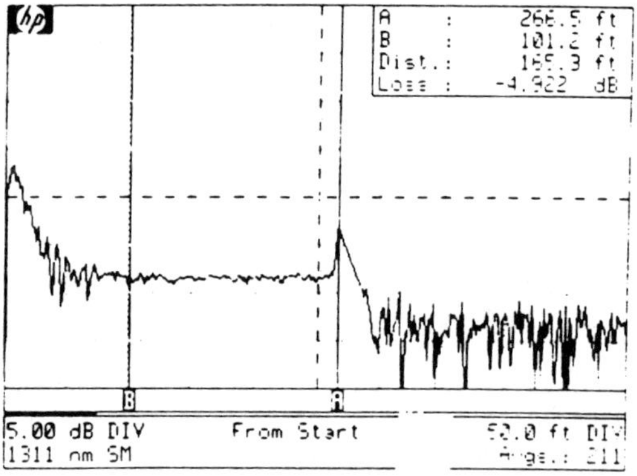

Figure 3. Screen Print From The Hewlett-Packard OTDR Showing Basic Characteristics

As seen, the fiber being tested is 266.5 feet in length. This is apparent since the end of the fiber produces a Fresnel reflection located at point A of the figure. To the left of point B is the "dead zone" of the fiber due to the input connector attached to the OTDR. Between points B and A is the readable section of the fiber. To the right of point A (the end of the fiber) is the noise floor of the OTDR. In this area there is no fiber to sample so the generated noise is not averaged out.

In Figure 4, a point of high attenuation (a bend) was placed 103.5 feet from the input end of the fiber.

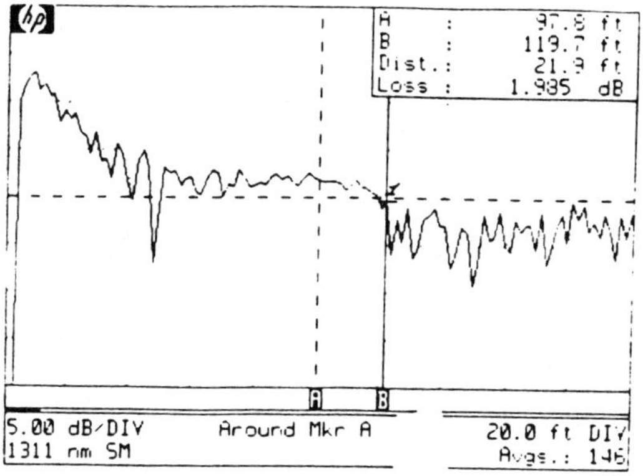

Figure 4. Display Showing A Point of High Attenuation

To the left of the point, the mean of the line is linear and horizontal as a result of minimal light lost from the unbent portion of the fiber. To the right of the point, the trace shows the noise floor. Since the noise floor is present immediately after the bend, the bend acts to attenuate the light fully and it can not continue through the fiber. The Fresnel reflection for the end of the fiber was not detectable since the fiber was attenuated fully at the bend. The OTDR evaluates the point at a length of 119.7 feet from the input end of the fiber giving an error of 13.5 %.

Ando OTDR

The Ando Model AQ-7130 OTDR was equipped with a 1310 nm singlemode plug-in module (AQ-7133) capable of detecting both Rayleigh backscatter and Fresnel losses. The 1300/1550 nm SM communications fiber tested with the HP OTDR was also used in this test. The equipment was setup to take 10 average counts while in the START mode. Note that the number of averages of the Ando can not be compared numerically to the Hewlett-Packard. Assuming equal clock speeds for the different brands, the Ando was averaged approximately 3 times longer.

OTDR Comparisons

Since both brands of equipment were tested on the same fiber, the two figures are overlaid for comparison in Figure 5. 2 dB/km attenuation was added to each point on the Ando OTDR's results to aid in showing the trace comparisons.

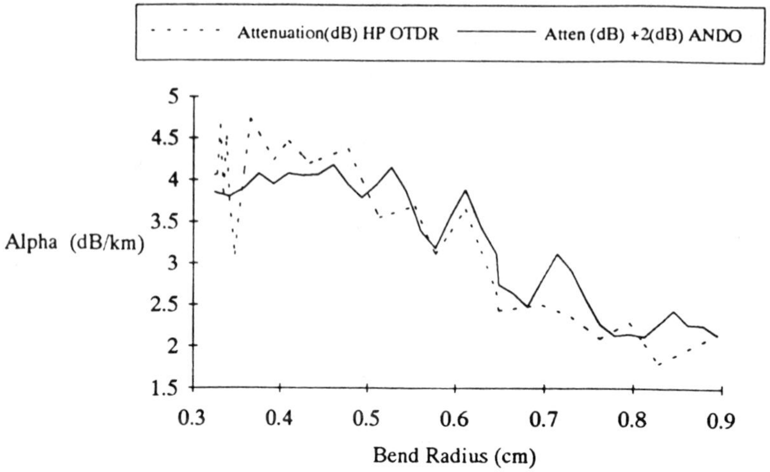

Figure 5. α Attenuation vs. Bend Radius For The HP and The Ando OTDRs

As seen, the mean of each trace is approximately parabolic. This shows that the smaller bend radius region would produce the greatest sensitivity to strain induced radius changes since the slope is the largest. The fluctuations from the mean should be noted at this time and will be discussed later in the paper. From the figure we also see that the traces compare closely with each other in attenuation variations as well as in fluctuations. This is expected since the same fiber was tested and both pieces of equipment use the same technology.

Opto-Electronics Photon Counting Reflectometer

The Opto-Electronics Fresnel Detecting Reflectometer was equipped with a 850 nm MM plug-in module. An 850 nm MM optical fiber was tested by comparing the Fresnel end reflection for a change in the bend radius. The equipment was not able to detect Rayleigh backscatter since the data sampler/processor was not setup to take data at a very high rate, i.e. tests were performed in Fresnel detection mode only. The attenuation for the bends of varying radius was detected by taking a two point attenuation value. The points were placed before and after the location of the bend so the attenuation due to the bend could be made. The results of the testing are shown in Figure 6. Although not shown, it should be noted that the same curve can not be obtained from repeated tests. The relationship between attenuation and bend radius changes with different tests even though the setup is the same.

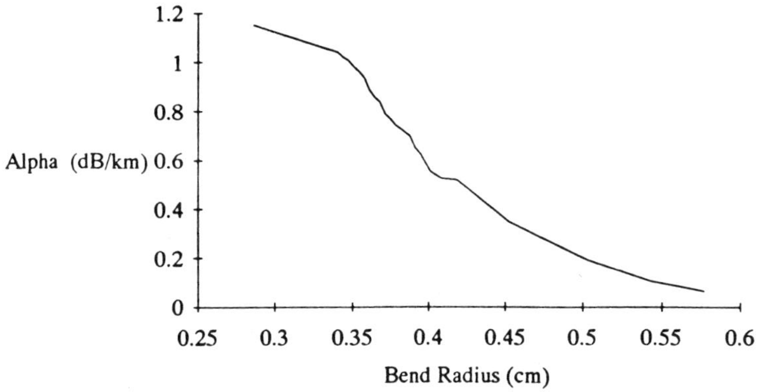

Figure 6. α Attenuation vs. Bend Radius For An 850 nm MM Fiber Using The PCR

As seen, the trace is much smoother than the ones using the typical OTDRs. This is mainly due to the use of multimode fiber instead of singlemode. The traces for the OTDRs and for the PCR can not be compared graphically since the test setups were different.

Continuous Wave Through Transmission

From the OTDR's and PCR's results of attenuation versus fiber bend radius change shown above, it is obvious that there are some underlying physics that effect the correlation. Since the tests performed involved using singular bends instead of multiple bends, the CWTT method of detection could be used to more accurately describe the effects of radius change on attenuation. This is because the signal-to-noise ratio is much higher for CWTT so details are amplified.

For the CWTT testing, a Seastar temperature stabilized laser source was used. The Seastar laser system employed a pre-pigtailed laser diode stabilized by a heat sink and a temperature controller and powered by a laser diode driver. Using this source, an O/E converter, and a LeCroy sampling oscilloscope, tests were performed on 850 nm singlemode and multimode fibers.

Using CWTT with the SM 850 nm fiber, three separate tests were performed and combined on the same graph as shown in Figure 7.

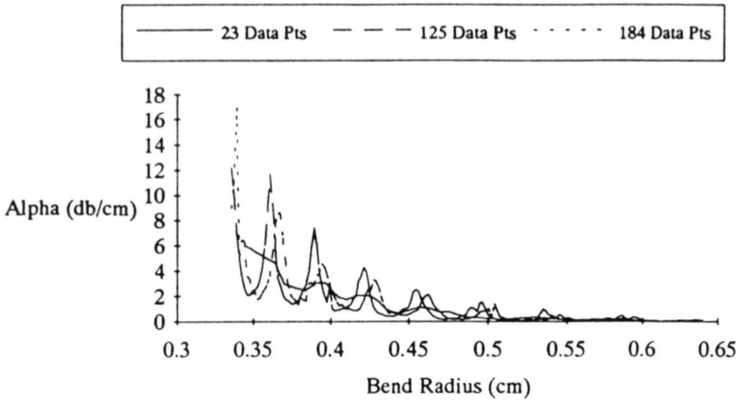

Figure 7. α Attenuation per Bend Length vs. Bend Radius for a 850 nm SM Fiber and a 830 nm Laser Diode Using CWTT

From the figure, the fluctuations seen in the OTDR testing are present. As the number of data points were increased, these fluctuations, or whispers, become more pronounced. The multiple tests conclude that the whispers are a repeatable function of bend radius change. Although the tests performed with the OTDRs used 1300 nm SM fiber and those with CWTT were 850 nm SM fiber the whisper modes were still present. This leads to the belief that regardless of the wavelength, singlemode fibers exhibit the whisper mode. The whisper mode issue will be discussed in the next section. For a comparison to the PCRs results shown in Figure 6, tests using the CWTT method were performed on an 850 nm MM fiber. The results of this test is shown in Figure 8.

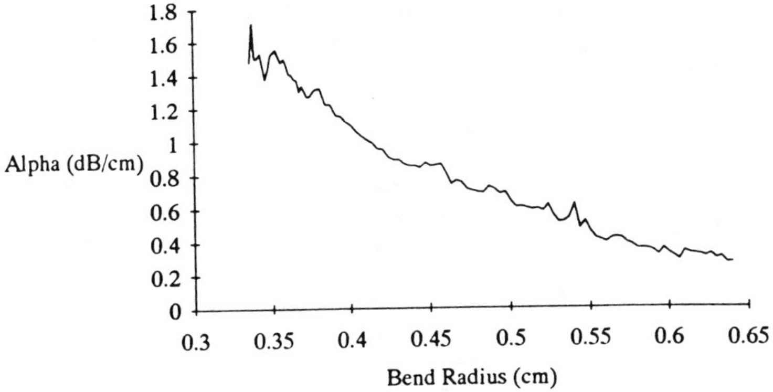

Figure 8. α Attenuation per Bend Length vs. Bend Radius for a 850 nm MM Fiber With a 830 nm Laser Diode Using CWTT

Comparing Figure 6 to Figure 8 we see that the traces are almost identical. The number of data points taken was much larger for the CWTT than for the PCR so slight fluctuations in the trace are more pronounced. If the results for any test shown for singlemode and multimode fibers are compared, it is seen that the multimode fiber exhibits little if any whisper mode fluctuations. This was further proven when multiple tests were performed on the same MM fiber and the fluctuations were not stable or repeatable. Initially it was thought that MM fiber would provide the best fiber optic strain sensor, but this decision was incorrect. The multimode fiber possesses a simpler relation, but repeatability, stability, and guide mode simplicity reduce the advantages of the simple correlation. With the singlemode fiber, the whisper mode effects are present, but they are stable and repeatable. This makes it harder to develop a mathematical function for the sensor, but once found, reduces the error in measurement and produces a better transducer.

The Whisper Mode Issue

From the tests performed on the singlemode fiber using the OTDRs and CWTT it is seen that the whisper mode issue is an important factor in the correlation between attenuation and fiber bend radius change. The whisper modes are apparently caused by transition losses encountered when the optical fiber makes the change from a straight fiber to one of a specified radius. Due to the transition, power is transferred from the guided HE_{11} mode to radiating modes which can also couple back to the guided mode[2]. Therefore it is concluded that the transition losses cause the power to oscillate as the radius is changed. The pure bend losses account for the exponential decay of the trace.

Testing of the Graphite Tube with the Embedded Sensor

The filament wound graphite tube previously mentioned was used as the sample structure for the optical strain sensor testing. The tube is 12 inches in length with an outside diameter of 3 inches. The wall thickness is 3/32 of an inch composed of the five graphite layers and one transducer layer. A 1300 SM communications fiber was embedded with a serpentine orientation in the tube. The new optics connector was used as the exit terminator. The layout of the transducer in the tube was previously shown in Figure 1.

The tube will be tested using a three point bend apparatus with the center loading directly above and on the opposite side of tube to the strain sensing bend of the transducer. A schematic of the apparatus with the bottle orientation is shown in Figure 9 along with a free body diagram of the forces.

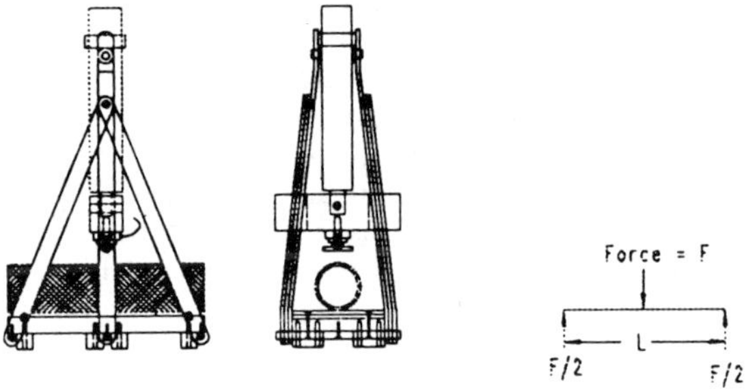

Figure 9. Schematic of The Three-Point Bend Fixture with Tube Orientation Shown

From static analysis, the moment at the center of the tube is:

$$M = FL / 4 \tag{2}$$

where: F = Force applied to tube,
L = Length from bottom load points.

The stress due to bending for a beam is:

$$\sigma = (M*c) / I \tag{3}$$

where: M = Bending Moment,
c = Distance from neutral axis to the point in question,
I = Moment of Inertia.

Substituting the values for the tube test and knowing that the Young's modulus is the ratio of stress to strain, the relationship for strain as a function of load is found. This strain relation is:

$$\varepsilon = (16FLD_o) / (E\pi(D_o^4 - D_i^4)) \tag{4}$$

where: D_o = Outside diameter of tube,
D_i = Inside diameter of tube,
E = Young's Modulus.

From Reference[3] it is seen that strain is also related to the amount of curvature applied to the structure:

$$(1 / \rho) = (\varepsilon / c) \tag{5}$$

where: ρ = The radius of curvature of the structure.

From Reference[1] the relationship of structural bend radius change to serpentine optical fiber bend radius change was found to be:

$$R_{fiber} = a / [(1 + (a/\rho)^2)^{1/2}] \tag{6}$$

where: a = Initial serpentine bend radius.

Using Equations (2) through (6), the amount of strain, the structural bending, and the fiber's radius change can be found based on the amount of loading applied to the structure.

A test of attenuation versus bend radius was performed on the same 1300 nm communications fiber embedded in the graphite tube. This was done using CWTT in order to obtain a theoretical relation between attenuation and bend radius. Using this correlation, a relation between strain and attenuation can be found using Equations (2) to (6). The results of the test are shown in Figure 10.

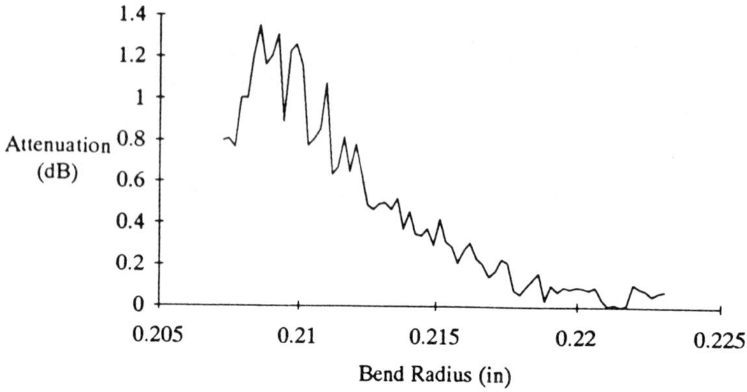

Figure 10. α Attenuation vs. Bend Radius For the 1300 nm SM Fiber Using CWTT

These results will be compared with the experimental results obtained from testing the graphite tube with the optical fiber transducer embedded. Again using Equations (2) to (6), a relationship between strain and experimental attenuation can be found. As long as the initial fiber radius is known, it is possible to obtain simple correlations between fiber bending and attenuation through experimentally obtained results. Without the knowledge of the initial serpentine pre-bend, the calibration curve obtained by CWTT can not be used since several different bend radii have the same attenuation level. Figure 11 shows the theoretical and the experimental results for the strain sensor.

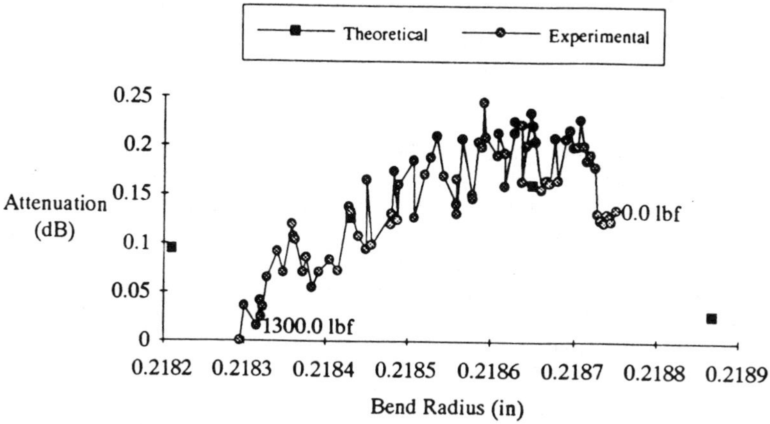

Figure 11. Theoretical and Experimental Results for Attenuation vs. Radius Change

Comparisons of Theoretical and Experimental Data From Graphite Tube Test

The experimental and theoretical data in Figure 11 show several comparisons. Although only four theoretical data points are witin the experimental radius change region, the maximum attenuation values are around the same bend radius. Also the mean of the experimental curve has the same slope sign changes as the theoretical data. From this test it is seen that only part of one whisper fluctuation was involved in the relationship between attenaution and bend radius. Even though the correlation contains the peak of the whisper mode, a relatively simple polynomial model can be found that mathematically relates the attenuation to the radius change.

Conclusions

From the tests performed with the different types of equipment the following advantages and disadvantages were found. The CWTT method produces the highest degree of accuracy and the fastest measurement rate due to the large signal-to-noise ratio, but lacks the ability to differentiate between multiple points of attenuation on a single fiber. The typical OTDRs have the ability to assign location values to points of attenuation on a fiber, but are limited to a minimum of centimeter spacing between location points. Other disadvantages of the OTDRs are: the need of large lengths of optical fiber before discontinuities to avoid dead zones and the increase in data processing time and reduction of accuracy due to the low signal-to-noise ratio. The PCR achieves the high resolution, minimal attenuation point spacing of a millimeter, and the significant reduction in dead zone not obtainable by the typical OTDRs. The PCR has a disadvantage in that the processing time is even longer than the OTDRs.

The requirements of using fiber optics as strain sensors determine the weight of the advantages and disadvantages of the different types of equipment and thus dictate which type is the best suited for an application. The requirements of the equipment of the optical strain sensors are that they must: be able to show multiple points of attenuation on a single fiber, return accurate and repeatable signals, be sturdy and easy to use, and show an acceptable resolution spacing between points of attenuation. The CWTT method provides the accuracy, but lacks the ability to distinguish between multiple points of attenuation. The OTDR can locate attenuation points, but requires long lengths of fiber before splices which makes sensor delicate and hard to physically apply to a structure. Although the PCR takes the longest to process data, it meets the optical sensors requirements. For these reasons, the PCR is chosen as the best type of equipment to be used with fiber optic strain sensors.

The protection of the optical fiber as it exits a part is another necessity for the design of a quality optical strain sensor. This is accomplished by using the fiber exit/terminator shown in Figure 2. Using this type of connection reduces the difficulty in sensor mounting, since the bare fiber is not exposed and allows for connection and disconnection of optical sensing equipment. This form of exit device dictates the use of the PCR type equipment since it is the only type capable of detecting zones of attenuation immediately following fiber splices or connections.

The tests of the different equipment types also determined which fiber type is best suited for strain sensing. The multimode fibers had a simple relation between attenuation and fiber bend radius, but lacked the stability, repeatability, and guide mode simplicity required for a quality strain sensor. The singlemode fiber produced whisper modes which added complexity to the relation between attenuation and fiber bend radius, but the results were stable and repeatable. Since typical structure radius changes are small compared to the radius change required to make the attenuation level change beyond one whisper, it will be a relatively simple task to correlate attenuation to fiber radius change. As seen in Figure 11, the loads applied to the graphite tube produced a correlation of attenaution versus radius change which varied over less than one whisper peak. Without the whisper modes, the sensitivity would be greatly reduced. For this reason, the SM fiber has a higher sensivity to bending than MM fiber.

From the tests performed on the graphite filament wound tube it was shown that knowledge of the pre-bend radius of the serpentine fiber transducer is essential in producing a strain sensor. Without the exact initial radius, an incorrect calibration between attenuation and strain for the sensor could be found since there are multiple bend radii which have the same level of attenuation.

In summary, these results show that an acceptable type of equipment has been found to be used with optical strain sensors, along with the proper fiber type, fiber orientation, and sensor exit terminator. However there is still a need to accurately determine the pre-bend radius of the serpentine fiber which makes up the transducer. Without this knowledge, accurate calibration of the sensor is a difficult task.

References

[1] A. C. Demirdogen, J. R. Houghton, A. K. Tay, D. A. Wilson, "Serpentine Optical Fiber Bending Transducer for Distributed Strain Measurements in Smart Structures," ASTM Symposium on Damage Detection and Quality Assurance in Composites, November 13, 1990. San Antonio, Texas. Also STP No 1128, 1992, pp. 55-71.

[2] W. A. Gambling, H. Matsumura, C. M. Ragdale, "Curvature and microbending losses in single-mode optical fibres," Optical and Quantum Electronics 11 (1979) PP. 43-59

[3] G. R. Buchanan, Mechanics of Materials, Holt, Rinehart and Winston, Inc., NY, 1988

III - EMBEDDMENT IN CONCRETE STRUCTURES

EVALUATION OF FIBER OPTIC BRAGG GRATING STRAIN SENSOR IN HIGH STRENGTH CONCRETE BEAMS

M. H. Maher[1], A.M. ASCE and E. G. Nawy[2], F. ASCE

Abstract

Fiber Optic Bragg Grating (FOBG) sensor was tested on large scale concrete beams for measurement of strain due to flexural deformations. The sensor, which has recently been developed at the United Technologies Research Center (UTRC) for monitoring of strain in structural composites, was tested with the objective of direct and quantitative measurement of load induced strain. The experimental results showed that the FOBG sensor can be used for monitoring strain in concrete structures through remote sensing. A number of issues such as sensor bonding to structure, sensor placement, sensor arrangement (embedded or exposed), and instrumentation on large concrete structures were addressed in this study.

1. Introduction

Bragg grating were first discovered by Hill et al. [1978] in standard telecommunications optical fiber. The gratings were formed by an optically induced refractive

---

1 Assistant Professor, Dept. of Civil and Environmental Engineering, Rutgers University, Piscataway, NJ 08855-0909.

2 Distinguished Professor, Dept. of Civil and Environmental Engineering, Rutgers University, Piscataway, NJ 08855-0909.

index change caused by the interference of two counter propagating waves in the fiber. The first glass composition to exhibit this property was germanium doped silica but since then many other glass composition have been shown to have similar properties. Meltz et al. in 1989 made it possible to make useful devices through a direct UV illumination of the fiber core from the side of the optical fiber, the transverse holographic method. The physics of the refractive index change is not completely understood but it is believed to be a photoinduced redistribution of color centers in the glass.

Since the introduction of useful devices, FOBG have been used to build laser cavities, Fabry-Perot interferometers, and sensors. Bragg grating in single mode fibers have been used to measure vibration, strain and temperature. Bragg grating in two mode optical fiber has also recently been shown to satisfy certain sensing applications. High sensitivity measurements of strain have been shown by monitoring the phase shift of the Bragg reflection.

A FOBG is a longitudinal periodic variation of the index of refraction in the core of an optical fiber as shown in Fig.1. The spacing of this periodic index variation is determined by the wavelength of the light that is to be reflected. Only light which has a wavelength that is twice the optical path (spacing) is reflected. This results in a very narrow spectral line, approximately 10-10m wide reflection band. The narrowness of this band provides good resolution during measurements. The condition that must be satisfied to get a reflection is

$$\lambda_{Bragg} = 2\ n\ D \ldots\ldots \quad (1)$$

where λ_{Bragg} is the reflected wavelength of the light; n is the average index of refraction in the core of the fiber; and D is the spacing of the periodic refractive index variation. As strain is applied to the fiber the spacing of the periodic variation of the refractive index is elastically deformed. In addition to mechanical deformation of the grating the average index of refraction is also changed through the strain optic effect. These two effects work together to cause a shift in the bragg wavelength reflected for an applied strain. By differentiating the Bragg condition expression and substituting an expression for the strain optic effect one can express the fractional change in wavelength as a constant (K) times the applied strain (ε).

$$\Delta\lambda / \lambda = K\ \varepsilon \ldots\ldots\ldots\ldots\ldots (2)$$

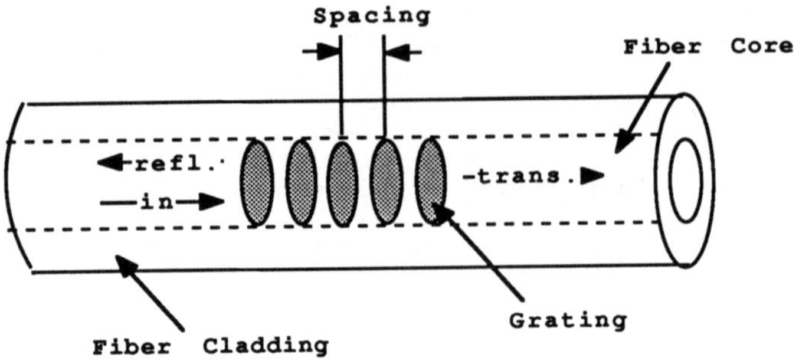

Fig.1.- Schematic diagram of gratings on a FOBG sensor

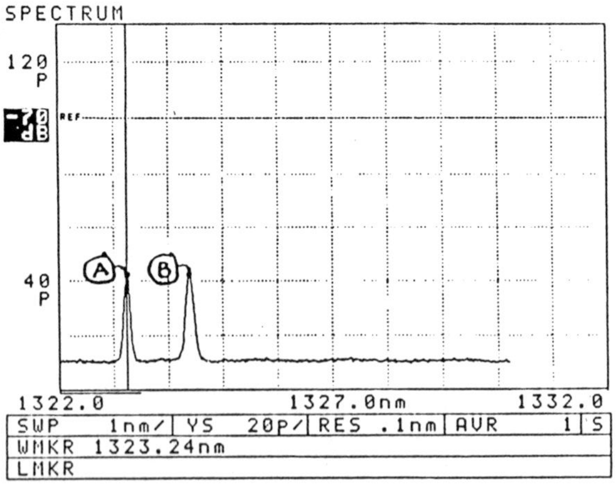

Fig.2.- Typical wavelength shift (from point A to point B) as a result of strain on FOBG sensor (Maher et al., 1992)

In the simplest configuration the spectral shift of the Bragg reflection corresponds to an applied strain or temperature change. This can be measured by using a broad band LED source at 1300 nm launching light into a 3dB coupler with one leg of the coupler connected to the fiber with the FOBG. The return leg of the coupler is then connected to a spectrum analyzer which monitors the spectral shift of the reflection band (Fig.2). A schematic diagram showing the system instrumentation is presented in Fig. 3.

The FOBG system was introduced by the investigators at Rutgers for the first time since its recent development for instrumenting reinforced concrete beams. It has shown good performance as a strain sensor embedded in concrete structural elements. Preliminary results of the first phase of the research at Rutgers, which included the testing of the FOBG sensor in both small and large scale concrete structures, has shown good potential for this sensor and has warranted more large scale tests (Maher-Nawy et al., 1992; Prohaska-Snitzer et al., 1992).

2. Objective

Based on the preliminary findings of the research done at Rutgers, a number of additional large scale model beam tests were recently conducted to provide more data on the performance of the FOBG sensor. This paper describes the results of these tests conducted so far, and the conclusions made based on the findings.

3. Experimental Program

Fiber Optic Bragg Grating (FOBG) sensors were tested in large scale concrete beams for measurement of strain. Third-point loading tests were performed on 10 ft. (3.05 in.) span beams having a cross section 10 in. x 12 in. The beams were made from high strength reinforced concrete with both embedded FOBG and electrical strain gauges attached to the reinforcing bars. Fiber Optic Bragg Grating sensors were also tested in an exposed position, attached to the underside of these beams with the objective of evaluating the sensor's performance on existing structures. A schematic diagram showing the cross sectional properties of the tested beams is presented in Fig.4.

The mix design for the concrete beam specimens gave a 7-day compressive strength of approximately 11,000 psi by using high early strength Type-III cement and silica-fume. the water-cement ratio used was 0.28. the coarse

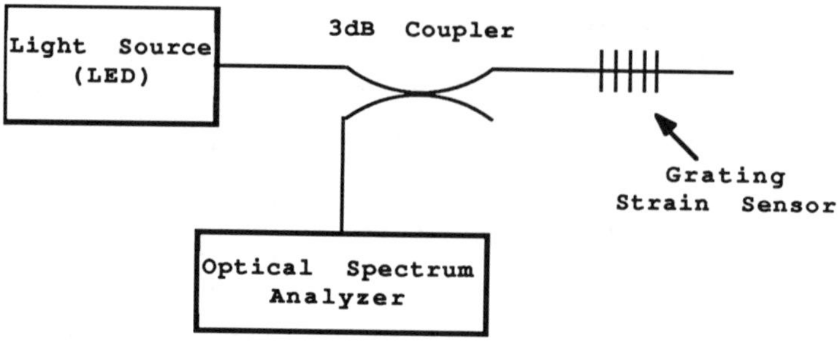

Fig.3.-Schematic of instrumentation for FOBG sensor (Maher et al., 1992)

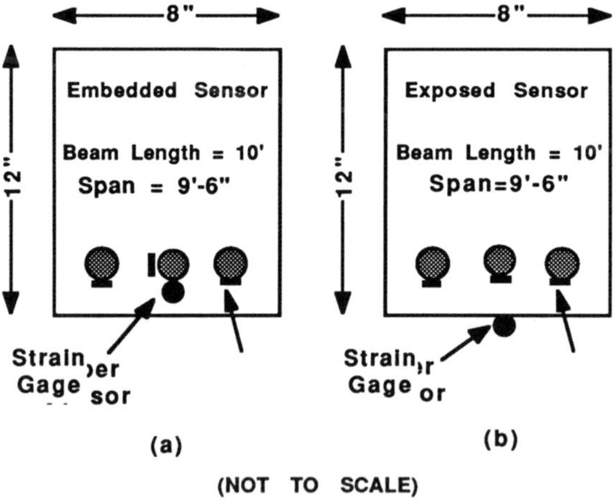

Fig.4.- Schematic Diagram of the concrete specimen cross-sections (1ft = 0.3048 m)

aggregate was crushed stone of 3/8 in.(0.95 cm.) maximum size; the fine aggregate was a natural local sand. Slump was about 6-7 in. The reinforcing steel bars used were #3 and #5 size bars with yield strengths of 58,000 psi (400 MN/m^2) and 79,000 psi (544 MN/m^2), and ultimate strengths of 80,645 (556 MN/m^2) and 109,677 (544 MN/m^2) psi, respectively.

3.1 Fiber Optic Bragg Grating Sensors Tested: In all the embedded cases tested, the FOBG sensors were attached to the reinforcing bars in the concrete beam by placing them in a small V-groove cut on the bar and mounting with epoxy. Ideally bonding of the sensor to the bar should provide good strain compatibility between the two. However, due to the low elastic modulus of epoxy in comparison to that of glass and steel, some loss in strain can occur during the strain transfer process. In order to evaluate the strain transfer process, the FOBG sensor to the reinforcing bar attachment was modeled by finite element (FEM). Results of the evaluation showed that there was some loss in strain transfer from the reinforcing bar to the FOBG sensor, mainly due to the lower modulus of epoxy in comparison to those of steel and the FOBG sensor, which was glass. The loss, however, could be minimized by choosing epoxy compounds with high elastic modulus and reducing the thickness of epoxy film between the sensor and the reinforcing bar surface (Maher et al., 1992).

In the case of the exposed FOBG sensor attached to the underside of the concrete beam, the sensor was wrapped and stretched between two steel cylinders (1" in diameter) which were mounted on the underside of the concrete beam. During the three point bending test, the FOBG sensor measured the tensile strain induced on these lower beam fibers. In both cases of embedded and exposed sensors, control electrical strain gauges were used for verification of strain measurement correlation.

An optical spectrum analyzer (ANDO AQ-6310B) was used for measurement of strain induced wavelength shift on the FOBG sensor.

4. Results

4.1. Flexural Tests on FOBG sensor embedded in 10" (0.25 m) x 1'-0" (0.30 m) x 10'-0" (3.05 m) concrete beams: Results of the FOBG sensor measurement of strain in 3 large scale concrete beam models is presented in Figs.5, 6 and 7. As shown in the figures, good agreement was observed between measurements obtained from the FOBG sensor with those obtained from electrical strain gages. In this test, the FOBG sensor was attached to one of the reinforcing bars for direct measurement of strain. A number of strain gauges were placed on the reinforcing bars for verification of FOBG sensor measurement.

4.2. Flexural Tests on FOBG sensor exposed on 10" (0.25 m) x 1'-0" (0.30 m) x 10'-0" (3.05 m) concrete beams: Results of strain measurement obtained from the exposed FOBG sensor attached to the underside of the concrete beam models is presented in Figs.8 and 9. The FOBG sensor measurement was verified by a number of embedded strain gauges attached to the reinforcing bars, and a set of LVDTs placed on the beam. Results of the tests showed good agreement between the sensors and the strain gages, taking into account the adjustments due to the difference in distance along the beam depths of the two types of strain measurement. The location at which the sensors were placed was 3 in. (76 mm) below the level of the strain gauges.

5. Conclusion

Based on the results of the ongoing research at Rutgers, it can be concluded that the FOBG sensor could become an effective tool for direct and quantitative measurement of strain in concrete structures including remote sensing. The specific conclusions reached in this study are as follows:

1. The fractional wavelength shift in FOBG sensor attached to the reinforcing bar of a large concrete beam showed good correlation with tensile strain measurements from control strain gages.

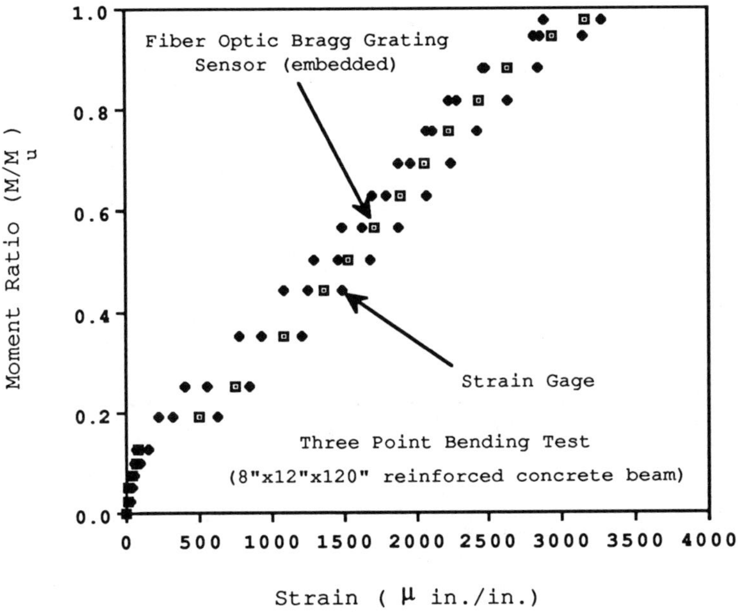

Fig.5.- Moment Ratio vs. tensile strain on a FOBG sensor embedded in a large scale concrete beam (Maher et al., 1992)

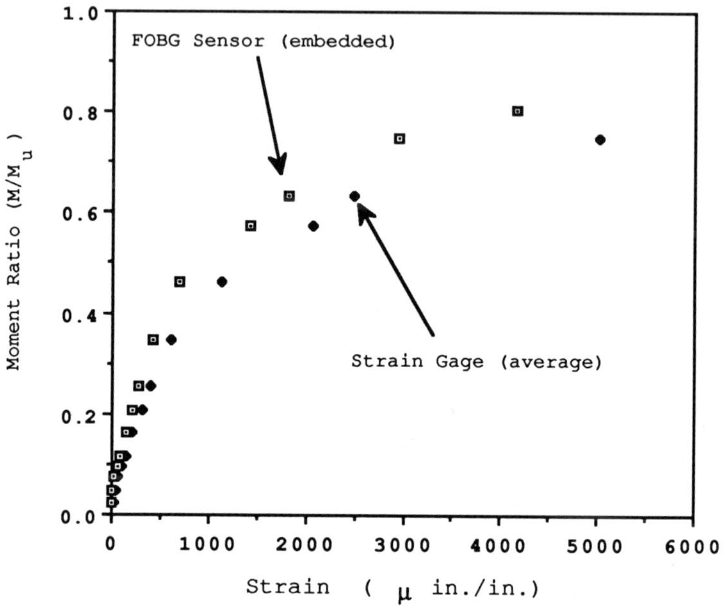

Fig.6.-Moment ratio vs. tensile strain on a FOBG sensor embedded in a large scale concrete beam

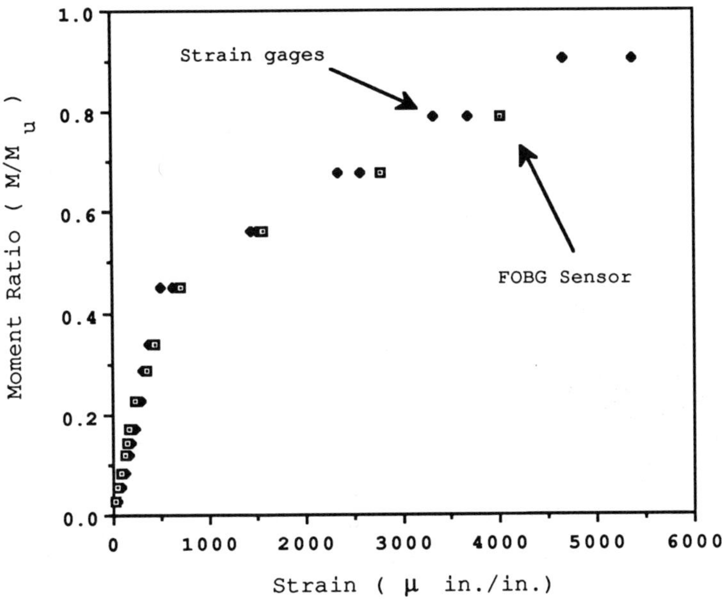

Fig.7.- Moment ratio vs. tensile strain on a FOBG sensor embedded in a large scale concrete beam

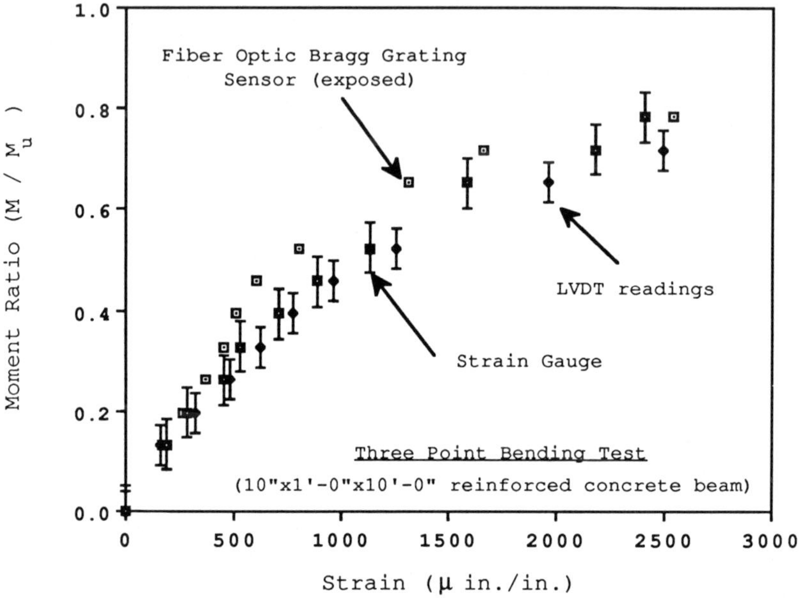

Fig.8.- moment ratio vs. tensile strength on FOBG sensor exposed on the underside of the large concrete beam (Maher et al., 1992)

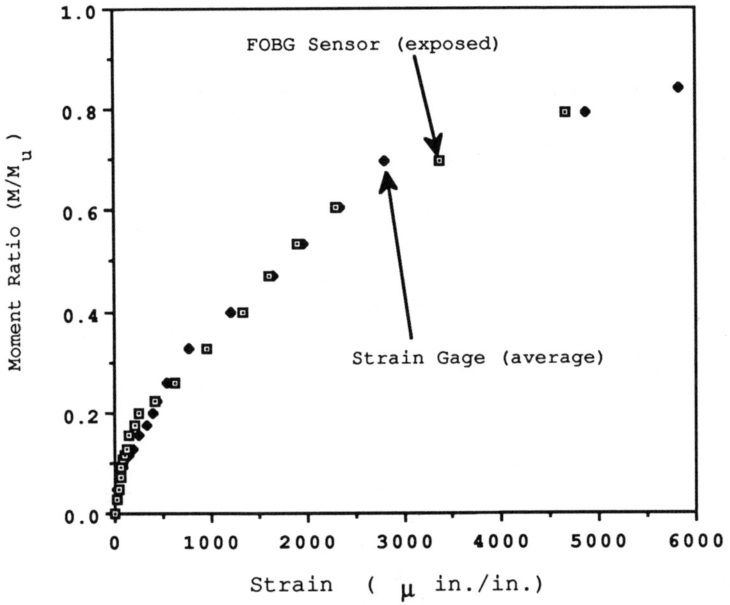

Fig.9.-Moment ratio vs. tensile strain on a FOBG sensor exposed on a large scale concrete beam

2.The fractional wavelength shift in Fiber Optic Bragg Grating (FOBG) sensor attached to the underside of a large concrete beam (exposed) showed good correlation with tensile strain measurements on the embedded reinforcing bar, and LVDT measurements outside the beam.

3.Good bonding of the sensor to the structure is crucial for maximum transfer of strain between the two. High elastic modulus epoxy, applied in thin layer between the sensor and the structure, provides the optimum bond.

4.Future research issues need to include sensor ruggedization, refinement and simplification of the instrumentation, and large scale field implementation and remote sensing.

References

Hill, K. O., Fugii, Y., Johnson, D. C. and Kawasaki, B. S., " Photosensitivity in Optical Fiber Waveguides," Appl. Phys. Lett. Vol.32, No. 10, May 1978, pp.647-649.

Meltz, G. and Morey, W. W., " Proc. OFC'90 Conference, San Francisco, 1990, p. 44.

Meltz, G., Morey, W. W. and Glenn, W. H., " Proc. OFC'90 Conference, San Francisco, 1990, p. 24.

Meltz, G., Morey, W. W. and Glenn, W. H., "Formation of Bragg Gratings in Optical Fibers by a Transverse Holographic Methods," Optical Letter Vol. 14, 1989, p. 283.

Maher, M. H., Chen, B., Prohaska, J. D., Nawy, E. G. and Snitzer, E., "Fiber Optic Sensor for Non-Destructive Evaluation of Strain in Concrete Structures," Submitted for Publication, New Experimental Techniques for Minitoring Structural Performance, ACI Convention, Vancouver, British , March 1993.

Morey, W. W., Dunphy, J. R. and Meltz, G., "Multiplexing Fiber Bragg Grating Sensors," Proc. SPIE Conference, Vol. 1586, Sept. 1991, pp. 1586-1582.

Morey, W. W., Meltz, G. and Glenn, D. H., "Fiber Optic Bragg Grating Sensors," Proc. SPIE, Vol. 1169 Fiber Optic and Laser Sensors VII, 1989, pp. 98.

Prohaske, J. D., Chen, B., Maher, M. H., Nawy, E. G. and Snitzer, E., "Fiber Optic Bragg Grating Sensor in large Scale Concrete Structures," Proc., SPIE, Vol. 1798, 1992.

Notations

λ_{Bragg} = Reflected wavelength of the light,
n = average index of refraction in the fiber core,

D = spacing of the periodic refractive index variation,
$\Delta\lambda$ = shift in wavelength,
K = constant,
ε = tensile strain (in./in.).

Acknowledgement

The study described in this paper was supported by National Science Foundation grant No. MSM-9018572. The authors wish to thank Dr. W. W. Morey and Mr. G. Meltz of the United Technologies Research Center, Connecticut, for providing the FOBG sensors, Prof. M. J. Matthewson of Rutgers' Fiber Optic Material program (FOMRP) for facilitating the control tests of the fibers. Prof. E. Snitzer, distinguished professor, FOMRP, for his active participation and support, John Prohaska of FOMRP together with Ben Chen and Ms. K. Latimer of the Dept. of Civil and Environmental Engineering for assisting the authors in concrete tests. This research was initiated and conducted in the Concrete Research Laboratory of Rutgers-The State University of New Jersey.

Dynamic Testing of Concrete with Fiber Optic Sensors

Dryver R. Huston[1]
Peter L. Fuhr[1]
Timothy P. Ambrose[1]

Abstract

This paper describes the use of embedded multimode fiber optic vibration sensors in concrete beams, slabs, and bridges. The embedding process used to place the optical fibers into the concrete as it was formed is described. Measurement procedures and the results obtained from vibration tests are presented.

Introduction

The use of fiber optic sensors for the monitoring of civil structures is currently undergoing a close examination to determine their applicability. Some of the potential advantages of fiber optic sensors are: high sensitivity, high bandwidth, multiplexing capabilities, reduced weight and size, immunity to electromagnetic interference, and in certain configurations, the ability to sense spatially distributed quantities. As a result, a wide variety of fiber optic sensors have been developed (Udd, 1991; and Krohn, 1992). Included among these sensors are those that measure standard mechanical quantities such as pressure, temperature, strain, displacement and vibration. Fiber optic sensors have already been deployed successfully in the monitoring of

[1]College of Engineering
University of Vermont
Burlington, VT 05405

composite structures such as aircraft, forming what has been termed "smart skins" (Udd, 1991). Fiber optic sensors have also been used in damage assessment of structures using strain measurements, acoustic response, and fiber breakage (Udd, 1991 and Measures et al. 1988).

Some of the reported uses of fiber optic sensors in concrete include: the monitoring of tension in post-tensioning strands (Wolff, 1990); crack detection in laboratory beams (Huston, et al. 1992); and the measurement of air entrainment levels in fresh concrete (Ansari, 1991). A distributed multimode fiber optic vibration sensor has been deployed previously on steel structural members (Spillman, et al. 1989; and Huston, et al. 1991). In this study the distributed multimode fiber optic vibration sensor is embedded and tested in various reinforced concrete members.

Statistical Fiber Optic Vibration Sensing

Fiber optic sensors may be classified as either extrinsic or intrinsic. Extrinsic sensors use the fiber as a light pipe or light conductor which carries light from a source to a remote transducer, where it is modulated, and then carried back to a detector. Intrinsic sensors use the modification of the light traveling through the fiber by external environmental effects as the transduction mechanism. Transduction in the intrinsic fiber sensor requires the perturbing signal to influence the manner in which the light in the fiber propagates. Such sensors typically rely on interference between the optical modes propagating within the fiber. Claus et al. (1989) demonstrated vibration sensing effects using a low number of optical modes with step index optical fiber sensors. Spillman et al. (1987) used a highly multiple modal domain intrinsic fiber optic sensors. The sensing technique performs a statistical analysis of the speckle pattern output from an optical fiber undergoing an external perturbation. This sensor is referred to as a Statistical Mode Sensor (SMS). The advantage of the statistical sensing scheme lies in the simplicity of the sensor components. The highly multiple mode interferometric sensor coupled with analog statistical analysis that forms the basis of the SMS vibration sensor used in this study.

SMS Fiber Sensor Theory

When coherent laser light is injected into a multimode optical fiber, the light travels down the fiber via many different paths resulting in a grainy interference pattern, composed of M individual speckles, being projected from the end of the fiber. When the fiber is perturbed, this speckle pattern changes slightly with the total intensity remaining constant. The intensity of the Nth speckle intensity, I_n, can be expressed as

$$I_n = A_n\{1 + B_n[\cos(\delta_n) - F(t)\phi_n \sin(\delta_n)]\} \tag{1}$$

where A_n is the amplitude of the nth speckle determined from mode self-interaction; B_n is the steady state mode interaction of the nth speckle; ϕ_n is a random phase factor of the nth speckle; δ_n is the phase change of the light traveling through the fiber; and F(t) is an external fiber perturbation.

For the system configuration used in this investigation, shown in Figure 1, a photodetector is illuminated by a subset of the available speckles. Therefore the photodetector output corresponds to an integration or summing of the speckle subpattern. Ignoring the steady-state DC response, the output signal is proportional to the intensity variation within the summed speckle subpattern ($m \ll M$), ΔI, namely

$$\Delta I = [\sum_{n=1}^{m} C_n \sin(\delta_n)] F(t) \tag{2}$$

where C_n is the amplitude of the nth speckle; M is the total number of output speckles; and m is the number of speckles detected. It has been shown (Spillman et al., 1989) that the bracketed term sums to a constant, K, which is dependent on experimental parameters. In the case of a vibrating perturbation force

$$F(t) = F_0 \sin(\omega t) \tag{3}$$

the summed speckle signal, ΔI, contains the vibration frequency component.

The implementation of the speckle pattern analysis is based on a statistical analysis of a signal resulting from photodetection of a piece of the output pattern of the fiber. Subsequent electrical signal processing and filtering is then performed. As illustrated in Figure 1, statistical intensity processing is achieved optically by spatially filtering the speckle pattern, i.e. illuminating a transparency mask. The mask passes only a portion of the entire speckle pattern which illuminates a photodetector and satisfies the $m \ll M$ criterion. By using a simple amplifier and high pass filter, which eliminates the DC portion of the optical field, a stable and

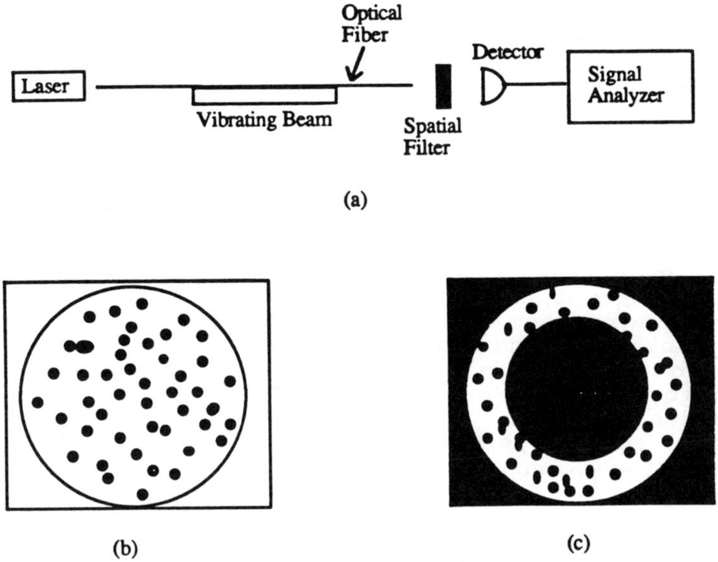

Figure 1. Fiber optic sensor: (a) configuration for measuring vibrations, (b) closeup of the time-varying speckle pattern, and (c) transmission of only a subsection of the total speckle pattern by annular filter.

repeatable signal is found. The resultant electrical signal varies linearly with respect to the vibratory perturbation field of the sensor. A detailed theoretical description of the performance of the sensor if given in Spillman et al. (1989). The sensor bandwidth is limited by the speed of the photodetector and the attenuation of the elastic waves as they traverse the jacketing layers of the fiber. The system used in these experiments was fabricated from inexpensive components and maintained a relatively flat band response from less than 1 Hz to in excess of 10 kHz.

Concrete Beam Tests

The first tests involved fabricating and then vibrating a set of small reinforced concrete beams with embedded SMS sensors. Four 1 meter long by 12.5 cm by 12.5 cm concrete beams, reinforced with two No. 4 rebars were fabricated using a commercial premixed concrete with gravel aggregate. Four multimode fiber loops and two No. 4 reinforcing bars with a measured yield strength of 372 MPa were placed into each form prior to pouring the concrete. Eight holes were drilled in one end of the form to allow the exit of the optical fibers and the reinforcement rods. The optical fibers were fitted with standard modular SMA connectors. Two of the fibers were wrapped around the rebar then covered with concrete aggregate. The other fibers were placed roughly in the middle of the beam, fashioned into a U-shape, then also covered with concrete. The concrete was vibrated manually, smoothed and allowed to cure for a minimum of 28 days. The fibers were then routed back to the laser source and detector. A disadvantage of this setup is that the fiber is a sensitive to mechanical perturbations anywhere along the length of the fiber, including that which is not embedded in the beam. The portion of the fiber external to the concrete was rigidly mounted to the laboratory table to minimize the transduction of spurious vibrations. Thus, in this arrangement, it was not possible to ascertain exactly which portion of the beam was vibrating, only that vibration along the fiber was occurring.

The concrete beams were subjected to a series of vibration tests. A concrete beam was mounted on rubber supports that were placed 0.22 m from the ends. This mount simulates a free-free boundary condition for the first flexural mode. The beam was vibrated by an electromechanical shaker that was pressed onto the top of the beam at midspan. Vibrations were measured with an accelerometer and with the SMS transducer. The setup is shown schematically in Figure 2.

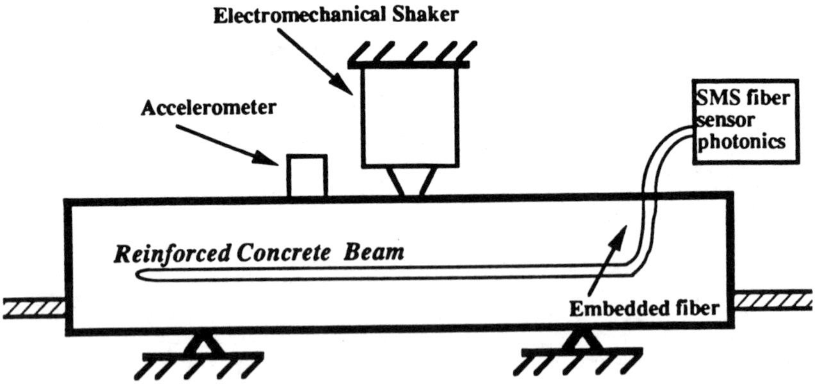

Figure 2. Schematic of beam vibration test.

The first set of tests involved driving the shaker with a sinusoidal signal. Figure 3. is typical of the results where the spectra of the accelerometer and the SMS transducer are very similar except that the SMS transducer carries a signal at the 60 Hz line frequency. The next set of tests used white noise as the driving signal. The frequency response function between the accelerometer and the SMS transducer is shown in Figure 4.

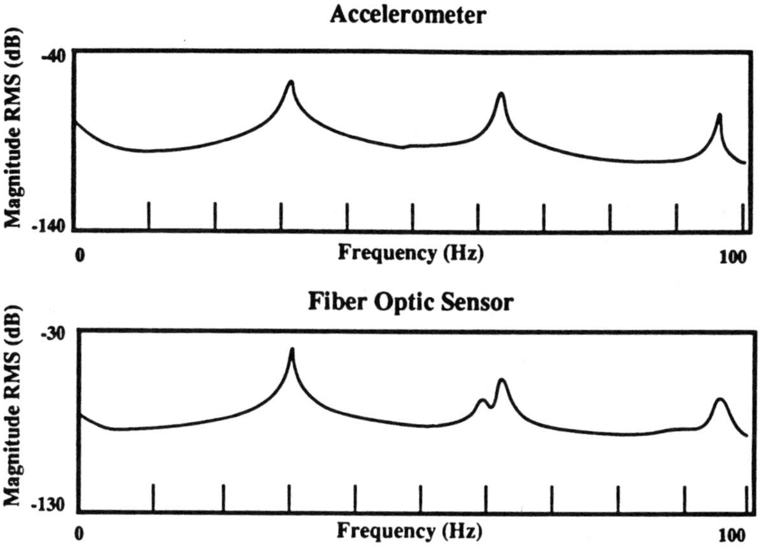

Figure 3. Spectra of the accelerometer and SMS transducer for a sinusoidal driving signal.

<u>Vibration Isolation Slab Test</u>

The Stafford Medical Research Building is a 65,000 sq. ft., 5 story, reinforced concrete building located on the campus of the University of Vermont, Burlington, VT, that has a vibration isolation slab in the basement. As part of a fiber optic installation project for the entire building, the isolation slab was fitted with an array of multimode fibers for use with the SMS sensors, Figure 4. The installation techniques for these fibers are described in detail in Fuhr et al. (1992) and Ambrose, et al. (1992).

The SMS transducers were evaluated with a set of impact-response tests in which the slab was struck with an instrumented rubber-tipped sledge hammer and the response of the slab in the form of an elastic wave was measured with the SMS transducer. The time of travel for the elastic wave versus position from the fiber is shown in Figure 5.

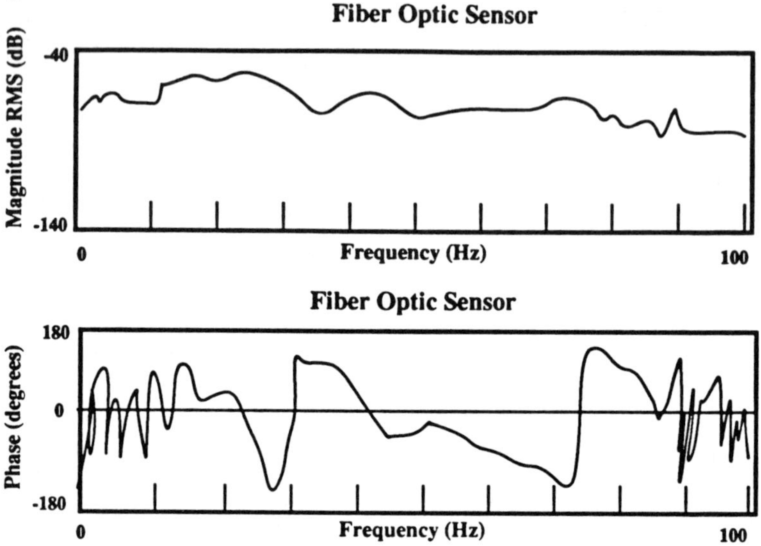

Figure 4. Frequency response function between the accelerometer and SMS transducer for the white noise driving signal tests.

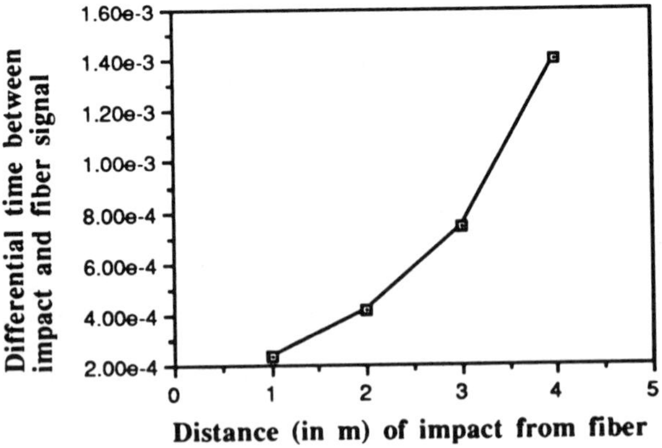

Figure 5. Elastic wave travel time versus distance from the fiber.

Middlebury, VT Railroad Bridge

A post-tensioned reinforced concrete railroad bridge was instrumented with an array of fiber optic sensors as it was constructed during the summer of 1992. The bridge has a total span of 14.1 m. 17 fiber sensors were installed in the bridge including four plastic core fibers. Figure 6 shows the fiber layout in the bridge.

The performance of the SMS transducers were assessed with an impact-response test that was very similar to the vibration isolation slab. A typical transient response that was recorded at the site is shown in Figure 7.

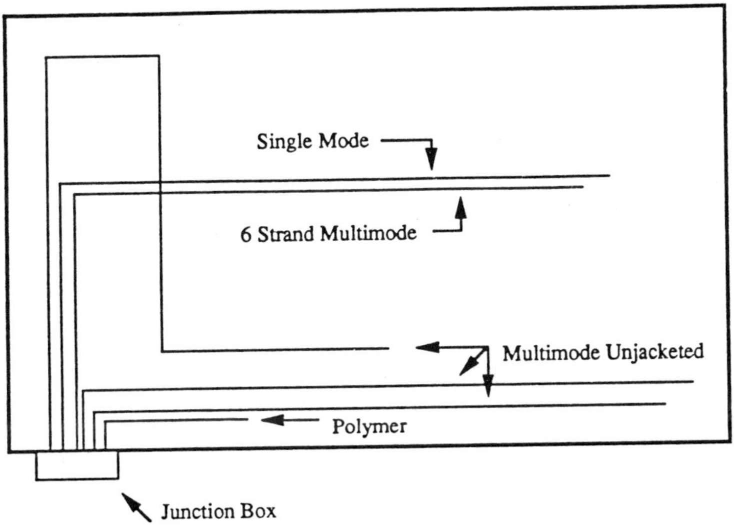

Figure 6. Fiber layout in the Middlebury, VT railroad bridge.

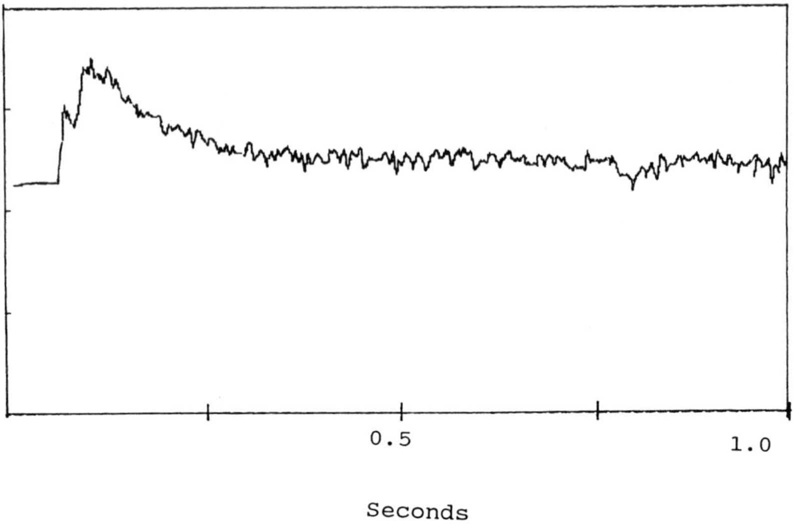

Figure 7. Transient time history recorded from SMS transducer in the Middlebury bridge.

Summary

While the use of fiber optic vibration sensing may be limited, it is envisioned that other potentially more exotic embedded sensors may allow further health monitoring and interrogation of concrete structures.

Acknowledgements

This work was supported by NSF grants no. MSM-8806126 and BCS-9024390 and optical fiber donations by the Ensign-Bickford, Northern Lights Cable, Siecor, and Hoechst-Celanese Corps.

References

Ambrose, T.P., Fuhr, P.L., Huston, D.R., and Kajenski, P.J. (1992). "Lessons Learned in Embedding Fiber Sensors in Civil Structures", SPIE OE/Fibers Conf., Boston, MA, Sept. 1992.

Ansari, F. (1991). "Rapid in-place air content determination in fresh concrete," Concrete International, pp. 39-41, January.

Claus, R.O., Bennet, K.D., Vengsarkar, A.M., and Murphy, K.A. (1989). "Embedded optical fiber sensors for materials evaluation," Jnl. of Nondestructive Evaluation, Vol. 8, No. 2, pp. 371-381.

Fuhr, P.L., Huston, D.R., Kajenski, P.J., and Ambrose, T.P. (1992). "Performance and Health Monitoring of the Stafford Medical Building Using Embedded Sensors", Smart Materials and Structures, Vol. 1, 63-68.

Huston, D., Fuhr, P.L., Beliveau, J.-G., and Spillman, W.B. (1991). "Structural Member Vibration Measurements Using a Fiber Optic Sensor", Jnl. of Sound and Vibration, 149(2), 348-353.

Huston, D.R., Fuhr, P.L., Kajenski, P.J., and Snyder, D. (1992). "Concrete Beam Testing with Optical Fiber Sensors," Proc. of the ASCE Minisymposium on the Nondestructive Testing of Concrete, F. Ansari, ed., San Antonio, TX, April 1992.

Krohn, D.A. (1992). Fiber Optic Sensors - Fundamentals and Applications, 2nd ed., Instrument Society of America, Research Triangle Park, NC.

Measures, R.M., Glossip, N.D.W., Lymer, J., and Tennyson, R.C. (1988). "Structurally integrated fiber optic damage assessment system for composite materials", Proc- SPIE Vol. 986, pp. 120-124.

Spillman, W.B., Jr., Kline, B.R., Fuhr, P.L., and Maurice, L. (1989). "Statistical Mode Sensor for Fiber Optic Vibration Sensing Applications", Applied Optics, Vol. 28, No. 15.

Spillman, Jr., W.B., Kline, B.R., Maurice, L.B., and Fuhr, P.L. (1989). "Statistical-mode sensor for fiber optic vibration sensing uses," Applied Optics, 28, 3166-3176.

Udd, E. (1991). Fiber Optic Sensors: An Introduction for Engineers and Scientists, J. Wiley, New York, NY.

Wolff, R. (1990). "Applications with fiber sensor system for monitoring prestressed concrete structures", in Structures Congress Abstracts, R. Corotis and B. Ellingwood, eds., ASCE, New York, pp.94-95.

Applications of Embedded Optical Fiber Sensors for Non-Destructive Testing of Concrete Elements and Structures

A. Méndez*

ABSTRACT

In this paper an analysis is presented on the use of embedded optical fiber sensors in concrete elements and structures for the non-destructive measurement of internal stress and strain as well as for the assessment of structural integrity. A discussion of the fundamental materials and micro-mechanical aspects regarding the fiber/matrix interaction is given. In addition, a summary of the experimental results obtained to date is made along with the applications sought for this new technology.

INTRODUCTION

Since the infrastructure of civil engineering works around the world is in state of deterioration due to the aging of the materials, excessive use, overloading, weathering, lack of maintenance and proper inspection. It has become increasingly important in the last few years to determine the safety of a structure by the NDE of its strength and integrity. This assessment is essential for the repair, retrofit, rehabilitation, life extension or replacement of the structure in question. Furthermore, it would be very useful to develop means for the feedback and control of the state of health of a structure.

The field of civil engineering, although highly advanced in materials and structural analysis and design, is comparatively deficient in its evaluation techniques. Many of the methods used are destructive. Various non-destructive methods for testing concrete have been developed (Malhotra, 1976). However, not all of these methods are very reliable and, in some instances,

*Laboratory for Lightwave Technology, Brown University, Providence, RI 02912

they can only be used on a comparative qualitative basis. NDE techniques being studied rely on the absorption, transmission and scattering of X and γ rays; response to nuclear activation; electrical resistance; projectile penetration; mechanical and ultrasonic pulse velocity; vibration; magnetization and several others.

Optical fibers, because their small size and lightweight, offer the possibility to be embedded within cement or concrete without affecting their properties and used as sensitive, but rugged, transducers of mechanical perturbations. The basic principle behind embedded fibers to characterize the state of a composite material is that light sent through the fiber has either its intensity, phase, wavelength or polarization altered by changes in the mechanical and thermal states of the surrounding host. In this respect, fibers have the ability to provide high resolution temperature and strain measurements, detect the onset and growth of cracks, as well as to monitor creep and thermal stresses. In addition, the actual location of a fault can be determined by interrogating the back-reflected signal coming from the fiber, using optical time-domain reflectometry (OTDR) techniques. All this information can ultimately be used to provide "real-time" information on the state of a specific concrete element or structure by means of a built-in damage detection and evaluation system based on a grid of optical fiber sensors embedded within the structure during its construction.

The use of optical fiber sensors embedded within composites for the measurement of internal strain and the detection of structural damage in aerospace applications has proved to be an effective non-destructive evaluation (NDE) technique and become the subject of substantial research (udd, 1988). Recently however, the focus of the applications has also centered around the field of civil engineering with aims at exploring the possibilities of embedding, prior to curing, optical fiber sensors into reinforced concrete elements and structures such as buildings, bridges, dams and tanks for the NDE of structural integrity and the measurement of the internal state of stress. The first proposal on the application of optical fiber sensors embedded in concrete buildings and structures was given by Méndez et al.(1989), who described the potential applications of this approach along with a study on the fundamental issues regarding its practical implementation. Subsequently, a small number of researchers in the U.S. and Europe have been working on different ways to measure stress and strain in concrete using optical fibers as well as to determine the health condition in a building.

However, in order to implement these ideas, a fundamental understanding of the factors involved in the embedding process in concrete is needed. In this paper, an analysis of the materials and micro-mechanical issues of the fiber/matrix interaction is presented. Different schemes for the internal measurement of stress and strain with fiber sensors will be described. In addition, a summary of the experimental results obtained to date will be given.

OPTICAL FIBER FUNDAMENTALS

The field of fiber optics has undergone a tremendous growth and advancement over the last 25 years. Initially conceived as a medium to carry light and images for medical endoscopic applications, optical fibers were later proposed in the mid 1960's as an adequate information-carrying medium for telecommunication applications. Ever since, optical fiber technology has been the subject of considerable research and development to the point that today lightwave communication systems have become the preferred method to transmit vast amounts of data and information from one point to another. Among the reasons why optical fibers are such an attractive are their low loss, high bandwidth, EMI immunity, small size, lightweight, safety, relatively low cost, low maintenance, etc.

At the heart of this technology is the optical fiber itself. A hair-thin cylindrical filament made of glass that is able to guide light through itself by confining it within regions having different optical indices of refraction. A typical fiber structure is depicted in Fig. 1. The central portion, where most of the light travels, is called the *core*. Surrounding the core there is a region having a lower index of refraction, called the *cladding*. From a simple point of view, light trapped inside the core travels along the fiber by bouncing off the interfaces with the cladding, due to the effect of the total internal reflection ocurring at these boundaries. In reality though, the optical energy propagates along the fiber in the form of waveguide modes which satisfy Maxwell's equations as well as the boundary conditions and the external perturbations present at the fiber (for a more detailed discussion on optical fibers and their fabrication see Miller and Chynoweth).

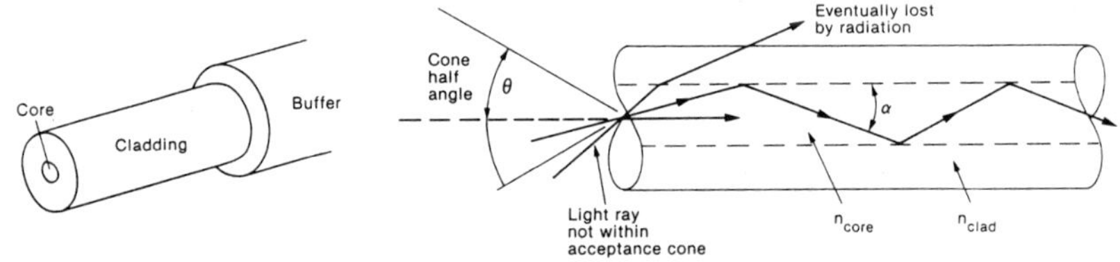

Figure 1: Schematic of an optical fiber.

Optical fibers are also attractive for other applications such as in sensing, control and instrumentation. In these areas, optical fibers have made a significant impact and are being the subject of substantial research over the last few years. In general, for these applications fibers are made more susceptible and sensitive to the same external mechanisms against which fibers were made to be immune for their effective operation in telecommunications. In its simplest form, an optical fiber sensor is composed of a light source, optical fiber, sensing element and a detector (see Fig. 2). The principle of operation of a fiber sensor is that the sensing element S modulates some parameter of the optical system (intensity, wavelength, polarization, phase, etc.) which gives rise to a change in the characteristics of the optical signal received at the detector. The fiber sensor can be either an *intrinsic* one, if the modulation takes place directly in the fiber, or *extrinsic*, if the modulation is performed by some external transducer (for a more detailed account of optical fiber sensors and their operation consult Culshaw and Dakin).

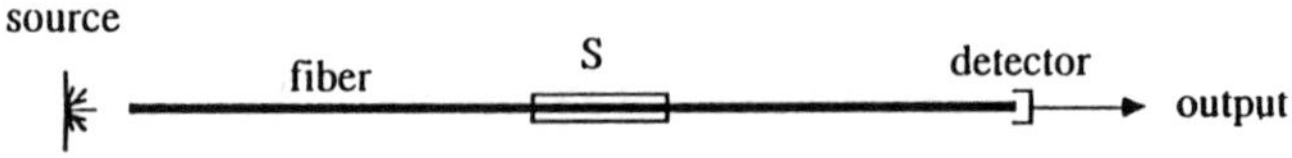

Figure 2: Basic fiber optic sensor

To date, several different types of fiber sensors are commercially available to measure parameters such as pressure, temperature, refractive index, displacement, gas concentration, etc. Others are a high stage of development allowing for a reliable and accurate measurement of stress and strain, electric current, vibration, sound, flow, etc. In the sections to follow we will explain how optical fibers can and are being used to measure diverse parameters of interest for geomechanical applications.

EMBEDDING OF OPTICAL FIBERS IN CONCRETE

In order to perform properly the function of internal strain/temperature sensor, several issues regarding the embedding of the optical fibers and the phenomena at the fiber/concrete interface must be investigated. For instance, the interfacial properties determine the efficiency of load transfer to the embedded fiber, as well as having a strong effect on the reliability and reproducibility of the sensor's response over time. Furthermore, an understanding of the micromechanics in the interphase region is of vital importance in the interpretation of what the sensor is measuring and how does this measurement relate to the measurand of interest in the composite.

In addition, the mechanical and chemical durability of the silica glass optical fibers in the harsh cement environment must be addressed. This presents two major problems. First, the introduction, without damage, of the fibers into the plastic, aggregate-filled and mechanically vibrated environment of the cement/concrete paste. Secondly, the water-rich and highly alkaline environment of a cement paste will have a tendency to attack glass fibers. Next, we will discuss some of these ideas in more detail.

- *Chemical Attack*

It is well known that water and hydroxyl ions react with the surfaces of glass fibers reducing their strength by promoting accelerated crack growth under stress (Wang et al., 1979). However, water is only present in cement and concrete mixes during the first few days of curing, since most of it is lost by evaporation and chemical reaction. Furthermore, depending on its characteristics (Fintel, 1985), concrete can be highly impermeable which explains in part why the reinforcing steel does not corrode once inside the concrete.

On the other hand, cement is strongly alkaline (with a pH of approximately 12) and will tend to corrode glass fibers (see Fig. 3). Fortunately, considerable research has been done on the use of glass fibers as a reinforcing material in portland cement (ACI, 1982). These studies have led to the development of alkali-resistant glass fibers (Majumdar, 1970), like the so-called Cem-Fil fiber (Proctor, 1975), characterized by having about 17% weight of zirconia (Z_rO_2) in its composition. Experimentally, it has been demonstrated that polymeric coatings on the fibers improve their chemical durability (Golosova et al., 1984 and Datta et al., 1986). Particularly interesting is the protection obtained by coating standard telecommunication silica-glass optical fibers with a polyetherimide coating (Kuwabura et al., 1989).

- *Interfacial Bonding*

Although an adequate bond might develop between fibers and concrete due to curing shrinkage (caused by the loss of moisture with a decrease in volume of about 0.05 percent), as well as by the mechanical keying and constriction caused by the formation and growth of crystals of hydrated calcium aluminosilicates, it is still not clear whether or not a good bond exists between glass fibers and concrete for an efficient load transfer. Therefore, experimental results such as

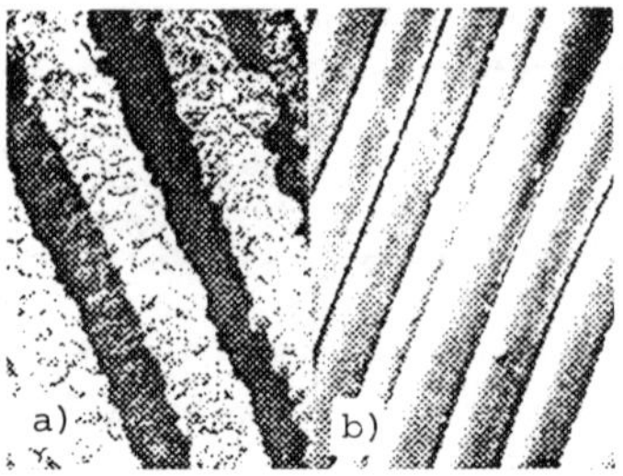

Figure 3: SEM pictures of a) Corroded glass fibers and b) Alkali-resistant fibers, after exposure to a cement environment. (Photo courtesy of Fiberglass UK Ltd.)

fiber pull-out tests are needed to determine the shear strength at the fiber/matrix interface.

The characteristics of the optical fiber coating play an important role as well. Good coating adhesion to both the concrete matrix and the fiber are necessary in order to prevent any delamination of the fiber from the coating, or worst yet, decoupling of the coated fiber from the concrete host. This is very important, for otherwise, any assessment made on the condition of the structure will be meaningless. Theoretically, Pak has determined that the shear transfer to the fiber, and hence the load transfer itself, can be controlled by the thickness and the modulus of the coating. It is thus shown that when the coating is softer than the matrix, a thinner coating will induce more load transfer; conversely, when the coating is stiffer than the matrix, a thicker coating will induce more shear transfer.

Some experimental studies on the coating performance of coated optical fibers embedded in epoxy resin revealed delamination at the coating/fiber interface in preference to the coating/epoxy resin interface (Haaksma et al., 1990).

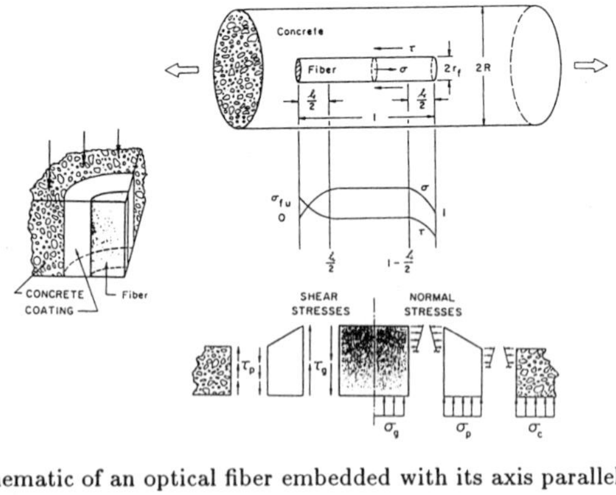

Figure 4: Schematic of an optical fiber embedded with its axis parallel to the applied load.

• *Fiber Orientation and Coating Effects*

Assuming that a perfect bond exists between fiber and coating as well as between coating and concrete matrix, the only parameters that affect the response of the embedded sensor are the fiber orientation with respect to the applied load and the mechanical properties of the layered matrix/coating/fiber region.

Consider the two most general cases. When the embedded fiber runs parallel to to the direction of the applied strain and when the fiber is perpendicular to it. This in turn results in axial and radial deformations, respectively, of the fiber. Fig. 4 depicts the case of a fiber parallel to the applied load. Although a good bond is assumed at the interface, notice how the stress seen by the fiber varies as a function of position along its length. There is, however, a region in which the stress is uniform all along. The minimum distance for which this occurs is called the critical length (l_c) of an embedded fiber. Physically what this means is that, in order to develop a strain in the fiber equal to that of the surrounding host, the gage length of the embedded section must be at least as long as the critical length for a given set of radii and elastic moduli. Nanni et al. have computed l_c to be approximately 20 times the fiber's diameter.

For the case of a fiber embedded with its axis perpendicular to the applied load (see Fig. 5), the actual strain present in the fiber is now dependent on the stiffness of the coating material as compared to that of the glass fiber. In other words, the fiber coating acts as a mechanical buffer that can drastically diminish the load transfer between the matrix and the fiber. The relation between strain in the fiber to strain in the concrete matrix is found to be (Nanni et al., 1991):

$$\epsilon_{fiber} = \epsilon_{concrete} \frac{2E_{coating}}{E_{fiber} + E_{coating}} \tag{1}$$

Therefore, for the perpendicular configuration, it becomes necessary to coat the fibers with hard coatings in order to insure adequate load transfer. Standard telecommunication-grade optical fibers are not suitable for this application since their soft primary coating is designed to absorb mechanical perturbations to prevent microbending losses.

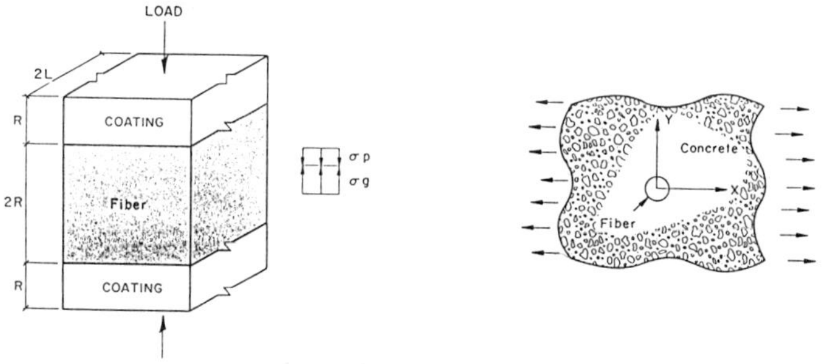

Figure 5: Optical fiber embedded with its axis perpendicular to the applied load.

• *Fiber Embedding and Placement*

One of the main difficulties associated with the practical implementation of these ideas consists in the introduction of the fibers into the uncured concrete mix. Furthermore, the placement of the fibers inside the concrete cannot be random, but rather it must comply with the requirements and objectives imposed by the type of measurement or detection one wishes to make with the embedded sensor.

Since about 70% of the total volume in a concrete mix is occupied by stone aggregates and other fillers, many obstacles will be encountered in the way of the fiber in any given direction. In addition, fibers will be subjected to severe mechanical abuse during the pouring process as a result of the pumping, vibrating and shoveling of fresh concrete. To overcome these difficulties different techniques for embedding fibers into concrete have been proposed (Méndez et al., 1989). In one of such techniques (see Fig. 6a), the optical fiber is protected by a metallic tube of a small diameter such as hypodermic stainless steel tubing. The tube is laid in the desired position within the formwork while concrete is poured on it. Once the filling operation has been completed, and before hardening starts, the tube is gently slid out leaving the fiber in place. This technique has the added advantage of promoting the formation of a smooth cement boundary at the surface of the fiber as a result of cement "bleeding". Fig. 6b illustrates a different approach where the sensing region of the fiber (gage length) is allowed to move freely inside a protective sleeve. Loads are transmitted to the fiber ends by means of sliding metallic fiber clamps. Fiber embedding experiments using cabled optical fibers protected by plastic jackets have also been tried (Huston et al., 1992). However, it is not clear whether such approach will result in an adequate transfer of load to the fiber.

All of the above problems, along with the unfavorable conditions that in general prevail at most construction sites have led to the development of pre-fabricated fiber-optic sensing panels (Rossi et al., 1989). In this way, fiber sensors can be carefully protected and placed inside special panels or frames which are subsequently transported and installed on site.

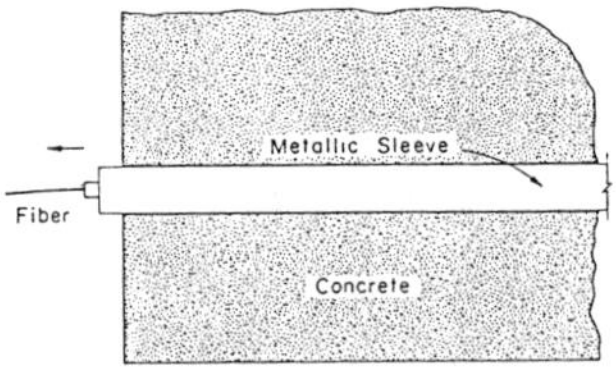

a)

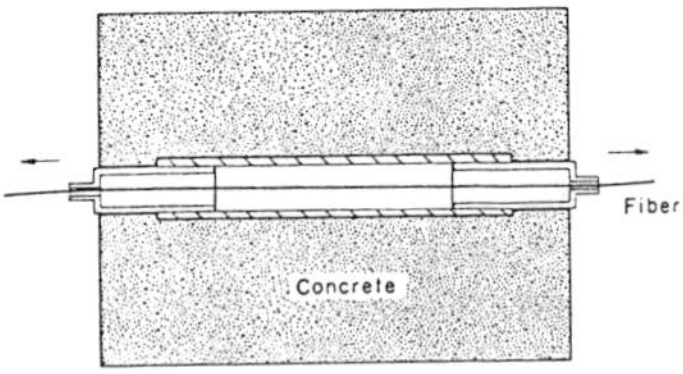

b)

Figure 6: Fiber embedding methods.

As mentioned before, an adequate placement of the fiber sensors is crucial for their effective use. Sensor location is strongly dependent upon application. Thus, what it might be an appropriate placement for a strain sensor, it might not necessarily be so for a crack or curing sensor.

To understand this better, consider the nature and behavior of a reinforced concrete (RC) structure. A RC structure is a continuous frame in which all of its members are rigidly connected to one another. Said members experience deformations under the action of tensile and compressive forces produced by the loads that the structure must bear. Associated with these deformations are internal stresses and strains that differ in character and intensity upon their location in the structure as well as on the nature of the loads. However, because concrete is strong in compression but weak in tension, tensile stresses and strains are of greater concern. Thus, fiber-optic strain sensors should be placed running parallel to the steel reinforcement where the tensile stresses are to be found. Fibers should also be placed next to stirrups, which are steel bars used as reinforcement against diagonal tensile stresses resulting from the interaction between horizontal and vertical shear stresses in the concrete.

For the case of detection and monitoring of cracks, fibers should be loop around an actual crack, or in the potential site for the formation of one. In this way, the fiber can measure the displacement of the crack opening without sacrificing its own integrity (Fig 7a). This is applicable to regions in the structure where cracks are to be expected, but that their size must be monitored to be kept within safe limits. Conversely, there are instances where the sole presence of a crack is cause of alarm, as is the case in prestressed concrete elements. In this type of application, fibers must be placed perpendicular to the probable direction of crack growth as seen in Fig. 7b to detect their presence (Rossi et al., 1989 and Ansari, 1992).

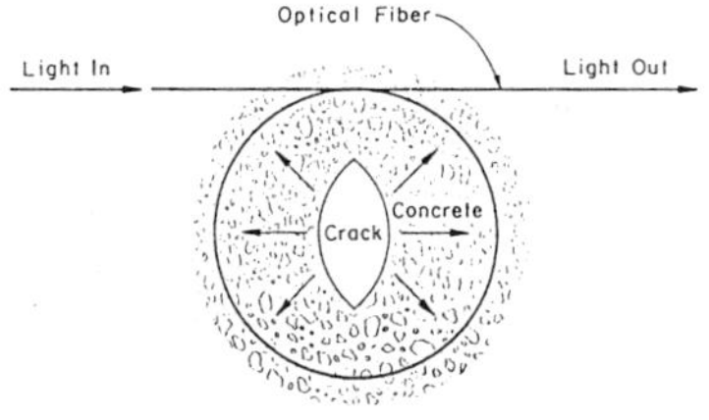

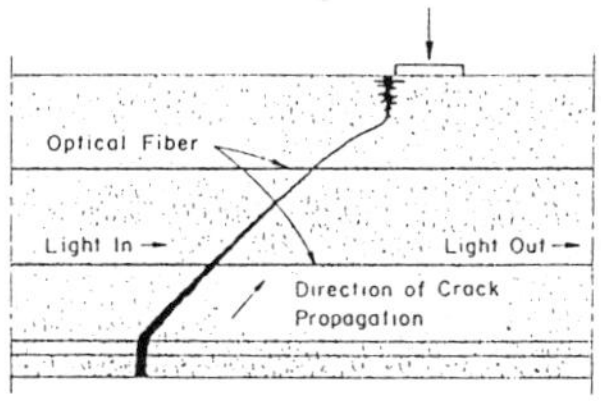

Figure 7: Placement of a fiber sensor for detection of:a) Crack aperture and b) Crack onset.

APPLICATIONS

The fundamental applications envisioned for this new technology are listed in Table 1 below. They can be grouped in three main areas, namely:

- Structural monitoring and damage evaluation
- Experimental stress analysis
- Management and control of systems and service installations

The first group explores the incorporation of single- and multi-mode fiber sensors within structural concrete elements such as beams, columns, arches, slabs and others, so that stress, strain, flexure, bending, curing, cracks and creep in concrete can be measured individually, as well as the deflection and bending of structures as a whole. In the next section, a summary of the experimental results obtained to date by various groups working in this field will be made.

Table 1: Applications of embedded optical fiber sensors in concrete

MATERIAL	STRUCTURAL	INSTALLATIONS
strength	stress	heating
curing	strain	cooling
creep	shear	pressure
shrinkage	bending	electric currents
uniformity	deflection	security
voids and defects	cracks	fire alarm
modulus of elasticity	thermal stresses	elevators
steel corrosion	damage detection	telecommunications

In the field of experimental stress analysis, fibers would make sensitive and versatile sensors for the measurement of mechanical characteristics of structural members in experimental studies. This might be particularly useful with structures and members having complex shapes for which analytical solutions can be difficult to obtain. An answer can be obtained by measuring experimentally the state of stress in a model. Another example would be to use a grid of fiber strain sensors embedded in the runways of airports to evaluate the stresses on the pavement during the landing and take-off of airplanes. The 2-D stress mapping obtained in this fashion would be helpful in the redesign and maintenance of such pavements.

Furthermore, embedded fiber optic strain sensors would enable structural engineers to compare between measured and designed values of stress, bending moments and deflection. From this information more accurate design factors can be determined which would make structures safer and more economical to build.

The third set of applications considers that building services and installations (such as heating, air conditioning, lighting, electricity distribution and consumption, security, fire alarms, etc.) can be operated more efficiently and economically using fiber optic sensors that monitor and measure the state of affairs of pertinent parameters. For instance, reading of the pressure, flow and temperature of water pipes could be used to regulate the temperature and distribution of running water by controlling valves, pumps and boilers as required. Similarly, the temperature in rooms, hallways and offices could be monitored and controlled automatically. Electric currents could be controlled and regulated according to demand. Fibers coated with special heat-sensitive coatings could be used as distributed heat sensors for use in fire alarm systems. And the list goes on. In this fashion, integration of all the information supplied by a network of fiber and regular sensors into a single processing center within the same facility would result in a "smart building".

PREVIOUS WORK

As we mentioned before, the potential of fiber optic sensors for the non-destructive evaluation and monitoring of concrete structures has just begun to be appreciated. To date, only a small number of studies report research activity connected with this field. However, it is expected that this will change in the near future. Next, we will describe briefly the latests developments in the field by area of application.

- *Thermal Stresses and Curing of Concrete*

Concrete curing is always accompanied by the liberation of heat from the chemical reactions which bring about the hardening of the portland cement. This heat of hydration causes a rise in the internal temperature of concrete in comparison to the exterior surfaces. As cooling begins, the outermost layers rapidly gain both strength and stiffness. Any restraint of the fine contractions will result in tensile stresses. These effects are not critical in regular cast concrete structures, but they are a major consideration in massive concrete. Evidently in this latter case, large amounts of heat need to be dissipated during the curing process if flawless concrete members are desired. In this respect, embedded fiber optic temperature sensors would make possible the measurement of the internal temperatures in massive concrete, allowing the control of the cooling rate during the curing process.

The ideas outlined above have already been successfully implemented in practice by a construction company in Japan which reports monitoring concrete curing in a distribution tunnel using a commercially available fiber-optic distributed temperature sensor (Photonics Spectra). In addition, using a different approach, Huston et al. have monitored the curing of concrete by measuring the light transmission of embedded multimode optical fibers inside a test beam.

Furthermore, information on the dynamics of heat transfer for an entire building or structure (caused by the thermal stresses due to changes in ambient temperature such as freezing and thawing, wetting and drying or heating and cooling) could be obtained by a network of embedded temperature sensors situated in strategic locations.

Although not a curing sensor, a fiber optic refractive index sensor has been developed by Ansari et al.(1991) to detect the air-bubble content in entrained fresh concrete . The fiber sensor detects changes in the reflected light intensity from a scanning probe tip as a consequence of the difference in refractive index between cement paste and air.

- *Crack Detection and Monitoring*

The presence of cracks in concrete is inevitable. Cracks are caused by the stressing of the steel reinforcement under normal load conditions, restrained drying shrinkage or by temperature-induced volumetric changes. The onset and growth of cracks can easily be detected by monitoring the intensity of the light transmission of strategically embedded fibers. While their location can be determined by means of OTDR techniques.

Rossi and LeMaou report the detection of cracks in concrete using an embedded multimode ($100\mu core$) optical fiber, which has its coating material removed in several small sections. A drop in the fiber's light throughput was observed whenever a crack propagated across any of the uncoated fiber sections. The method was applied to monitor the presence of cracks at different locations within the shell of a motorway tunnel. During their embedding, fibers were protected by metallic tubes which were subsequently removed after the concrete pouring took place.

Crack opening displacements in concrete beam specimens were measured by Ansari (1992) using circular fiber loops. The light transmission through the loop varied as a function of the transverse perturbation induced by the crack growth. However, it was concluded in this study that the main disadvantage of such approach is that the crack paths must be known a priori.

• *Bending, Deflection and Displacement*

Other mechanical effects of interest to civil engineers and architects are the deflections, bending and displacements of long-span members, frames and massive structures. In this area, Wolff and Miesseler measured the elongation and tensile stresses present in a $53m$ long bridge by embedding optical fiber sensors inside the deck. A total of 8 optical fibers were used (4 placed in the upper and 4 in the lower side of the concrete slab). Measurement of the fiber's elongation over time revealed information regarding the effects of temperature and creep in the bridge.

A different approach proposed measuring the loads acting on a concrete bridge by means of multimode optical fibers embedded inside elastomeric bearings fitted between the bridge's deck and supports (Caussignac et al., 1992). The magnitude of the loads are determined by the attenuation induced by the microbending loss in the fibers.

Displacements of the span joints in a concrete dam have been monitored using specially designed opto-mechanical arrangements (Holst and Habe). In one of its configuration, out-of-limit displacements are detected by the loss in light transmission through a multimode fiber when pressed by a spring-loaded movable ram. A second configuration allowed measurement of the relative joint displacements from the attenuation changes resulting from bending a small fiber loop mechanically connected to the joints.

• *Stress and Strain Measurement*

Knowledge of the state of stress and strain in a structure is vital for the prediction of its behavior. Mechanical strains present, from which stresses can subsequently be inferred, can be determined with relative ease by embedding fibers within critical regions. Dimensional and/or refractive index changes in the fiber can then be used to measure the strain transferred to the fiber.

Different fiber sensors are suitable for this purpose. Experimentally though, Nanni et al. have measured compressive strains inside cement test cylinders by measuring the interference between the two polarization components of the light traveling through a birefringent fiber embedded parallel to the applied load. Similarly, for the case of a fiber embedded perpendicular to the load, polarization maintaining fibers were used. In this case, the interference is produced by the phase difference between the two orthogonal polarization modes induced by the stress acting on the fiber.

The birefringence effect has also been used to measure the internal strain inside a concrete column. Tardy et al. developed a fiber optic transducer consisting of an uncoated low birefringence single-mode fiber squeezed between two metallic ribbons. The external load acting on the ribbon (plus the pre-load resulting from the ribbon's deflection) give rise to an induced birefringence that is easily measured by the interference obtained between the two polarization modes in the fiber. In this fashion, fiber sensors 14 cm long were sand-coated and embedded inside concrete specimens without special care. Sensor response during compression tests showed a relatively large induced birefringence as indicated by the reduced degree of polarization observed. This same technique has also been used to detect and measure the weight of vehicles in motion on a road (Teral, 1992).

Using an interferometric technique, Escobar et al. have measured strains during three-point flexural tests of concrete beams instrumented with surface-bonded and embedded single-mode fibers. The results obtained demonstrated the feasibility of embedding optical fibers within concrete elements without having to change normal casting procedures. Good linearity in the sensor's response was obtained, while no degradation in the fiber's response was observed after the casting process. Also, no detrimental aging effects were noticed due to the concrete environment for the teftzel-coated fibers after five years of embedding.

Kruschwitz et al. describe the use of extrinsic Fizeau fiber interferometers attached to and embedded within concrete specimens. The Fizeau cavities are formed by the flat ends of two optical fibers closely space and aligned inside a glass capillary. Any displacement between the fiber ends, due to applied strains or temperature changes, varies the gap separation which can be measured by the spectral shift of the interference fringes in the cavity. Sensor heads of this type have been embedded inside reinforced concrete beams. Sensors were placed in the cement matrix and attached next to the rebar. Strains measured under mechanical loading revealed a five percent difference between strains in the matrix and the steel reinforcement. This difference is attributed to several factors such as slippage between host and sensor, non-parallel alignment or experimental errors. Furthermore, similar sensors were epoxy-bonded to the lower side of the deck of a concrete bridge to measure the strains produced by a moving truck at different points along the span.

A more basic approach has been used by Huston et al. who have used embedded multimode optical fibers, with and without plastic jackets, for the load-to-failure testing of a reinforced concrete beam. In this scheme, the light transmission intensity in the fibers was monitored as a function of applied load. Experimental results indicate that the light intensity remains practically unaffected until a certain load is reached and the fiber breaks. At this point, the power intensity of the light is a function of the axial separation loss between the two fiber ends. Furthermore, it was observed that bare fibers were prone to breaking during the curing process.

• *Building Monitoring*

As we mentioned it before, a network of several fiber optic sensors embedded throughout different locations in a building could act as a "nervous" system for said structure, supplying information on its state of health. This combined with a computerized processing center could probably lead to a smart building capable of self interrogation and control.

The first step towards achieving some of the above ideas has been taken by Fuhr and coworkers who have embedded a variety of single- and multi-mode fiber optic sensors into a university campus building while it was under construction. The fiber sensors are intended to measure vibration, wind pressures, in-service loading, concrete creep, temperature, detect cracks, etc. while interrogated by a microcomputer. The information obtained from such a sensor grid will be used to study and monitor the building's in-service performance and overall structural health.

SUMMARY

Fiber optic sensing technology offers the possibility of non-destructive, in-situ, measurement of stress, strain, temperature and deformation of elements and structures made of concrete.

Novel applications in the analysis and evaluation of structural integrity as well as in the management and control of building services are envisioned.

The performance of embedded optical fiber sensors in concrete is affected by the mechanical and chemical durability of the glass fibers, interfacial bonding, fiber orientation and coating characteristics. An understanding of the material and micro-mechanical phenomena present at the interface region, as well as their connection to the measurand one wishes to measure in the host matrix, are essential in the development of durable sensors with accurate, reliable and repeatable response.

To date, only a small number of experiments using optical fibers in conjunction with concrete or cement are reported in the literature. Most of them focus on the measurement of stress/strain inside concrete elements using birefringent or microbend-sensitive fibers as well as Fabry-Perot cavities. Other sensors have been designed to detect the onset of cracks, measure entrained air content in fresh concrete, weight vehicle traffic, determine displacements in bridges and dams, monitor thermal stresses and curing of massive concrete.

Nonetheless, experimental research is needed to investigate, among others, adequate embedding techniques, the long-term stability and durability of the fibers inside concrete, coating adhesion to fiber and matrix, micro-mechanics of load transfer as well as practical measurement schemes. In addition, proper interfacing and collaboration is needed between researchers in the fields of fiber optics and civil engineering.

Although new and very young, this technology offers interesting possibilities for NDE applications and certainly promises to enhance the fields of fiber optics, smart structures and civil engineering.

ACKNOWLEDGEMENTS

This work was supported by NSF under grants MSS-8904872, Solid-Geo Mechanics Program and MSS-901793, Non-Destructive Evaluation Program. The author wish to thank Prof. F. Méndez for helpful discussions.

REFERENCES

ACI Committee 544, "State-of-the-Art Report on Fiber Reinforced Concrete", ACI Report No. 544-1R-82, 1982.

Ansari, F. and Chen, Q., "Fiber-Optic Refractive-Index Sensor for Use in Fresh Concrete", Appl. Op., Vol. 30, No. 28, pp. 4056-4059, 1991.

Ansari, F., "Real-Time Monitoring of Concrete Structures by Embedded Optical Fibers", in Nondestructive Testing of Concrete Elements and Structures, Proceedings of the ASCE, San Antonio, TX, April 1992, pp. 49-59.

Caussignac, J. M., Chabert, A., Morel, G., Rogez, P. and Seantier, J., "Bearings of a Bridge Fitted with Load Measuring Devices Based on an Optical Fiber Technology", Proc. 1st European Conference on Smart Structures and Materials, Glasgow, 1992., pp. 207-210.

Culshaw, B. and Dakin, J., Eds., Optical Fiber Sensors: systems and applications, Vol. II, 1989, Artech House, Norwood, MA.

Datta, A. B., Gupta, A. P. and Paul, A., "Alkaline Durability of Glass Fibre Containing S_iO_2, PbO and Al_2O_3", J. Matt. Sci., Vol. 21, 1986, pp. 2633-2642.

Escobar, P., Gusmeroli, V. and Martinelli, M., "Fiber-Optic Interferometric Sensors for Concrete Structures", Proc. 1st European Conference on Smart Structures and Materials, Glasgow, 1992., pp. 215-218.

Fintel, M., Editor, Handbook of Concrete Engineering, Van Nostrand Reinhold, 1985.

Fuhr, P. L., Huston, D. R., Kajenski, P. J. and Ambrose, T. P., "Performance and Health Monitoring of the Stafford Medical Building Using Embedded Sensors", Smart Mater. Struct. 1, pp. 63-68, 1992.

Golosova, L. V. and Varlamov, V. P., "The Interface in Portland Cement Matrix Reinforced with Fiber Glass with Long Hardening Times", Inorg. Matt., Vol. 20, No. 12, 1984, pp. 1782-1785.

Haaksma, R., Cehelnik, M. and Kerkhoff, J., "An Investigation of Optical Fiber Coating Performance in Embedded Sensing Applications", Mat. Res. Symp. Proc., Vol. 172, pp. 71-77, 1990.

Holst, A. and Habe;, W., "Fiber-Optic Intensity-Modulated Sensors for Continuous Observation of Concrete and Rock-Fill Dams", Proc. 1st European Conference on Smart Structures and Materials, Glasgow, 1992., pp. 223-226.

Huston, D., Fuhr, P., Kajenski, P. and Snyder, D., "Concrete Beam Testing with Optical Fiber Sensors", in Nondestructive Testing of Concrete Elements and Structures, Proceedings of the ASCE, San Antonio, TX, April 1992, pp. 60-69.

Kruschwitz, B., Claus, R. O., Murphy, K. A., May, R. G. and Gunther, M. F., "Optical Fiber Sensors for the Quantitative Measurement of Strain in Concrete Structures", Proc. 1st European Conference on Smart Structures and Materials, Glasgow, 1992., pp. 241-244.

Kuwabara, T. et al., "Polyetherimide Coated Optical Silica Glass Fiber with High Resistance to Alkaline Solution", OFS '89 Conference, Houston, TX, paper WA4.

Malhotra, V. M., Testing Hardened Concrete: Nondestructive Methods, American Concrete Institute, Monograph No. 9, 1986.

Majumdar, A. J., "Glass Fiber Reinforced Cement and Gypsum Products", Proc. Roy. Soc. Lon., No. A 319, 1970, pp. 69-78.

Méndez, A., Morse, T. F. and Méndez, F., "Applications of Embedded Optical Fiber Sensors in Reinforced Concrete Buildings and Structures", SPIE Proceedings, Vol. 1170, Fiber Optic Smart Structures and Skins II, pp. 60-69, September 1989.

Miller, S. E. and Chynoweth, A. G., Eds., Optical Fiber Telecommunications, Academic Press, New York, 1979.

Nanni, A., Yang, C. C., Pan, K., Wang, J. and Michael, R., "Fiber-Optic Sensors for Concrete Strain/Stress Measurement", ACI Mat. Jor., Vol. 88, No. 3, pp. 257-264, 1991.

Pak, Y. E.,"Longitudinal Shear Transfer in Fiber Optic Sensors", Smart Mater. Struct., Vol. 1, pp. 57-62, 1992.

"Fiber Optics and Curing Concrete", Photonics Spectra, January 1990, p. 22.

Proctor, B. A., "Glass Fibre Reinforced Cement", Physics in Technology, Vol. 6, 1975, pp. 28-32.

Rossi, P. and LeMaou, F., "New Method for Detecting Cracks in Concrete Using Fiber Optics", Materials and Structures, Research and Testing (RILEM), Vol. 22, No. 132, pp. 437-442, November 1989.

Tardy, A., Jurczyszyn, M., Caussignac, J. M., Morel, G. and Briant, G., "High sensitivity Transducer for Fibre-Optic Pressure sensing to Dynamic Mechanical Testing and Vehicle Detection on Roads", Springer Proceedings in Physics, Vol. 44, Optical Fiber Sensors, H. J. Arditty, J. P. Dakin and R. Th. Kersten, Eds., pp. 215-221, 1989.

Teral, S., "Vehicle Weighing in Motion with Fiber Optic Sensors", Proc. 1st European Conference on Smart Structures and Materials, Glasgow, 1992., pp. 139-142.

Udd, E., "Overview of Fiber Optic Applications to Smart Structures", Review of Progress in Quantitative Nondestructive Evaluation, Plenum Press, 1988.

Wang, T. T., Vazirani, H. N., Schonhorn, H. and Zupko, H. M., "Effects of Water and Moisture on Strengths of Optical Glass (Silica) Fibers Coated with a UV-Cured Epoxy Acrylate", J. Appl. Polmr. Sci., Vol. 23, pp. 887-892, 1979.

Wolff, R. and Miesseler, H., "Monitoring of Prestressed Concrete Structures with Optical Fiber Sensors", Proc. 1st European Conference on Smart Structures and Materials, Glasgow, 1992, pp. 23-29.

IV - FRACTURE MECHANICS

A Fiber Optic Sensor for The Determination of Dynamic Fracture Parameters in Fiber Reinforced Concrete

By:
Farhad Ansari[1] MASCE, Rajendra K. Navalurkar[2]

Abstract

A new fiber optic sensor is introduced, by which it is possible to directly measure the displacements associated with opening of cracks in cementitious composites provided their locations are known a priori. Hence, such sensors are most suited for experimental determination of crack tip opening displacements (CTOD) in notched specimens. The Development and calibration of this fiber optic sensor for embedment within cementitious composites is described. The fiber optic sensor developed in the present study can be employed in experiments involving both static as well as dynamic loads. Fiber reinforced concretes (FRC) have been extensively employed in particular applications to strategically sensitive structures such as nuclear containment and military installations. Development of means by which the structural response of FRC structures can be determined under dynamic loading conditions is important in prediction of residual capacity. It is therefore important to assess the fracture characteristics of FRC subjected to impulsive loading conditions. In the study reported herein, three-point-bend fiber reinforced concrete (FRC) specimens were tested under an impact velocity of 1.1 m/s (43 in/s). The range of specimens tested in this study included plain concrete as well as FRC reinforced with 0.5, 1.0, and 1.5 percent by volume of straight steel fibers.

---

[1]Professor, and director of Smart Structure Laboratory, Civil & Environmental Engrg. New Jersey Institute of Technology, Newark, New Jersey 07102
[2]Graduate Research Assistant, Smart Structure Laboratory, Civil & Environmental Engrg., New Jersey Institute of Technology, Newark, NJ 07102

Introduction

Researchers at the Smart Structure laboratory in the department of civil and environmental of NJIT are currently involved in several research projects, involving the utilization of fiber optic sensors for testing and monitoring of civil structures. Development of sensors for detecting and quantifying the fracturing process in civil engineering materials has been a continuing goal of the researchers at the smart structure laboratory. As shown in Figs.1 thru 3, monitoring fracture and crack propagation in concrete during a typical fracture test can be easily accomplished by way of embedded optical fibers. Optical fibers are specially attractive in applications to concrete structures where they can be directly embedded within the material during pouring and casting operations. Fiber optic sensors have been extensively employed for condition monitoring of structural components in advanced materials for use in aircraft and space vehicles (Butter and Hocker 1978, Claus et al 1988, Measures et al 1989). External perturbations such as strain, pressure, and vibrations induce changes in the phase, intensity, or the wavelength of lightwaves propagating through optical fibers. This change in one or more properties of light can then be related to the parameter being measured. Optical fibers are geometrically versatile and can be configured to arbitrary shapes. The "smart material" concept takes advantage of the geometric adaptability of optical fibers. In this technology, optical fibers are embedded within the structural material for the purpose of real-time damage assessment.

Determination of fracture mechanics parameters in concrete have been the subject of numerous investigations. However, most of the accumulated fracture data pertain to experiments which were performed under static loading conditions. Concrete structures may be subjected to impulsive loads due to earthquakes, wind gusts, and projectile impacts. Accurate prediction of structural response under impact loading conditions require material property data representing behavior at high rates of loading. One parameter of significant importance in determining fracture properties is crack tip opening displacement (CTOD). For instance, crack extension or fracture may be assumed to occur as soon as the crack opening displacement exceeds a critical value. Experimental determination of CTOD is difficult even under static loading conditions. Attempts have been made to determine CTOD from measurement of surface deformations using laser speckle interferometry (Ansari 1987, 1989). However, surface deformations are not entirely representative of deformations and cracking processes taking place within the cementitious composites (Gillespie et al 1984, Swartz et al 1984). Recent interest in the fracture properties of component materials in structures at high rates of loading, has led to a number of interesting studies (John and Shah 1986, Shah 1986). In most of these studies, surface strains were measured in order to characterize crack opening kinematics. But, for a material which exhibits extensive microcracking prior to fracture, strain measurements may not prove to be acceptable in determining fracture characteristics of the material. Thus, it is very important to develop a new experimental technique for the determination of crack tip opening displacements in cementitious composites.

Fig. 1 Monitoring crack propagation in a compact tension specimen by four embedded optical fibers.

Fig.2 Measurement of CTOD in a three-point-bend concrete specimen.

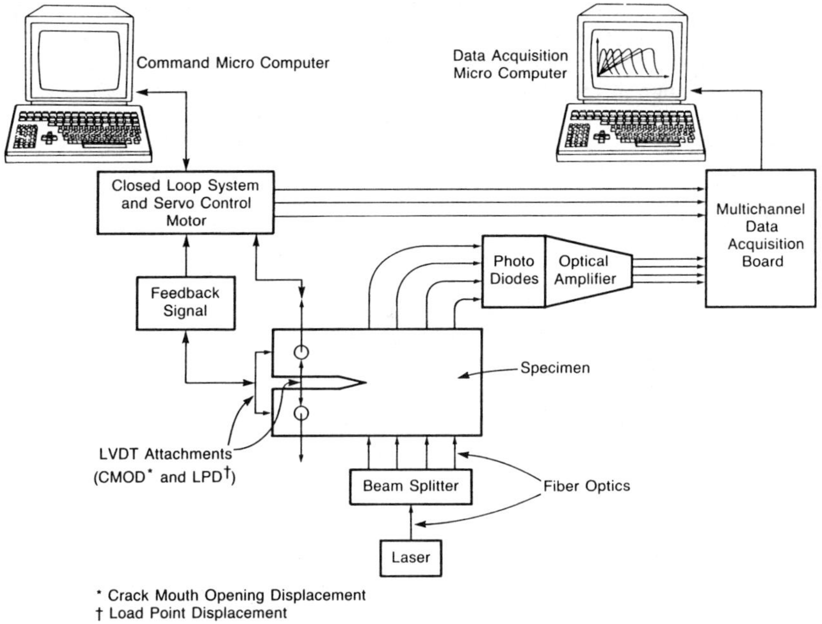

Fig.3 Instrumentation involved in a typical fracture test by intensity modulated fibers.

In the study reported herein, optical fibers were configured in such a way so that they could be embedded within fiber reinforced concrete (FRC) specimens for measurement of crack tip opening displacements. These specimens were then employed in a series of impact tests using an instrumented Charpy impact tester. By using optical fibers, it was possible to measure crack tip opening displacements under impulsive loading conditions. The concept of J-integral was employed in determining the tension softening (stress-CTOD) behavior of FRC at high rates of loading.

Fiber Optic CTOD Sensor

Transmission of light through optical fibers can be explained by the Snell's law and the concept of total internal reflection. Many modes of light rays travel through a typical multi-mode optical fiber. Each mode propagates at an angle larger than the critical angle of incidence at the core to cladding interface. These modes stay within the core, and therefore no loss in optical intensity occurs. Idealy, the optical power at the output end of the fiber remains equal to the transmitted power at the input level. If the fiber is bent at any point along it's length, the change in the angle of incidence will cause some of the light rays to escape out by way of the cladding (Fig. 4). Consequently, higher order modes which travel near critical angle of incidence enter the cladding and are lost (Gloge 1972).

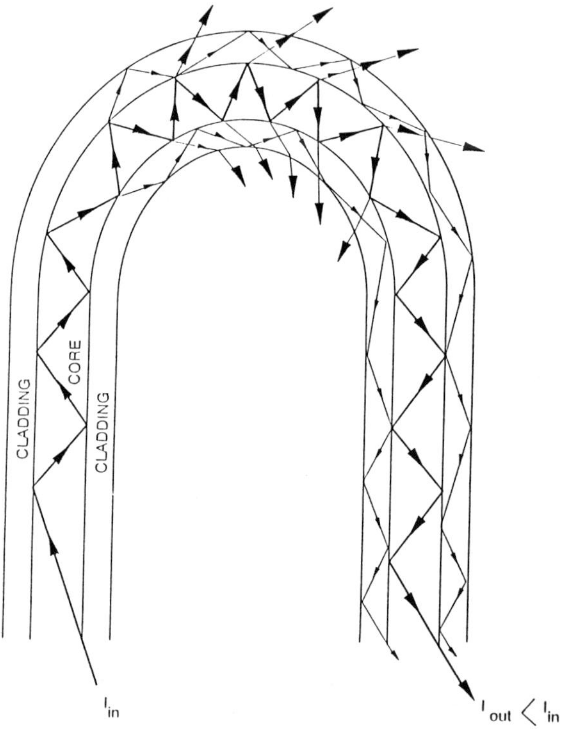

Fig.4 Light intensity perturbation in an optical fiber due to macro bend effect.

However, loss of higher order modes do not significantly influence the intensity of light at the output end of the fiber. Major portion of the energy associated with the electromagnetic field is carried through the fiber by way of the fundamental modes. Fundamental modes of light propagate at angles much larger than the critical angle of incidence. These modes refract into the cladding only when the curvature of the bend is significantly increased beyond a critical level. The critical curvature varies from fiber to fiber, depending on the diameter as well as the optical and mechanical properties of the optical fiber. Reduction in the number of fundamental modes is associated with significant intensity loss at the output end of the fiber. Since fundamental modes propagate at angles within close proximity of one another, minute increases in the bend curvature beyond the critical level bring about large losses in intensity (Marcus 1976).

The fiber optic sensor developed in the present study, measures the light intensity signal modulations due to crack opening displacements through a specific geometric arrangement shown in Fig. 5. The curvature due to circular bend positions the propagation angles of the fundamental modes near the critical level. Crack opening displacements result in further bending of the bent optical fiber and a large decrease in light intensity.

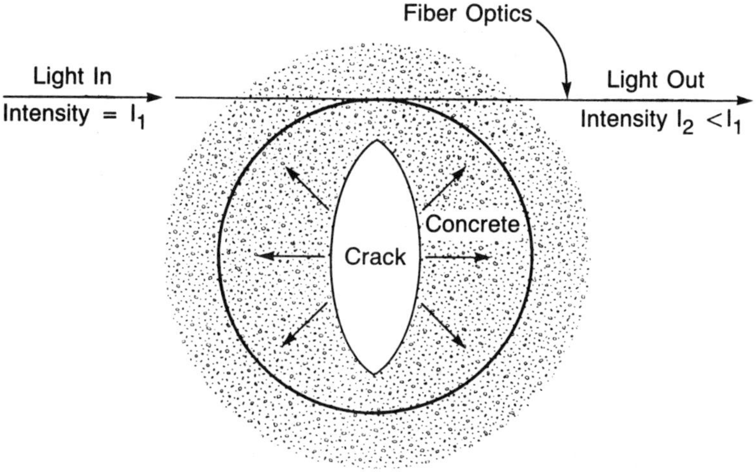

Fig. 5 Fiber optic CTOD sensor for concrete.

Correlation between the displacements and corresponding signal modulations are accomplished through calibration. The configuration shown in Fig.6 corresponds to the calibration process by which the light intensity loss measurements are related to CTOD values. Accordingly, the crack opening displacements through the thickness of a specimen are simulated by the separation of cylinder halves. Crack opening displacements along the crack profile in concrete are not uniform mainly due to heterogeneous nature of the material.

Therefore, the calibration similitude presented in this study corresponds to an average CTOD value along the crack profile.

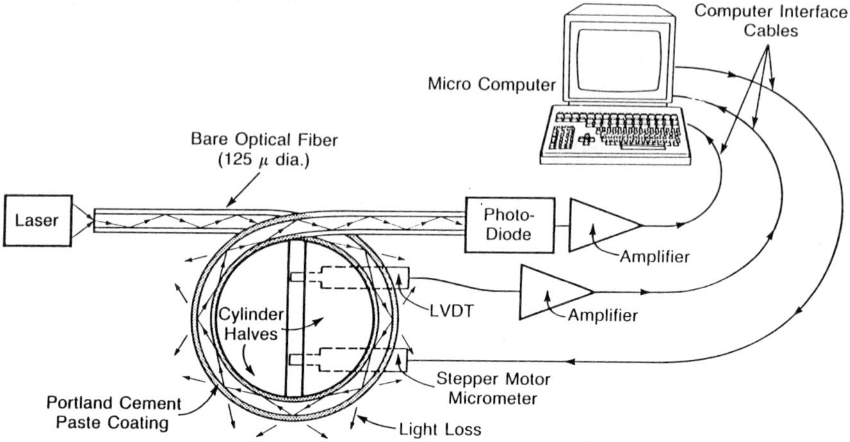

Fig.6 Instrumentation associated with the calibration of the fiber optic CTOD sensor.

The calibration process involves using portland cement paste for affixing the fiber optics around two cylinder halves of diameter 31.75 mm (1.25 inch). Trial and error experiments indicated that the loop diameter 31.75 mm for the particular optical fiber employed in this study provided sufficient intensity loss. Smaller loop diameters result in larger intensity drops and therefore higher sensitivity. However, for long term testing, bending of these fibers into loops of less than 25.4 mm may result in permanent damage to the fiber and unpredictable results.
A laser source is employed in launching a beam of light into the optical fiber from one end, and a photodetector receives the light at the other end of the optical fiber. A motorized micrometer positioner separates the cylinder halves, and an LVDT measures the separation. A data acquisition board is employed in receiving the light intensity loss, and data transmitted by the LVDT due to the separation of the cylinder halves. For the optical fiber tested in this study, a calibration constant of 0.045 Volts intensity loss per one milli-inch of displacement was obtained by calculating the slope of the line relating the modulated light intensity to the displacement. The cement paste coating employed for affixing the optical fiber to the calibration cylinder prevented slippage. Besides adhesion, the circular geometry of the optical fiber develops a stronger bond by way of anchorage. Slippage due to bond failure can be detected from increases in the light intensity output. During the calibration process light intensity signal continuously decreases due to separation of the cylinder halves. Loss of bond provides a momentary

pressure relief for the optical fiber, and results in an increase in intensity. Similarly, bond integrity between the optical fiber and surrounding matrix in specimens were examined through the examination of acquired light intensity data. Analysis of light intensity data indicated existence of perfect bond between the optical fiber and the matrix.

Precision of the calibrations were checked by performing a calibration test for each one of the tested specimen series. Calibration experiments were also performed for each new lot of optical fibers employed for embeddment within specimens. The scatter in calibration data was within seven percent of the mean value. This scatter is due to the interferometric noise associated with partial refraction of cladding modes back into the core (Yao and Asawa 1983). The main parameter which directly affects calibration results is the diameter of the circular fiber optic loop. Therefore, careful attention was given in forming the 31.75 mm diameter fiber optic loops for embedment in specimens.

Tension Softening Relations

The relationship between the traction across a crack plane and the separation distance of the crack faces in a cementitious composite is termed as the tension softening. It is a very useful relationship, since it represents the constitutive law between the tensile stress and Crack Opening Displacement. Li and Leung 1987, developed an experimental procedure in which the J-integral was employed in conjunction with compact tension specimens in order to determine the tension softening curves under quasi-static loading conditions. The J integral technique was also employed in the present study in order to check the applicability of the fiber optic sensor in studies related to cracking and fracture in cementitious composites. In the study reported here, the tension softening curves were indirectly obtained from three point bend tests under impact loading conditions. Theoretical basis for extraction of tension softening relations from the J-integral method is thoroughly described by Rice 1968, and Li and Leung 1987. In the following, some of the main features of the technique which are directly related to the experiments are explained:

The path independent J-integral may be interpreted as the change in potential energy per unit crack advance. Therefore:

$$J = -\frac{\partial V}{\partial a} \qquad (1)$$

where V is the potential energy, and a is the crack length. For a material which exhibits softening in tension, it can be shown (Li and Leung 1987) that Eq. (1) reduces to:

$$J(\delta) = \int_0^{\delta_c} \sigma(\delta)d\delta, \qquad (2)$$

Where $\sigma(\delta)$, and δ are the normal stress and the opening displacement at a point in the process zone. If δ_c is the material separation at which the load carrying capacity just vanishes, then $J = J_c$ is a critical value when CTOD reaches δ_c . In this case J_c is the complete area under the tension softening curve.

In the present study, the multi-specimen technique was employed in the evaluation of J . J can be evaluated from a pair of three-point-bend beam specimens identical in every respect except for a difference in midspan notch lengths. The area under the load-load point displacement curves for each one of the two specimens are evaluated. The difference between the two areas obtained in this way results in J according to:

$$J(\Delta) = \frac{A}{B(a_2 - a_1)} , \qquad (3)$$

where Δ is the load-point-displacement, A is the difference in area between the two load, and load-point-displacement curves. B is the specimen thickness, a_1 is midspan notch length for the specimen with shorter notch length, and a_2 is the midspan notch length for the specimen with longer notch length. The tension softening curve can then be determined by direct differentiation of Eq. (2):

$$\sigma(\delta) = \frac{\partial J(\delta)}{\partial \delta} \qquad (4)$$

Crack closing pressure is therefore determined by numerical differentiation of the experimentally determined J-CTOD relationship. A major drawback in using this technique is the representation of crack growth by the difference between the notch lengths. Therefore, J determined in this way corresponds to the energy level for crack initiation and not growth.

Experimental Program

Determination of a single tension softening curve requires testing two specimens. All specimens were 15 X 5 X 86 cm (6 X 2 x 34 inches), and tested on a span length of 81 cm (32 inches) under three-point bending. The mix proportions by weight of FRC used in this study were designed to be 1:2:2:0.50, corresponding to cement : sand : aggregate: and water ratio by weight. The amount of straight (unhooked) steel fibers used included 0, 0.5,1.0, and 1.5 percent by volume. Type I portland cement conforming to ASTM C 150, ASTM No. 2 grade river sand passing through sieve No. 8, and maximum coarse aggregate size passing 9.5 mm (0.375 inch) and retained on No. 4 sieve were used. Specimens were cast in plexiglass molds. Preparation of individual specimen involved a rigorous procedure for placement of the fiber optic sensor in the mold. The circular fiber optic loop was attached to the bottom of the mold by a loose transparent tape. The deliberate slack in the tape was incorporated so that the fiber could have a clear distance of approximately 10 mm (0.375 inch) from the bottom of the mold. The fiber optic loop was then sandwiched between two thin aluminum sheets for alignment. The aluminum sheets were held in position by a set of holding arms from the top of the mold. The mold was placed on a vibrator and fresh concrete was slowly poured in during vibration. The aluminum sheets were taken out once the mold was totally filled. Vibration was continued for another fifteen seconds to assure uniform dispersion of concrete through the loop. Fig. 7 depicts the location of the fiber optic loop in the mold prior to pouring of concrete.

Specimens were cured at room temperature for one month. Notches were cut with

Fig.7 Placement of the fiber optic sensor in the concrete mold.

a circular diamond saw. It was planned to produce notch lengths of 3.8 and 4.5 cm (1.5 and 1.75 in) for the pair of specimens in each set. However, actual notch lengths after sawing operations differed from the above-mentioned values and are given in table 1. Perhaps prenotched specimens would serve the purpose better as some of the specimens were destroyed by accidental cutting of the optical fibers during sawing operations. Three sets (2 per set) of specimens were cast per type, as designated by fiber content in table 1.

Instrumentation

A 25 mW Helium-Neon laser source provided the light source entering the optical fiber. Light intensity output is detected by a photodiode from the exit end of the optical fiber. At this point, the optical intensity is amplified by an amplifier, and is converted to digital signals through a multichannel high speed waveform digitizer (Lecroy 8210) at a sampling frequency of 1 MHz. The output of the digitizer is transferred to a microcomputer for analysis. A Tinius Olsen Charpy impact machine was modified and instrumented for testing FRC beams under three point bending loading conditions. All specimens were tested under an impact velocity of 1.1 m/s (43 in/s). Impact loads were recorded by an instrumented striker. Load cell sensing elements were made from 350-ohm resistance type strain gages in full bridge configuration. Mid point deflections were acquired by an LVDT at a point immediately next to the sawed notch. Triggering of the waveform digitizer was achieved by blockage of light to a phototransistor circuit at the time of striker's impact.

Table 1. Experimental Program.

Specimen type	Number of Specimens	Fiber content by volume	Avg. notch length (inch)
PC-A	3 (1)*	0	1.2
PC-B	3 (1)	0	1.5
FRC05-A	3	0.5	1.4
FRC05-B	3	0.5	1.6
FRC10-A	3 (1)	1.0	1.4
FRC10-B	3 (1)	1.0	1.6
FRC15-A	3	1.5	1.4
FRC15-B	3	1.5	1.6

1 inch = 2.54 cm

Results and Discussions

Typical CTOD data for specimens with various amounts of fiber reinforcement are compared in Fig. 8. As shown in Fig.8, addition of steel fibers results in lower CTOD values and longer fracture times. Crack tip opening displacement velocities are acquired by numerical differentiation of the CTOD -time relationships. For all the specimens tested, Crack tip opening displacement velocities increased at a constant rate throughout the fracturing process. Typical velocity - time relationships are shown in Fig.9. Crack Tip Opening Accelerations (CTOA) are given in Table 2. As shown in table 2, on the average, CTOA in plain concrete is three times larger than the concrete reinforced with 1.5 percent by volume of steel fibers.

Load vs. load point displacement values were employed in the evaluation of J

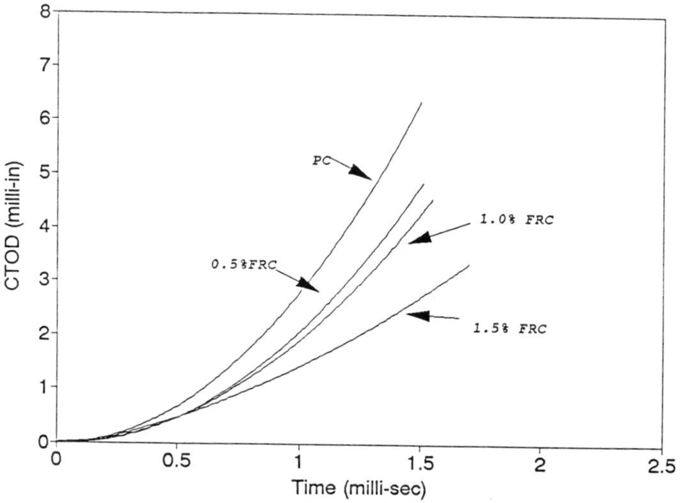

Fig.8 Typical CTOD response for specimen with different amounts of fiber.

according to the procedure explained earlier. As shown in Fig.10, J increases for increasing amounts of steel fiber. Higher percentages of steel fibers did not markedly increase the load carrying capacities of the specimen tested in the present study. The increase in J for increasing amounts of steel fiber, is attributed to the increase in ductility due to larger deformation capability prior to fracture. This can be illustrated by comparing the deformation axis in Figs.11 through 12 representing the load-deflection relationships in plain and fiber reinforced concretes respectively.

As it was pointed out earlier, the stress-separation relationships are directly derived from J-CTOD curves by eq. 4. Typical stress-CTOD curves for the four types of specimen tested are shown in Fig.13. The critical value of CTOD in these curves corresponds to the point of zero stress on the horizontal axis. This point corresponds to J_c which is determined as the difference in area between the load-deflection curves of the two identically similar specimens with unequal notch lengths, up to the point where these curves intersect. For the specimens tested in the present study, the critical value occurred on the descending branch after the peak load (Figs.11-12). Interpretation of stress-CTOD curves shown in Fig.13 require careful attention. These results were obtained under impulsive loading conditions, and the time duration to reach the critical CTOD's for plain concretes and specimens with lower fiber contents were lower than the specimens containing higher volumes of fiber.

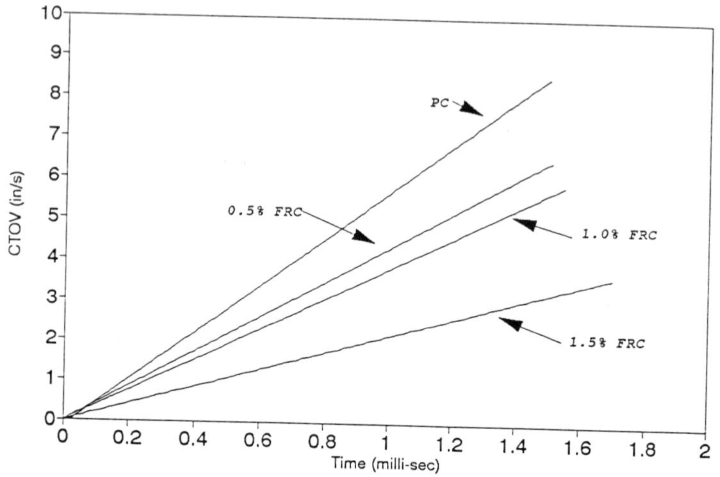

Fig.9 Comparison of the crack tip opening velocities for the various types of specimens tested.

Table 2 Average experimental results

Specimen type	J_c (lb/in)	Critical CTOD (milli-inch)	CTOA (in/s^2)
PC	1.30	6.85	5.80
FRC05	1.52	5.25	4.5
FRC10	1.72	4.23	3.75
FRC15	2.35	3.60	2.10

1 lb = 4.448 Newton
1 lb/in = 175.2 N/m
1 in/s^2 = 2.54 cm/s^2
1 milli-in = 25.4 micron

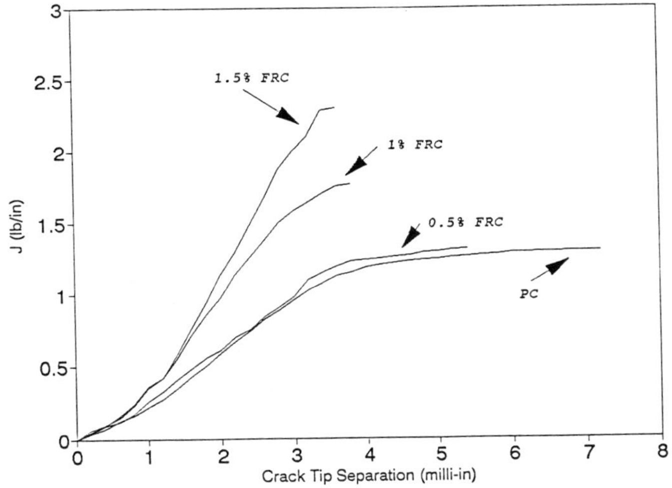

Fig.10 Typical J-CTOD relations.

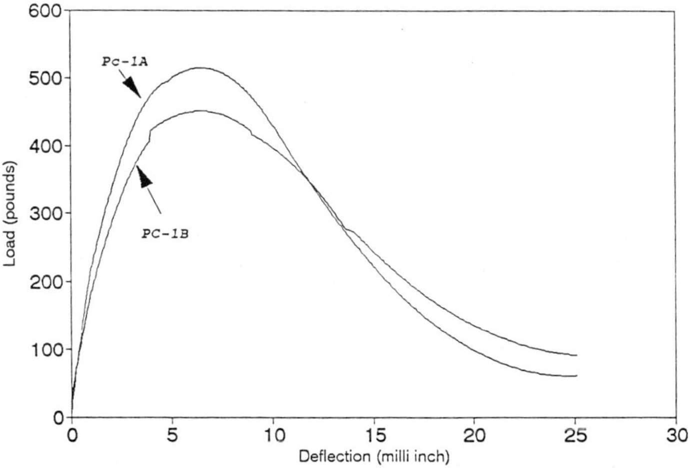

Fig.11 Load-deflectionrelations for a typical set of concrete specimens.

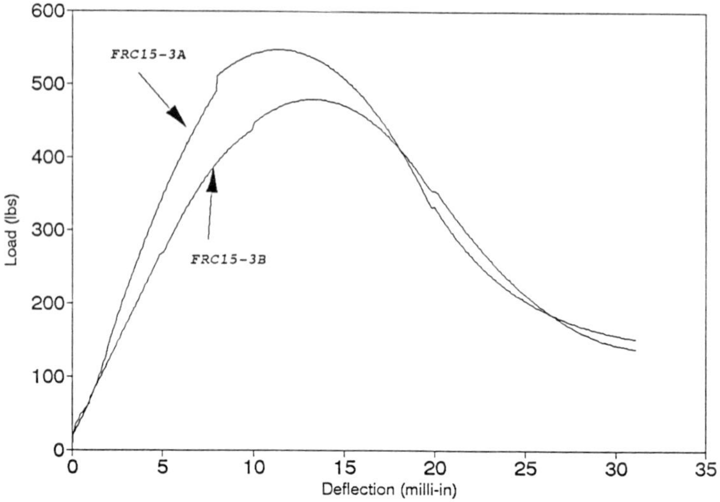

Fig.12 Load-deflectionrelations for a typical set of 1.5% FRC specimens.

Conclusions

1. A fiber optic sensor is developed which can be embedded within the material, and is capable of measuring the crack tip opening displacement provided that the crack location is known a priori.
2. The fiber optic sensor is potentially useful for condition monitoring of structures.
3. Besides CTOD's, using the fiber optic sensor, it is possible to obtain other useful information such as crack tip opening velocities and accelerations.
4. For the impact velocity used in the present study, the form of failure in FRC is governed by fiber cleavage and not slippage.
5. The flexural load carrying capacity of FRC specimens were not considerably higher than plain concrete. However, addition of steel fibers produced substantial increases in J_c which is due to increased crack closing pressure .
6. Under impulsive loading conditions, specimens containing larger quantities of steel fibers exhibited longer time durations to fracture, and smaller critical CTOD's. The crack tip opening velocities were also reduced as the fiber content increased.

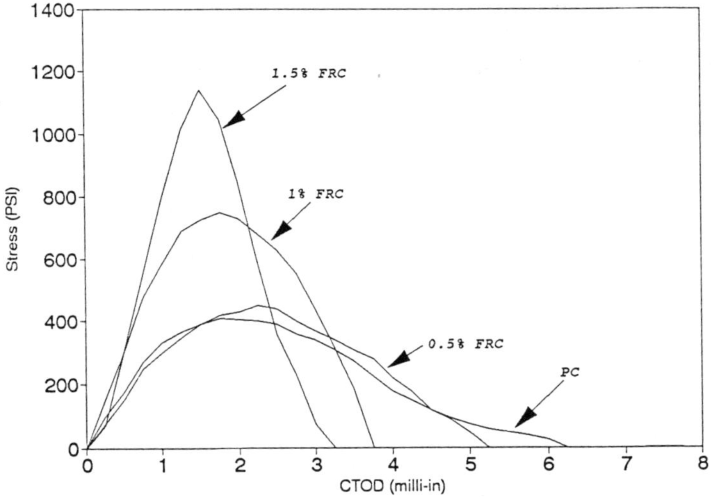

Fig.13 Typical stress-CTOD relations.

Appendix I.-References

Ansari,F. (1987),"Stress-Strain Response of Microcracked Concrete in Direct Tension," Mat. Jnl, ACI, 84(6), 481-490.

Ansari,F. (1989), "Mechanism of Microcrack Formation in Concrete ," Mat. Jnl, ACI,86(5),459-464.

Butter C.D.,Hocker G.B. (1978),"Fiber Optic Strain Sensor," Applied Optics ,17, 2867-2873.

Claus R.O.,Bennett K.D.,May R.G. (1988),"Optical Fiber Methods for the NDE of Smart Skins and Structures," Proc. Fall Conf., SEM, 19-24.

Gillespie R.C., Go C.G., and Swartz S.E. (1984),"Vertical Displacement Gage Fixture for Testing," Experimental Techniques, SEM, 8(5)

Gloge,D. (1972),"Bending Loss in Multimode Fibers with Graded and Ungraded Index," Applied Optics ,11,2506-2513.

John,R.,Shah, S.P. (1986),"Fracture of Concrete Subjected to Impact Loading," Cem. Conc. & Agg., ASTM, 8(1), 24-32.

Li V.C.,Leung K.Y.(1987),"Experimental Determination of the Tension Softening Relations for Cementitious Composites, Cement and Concrete Research,17, 441-452.

Measures R.M.,Glossop N.D.,Lymer J.,LeBlanc M.,West J.,Dubois S., Tsaw W., Tennyson R.C. (1989),"Structurally Integrated Fiber Optic Damage Assessment System for Composite Materials, " Applied Optics, 28, 2626-2633.

Marcuse,D. (1976),"Field Deformation and Loss Caused by Curvature of Optical Fibers," Journl Opt. Soc. Am.,66,311-320.

Rice,J.R.(1968), Mathematical Analysis in the Mechanics of Fracture," Fracture:An Advanced Treatise ,Ed. Liebowitz,H.,2,Academic Press,NY,191-311.

Shah,S.P. (1986),"Concrete and Fiber Reinforced Concrete Subjected to Impact Loading, " Cement Based Composites: Strain Rate Effects, MRS Symp. Proc., Eds. Mindess,S.,Shah,S.P., 6, 181-201.

Swartz,S.E.,Go,C.G. (1984), "Evaluation of the Validity of Compliance Calibration to Estimate Crack Lengths in Concrete Beams in Bending, "Experimental Mechanics, SEM, 24(2)

Yao, S.K.,Asawa,C.K., (1983), " Fiber Optical Intensity Sensors," IEEE, 1(3), 562-574.

Appendix II- Notations

A = difference in area between the two load, load point displacement curves.
a = crack length.
a_1 , a_2 = notch lengths.
B = specimen thickness.
J = Path independent J-integral.
J_c = Critical J.
V = Potential energy.
Δ = Load point displacement.
δ = crack tip opening displacement (CTOD).
δ_c = critical CTOD.
$\sigma(\delta)$ = crack closing pressure as a function of CTOD.

FIBER-OPTIC SENSORS AND FRACTURE MECHANICS

Arun Shukla[1], Raman Singh[2], Stephen V. Letcher[3], and Nadarajah Narendran[4]

ABSTRACT

This chapter presents a review on the studies done to test the applicability of fiber-optic sensors to fracture mechanics. The Mach-Zehnder, Michelson and Fabry-Perot interferometric sensors have been used, in both surface and embedded configurations, to measure strains near a crack tip. The strains are used to determine the opening mode stress intensity factor, K_I, using linear elastic fracture mechanics. Results show that all the three sensors can be used successfully to measure the stress intensity factor, with the Fabry-Perot sensor being found to be the most accurate.

INTRODUCTION

This chapter reviews the applicability of fiber-optic sensors to fracture studies. The use of fiber-optic sensors for smart-materials applications has been growing rapidly during the past ten years (Wiencko *et al.*, Valis *et al.*, Murphy *et al.*, Leonard, Measures *et al.*, Bruinsma *et al.*, Lee *et al.*, Udd, Liu *et al.*, and Measures). Incorporating optical-fiber sensors into a material during manufacturing opens an avenue for real-time health monitoring throughout the lifetime of the material. Smart-

---

[1]Professor, Department of Mechanical Engineering, University of Rhode Island, Kingston, RI 02881

[2]Graduate Student, Department of Mechanical Engineering, University of Rhode Island, Kingston, RI 02881

[3]Professor, Department of Physics, University of Rhode Island, Kingston, RI 02881

[4]Senior Research Engineer, Mechanical Technology Incorporated, 968, Albany Shaker Road, Latham, NY 12110

materials will not only have the capability to monitor the condition, but also will have active feedback control to take corrective action. Dielectric construction, geometric versatility, low weight, and small size make optical fiber sensors an attractive choice for embedded sensing applications.

The development of fiber-optic sensors first started in the mid seventies and many researchers have since entered this field and accelerated its progress (Giallorenzi *et al.*). In general, fiber-optic sensors are based on the detection of modulated light intensity, frequency, polarization, or phase of the optical beam propagating through the fiber. Fiber-optic sensors have been used in applications for sensing various physical parameters such as strain, temperature, pressure, vibration, etc. These sensors can be categorized as extrinsic and intrinsic sensors. An extrinsic sensor has an external sensing element that optically interacts with the measurand and the optical fibers serve as the light guides. In the intrinsic case, the measurand acts directly on the optical fiber and modulates the beam passing through. The latter is preferred for embedded sensing applications.

Fracture studies are important in predicting and preventing failure of machines and structures. Within the elastic domain, crack propagation in a material is governed by the stress intensity factor, K_I, which completely defines the stress field near a crack tip. Optical techniques of photoelasticity and caustics (Dally, and Rosakis *et al.*) are two of the most popular experimental methods that are used for obtaining this stress intensity factor in different materials. Both these methods require expensive, elaborate setups and also require that the material be polished or transparent. Therefore, these techniques have trouble with materials like ceramics, rocks or composites. Strain gages have also been used to study dynamic fracture (Shukla *et al.*, 1988). Strain gages have their own limitations in that they can only measure surface strains.

FIBER-OPTIC SENSORS

Three types of fiber-optic interferometric sensors, namely Mach-Zehnder, Michelson, and Fabry-Perot, were utilized in this study. Interferometric sensors usually operate in two regimes. In the first case the relative phase difference is less than 2π and the shift of the interference fringes is less than a full fringe. In the second case the phase shift is much larger than 2π and many fringes pass by the observation point. The extent of the shift of the optical fringes is related to the parameter being measured. In all the experiments discussed in this chapter the interferometers were operating in the latter regime.

Mach-Zehnder Interferometer

An experimental set-up of a Mach-Zehnder interferometer is shown in Fig.1. Light from a He-Ne laser is split into two beams and each is then coupled into an optical fiber. One fiber is exposed to the applied strain while the other serves as a

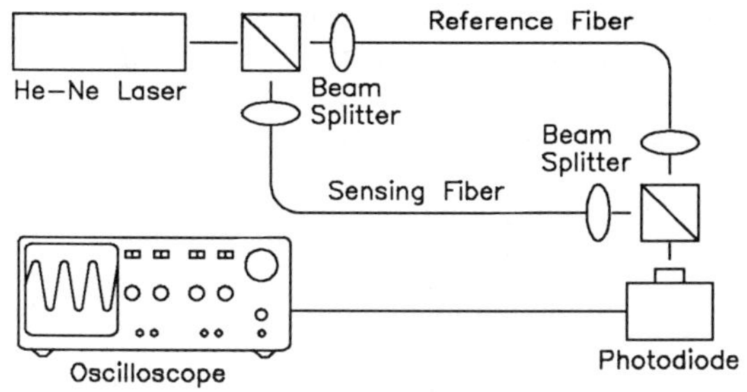

Figure 1. Experimental setup for the Mach-Zehnder interferometric sensor.

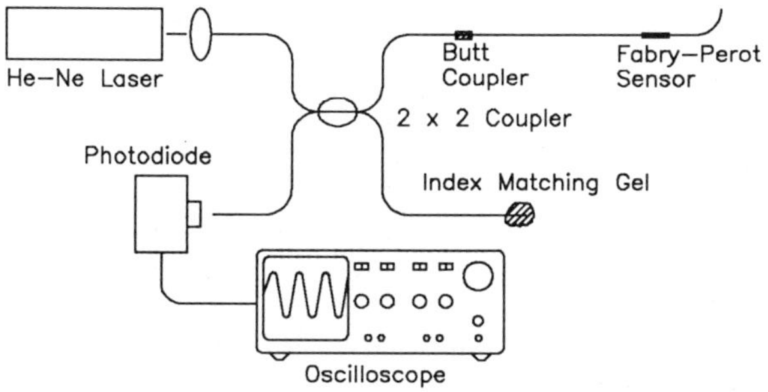

Figure 2(a). Experimental setup for the Fabry-Perot interferometric sensor.

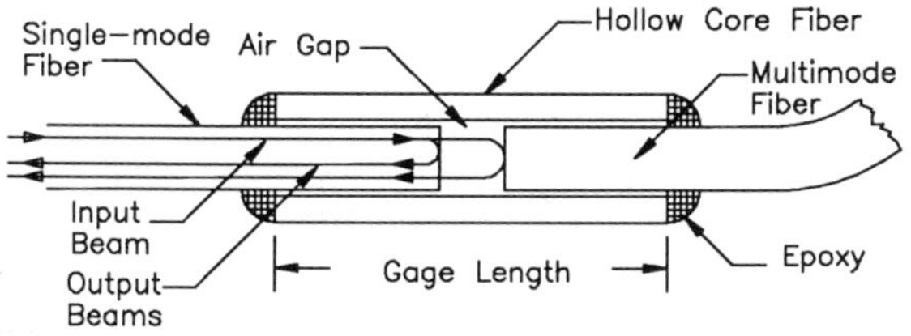

Figure 2(b). The extrinsic Fabry-Perot interferometric sensor.

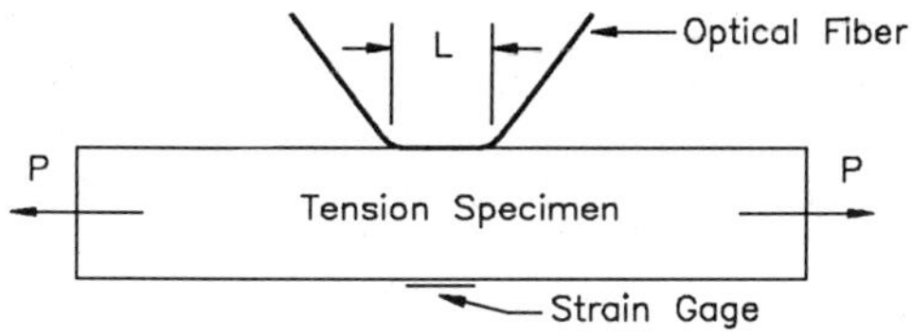

Figure 3. Tensile setup for calibrating attached fiber optic sensors.

reference path. The two beams are recombined and made to interfere. A strain applied to the sensing fiber causes a phase shift and a corresponding displacement of the fringe pattern. Fringes are detected by a photo-diode and the data is stored in a digital storage oscilloscope, from which it is transferred to a computer for further processing.

Butter and Hocker have shown, that for a surface mounted sensor subjected to an uniaxial strain, ε_x, the relative optical phase change between the sensing and the reference arm of the Mach-Zehnder interferometer is given by,

$$\delta\Phi = \left(\frac{2\pi\, n\, \varepsilon_x L}{\lambda}\right)[1 - c] \quad , \tag{1}$$

where

$$c = \left(\frac{n^2}{2}\right)\left[(1-\nu)p_{12} - \nu p_{11}\right] \quad , \tag{2}$$

and where λ is the vacuum wavelength of the optical beam passing through an optical fiber of length L and refractive index n, p_{11} and p_{12} are the strain-optic coefficients, and ν is the Poisson's ratio of the fiber core. The phase shift required to move one interference fringe past a given point is 2π; therefore, the number of fringes passing a given point caused by the strain ε_x can be written as

$$N = \left(\frac{\delta\Phi}{2\pi}\right) = \left(\frac{n\, \varepsilon_x L}{\lambda}\right)[1 - c] \quad . \tag{3}$$

Equation (3) can be rearranged to give

$$\varepsilon_x = DN \quad , \qquad D = \left(\frac{\lambda}{nL}\right)\left[\frac{1}{1 - c}\right] \tag{4}$$

Thus, the number of fringes passing a given point is proportional to the axial strain.

Recently Sirkis and Haslach extended Butter and Hocker's theory and derived a more complete phase-strain relationship for embedded interferometric optical fiber sensors. Mathews and Sirkis have shown that the relative phase retardation in an embedded optical fiber and the strain state along the fiber is given by,

$$\delta\Phi = \frac{2\pi n}{\lambda}\int_{\tau}\left[S_1 - \frac{1}{4}n^2[P_{2j} + P_{3j}]\, S_j\right]ds \tag{5}$$

where the strains are labelled as S_j, j = 1,....,6 in the order ε_{11}, ε_{22}, ε_{33}, ε_{23}, ε_{13}, ε_{12};

and S_1 is always taken as the normal strain component everywhere tangent to the optical fiber axis, and S_2 and S_3 are any perpendicular normal strains in the plane of the fiber cross-section. The path traversed by the fiber is τ and P_{ij} is the Pockel's strain-optic tensor, i,j = 1, 2, 3; and repeated indices mean summation. In their study, Mathews and Sirkis have shown that Butter and Hocker's theory fails for the cases when the stiffness of the embedded optical fiber is close to that of the host material.

Michelson Interferometer

The experimental setup for this sensor remains the same as for the Mach-Zehnder interferometer except that instead of recombining the two beams at the end they are both reflected back through the fibers by mirroring the fiber ends. The two reflected beams are combined and made to interfere. The light intensity output from this sensor can be related to the applied axial strain in the same manner as that for the Mach-Zehnder sensor. In this case the effective gage length of the sensor is optically doubled due to the fact that the sensing beam traverses the sensing zone twice.

Extrinsic Fabry-Perot Interferometer (EFPI)

Figure 2a shows the experimental setup for a Fabry-Perot interferometric sensor. The actual sensor, shown in Fig.2b, is a low-finesse Fabry-Perot cavity created by inserting two fibers with partially mirrored ends into a hollow-core fiber. Two separate end-reflections take place, as shown in Fig.2b. These two light beams combine to produce an interference pattern. A change in the "air gap", due to applied strain, causes the interference pattern to change and fringes are registered on the photodiode. The number of fringes N, can be related to the applied strain, ε_x, as,

$$N = \frac{\phi}{2\pi} = \frac{2L}{\lambda}\varepsilon_x \quad , \qquad \textbf{(6)}$$

where L is the gage length of the sensor and λ is the wavelength of light. The number of fringes passing a given point, N, is proportional to the applied axial strain, ε_x.

FRACTURE STUDIES

This section is divided into two parts. The first part describes the experimental procedure and results obtained by attaching the fiber-optic sensors to the surface of a fracture specimen and the second describes embedding the fiber-optic sensors in a fracture specimen to evaluate the stress intensity factors.

Surface-Attached Sensors

In order to evaluate the strain sensed by the optical fiber from the light intensity output of the interferometers, the proportionality constant D in Eq. (4) has to be determined. This constant can be calculated if the material parameters of the optical fiber, p_{11} and p_{12}, the strain-optic constants and the Poisson's ratio ν are known. But these quantities are not known accurately for the fiber used, thus it is necessary to calibrate the Mach-Zehnder and Michelson sensors. The EFPI does not require calibration because the constant D is only a function of the wavelength, λ, and the gage length L and these are known accurately.

Calibration Procedure

A tension experiment, as shown in Fig.3, was utilized to calibrate the Mach-Zehnder and Michelson interferometric sensors. The optical fiber was attached on one side of a tension specimen and a strain gage was bonded on the other side, at the same location. The specimen was monotonically loaded and the corresponding optical fringes were recorded. Figure 4 shows the variation of light intensity with time along with the corresponding strain gage output. The two outputs were used to obtain the calibration constant, D, as given in equation (4). The calibration procedure was repeated for various gage lengths to obtain D as a function of gage length, L.

Surface-Attached Fracture Experiments

A single edge notched (SEN) aluminum specimen was used with the attached fiber-optic sensors to determine the stress intensity factor. An initial crack of length, a, was saw-cut into the specimen as shown in Fig.5. The optical-fiber was attached to the specimen at an angle α to the crack direction and at a distance, r, from the crack tip. To ensure perfect transfer of strain from the specimen to the fiber, the plastic coating covering the fiber was removed. The angle α was 60° as explained later. The distance, r, was chosen so that no portion of the fiber was within half the plate thickness from the crack tip, where the three dimensional effects are dominant. The attached fiber formed the sensing arm of the Mach-Zehnder interferometer. The specimen was monotonically loaded, and the load and the corresponding number of fringes were recorded simultaneously. Sirkis has shown that only negligible amounts of shear and transverse strains are transferred to the fiber. Using the calibration constant, D, light intensity data was converted into axial strain, $\varepsilon_{x'x'}$. This procedure was repeated for six different crack lengths, a. The plots of strain, $\varepsilon_{x'x'}$, versus load, P, are shown in Fig.6. As expected, the strain varied linearly with load, P.

The strains obtained from the fiber optic sensor were used to calculate the stress intensity factor, K_I, for the crack tip. Dally and Sanford have shown that the axial strain at point P, in the rotated coordinate system x′y′ (see Fig.7), is given as,

$$\varepsilon_{x'x'} = \left(\frac{1}{2G}\right)\frac{K_I}{\sqrt{2\pi r}} - \left[\left(\frac{1-\nu}{1+\nu}\right)\cos\left(\frac{\theta}{2}\right) - \frac{1}{2}\sin(\theta)\sin\left(\frac{3\theta}{2}\right)\cos(2\alpha) + \frac{1}{2}\sin(\theta)\cos\left(\frac{3\theta}{2}\right)\sin(2\alpha)\right] + B_0\left[\left(\frac{1-\nu}{1+\nu}\right) + \cos(2\alpha)\right] , \tag{7}$$

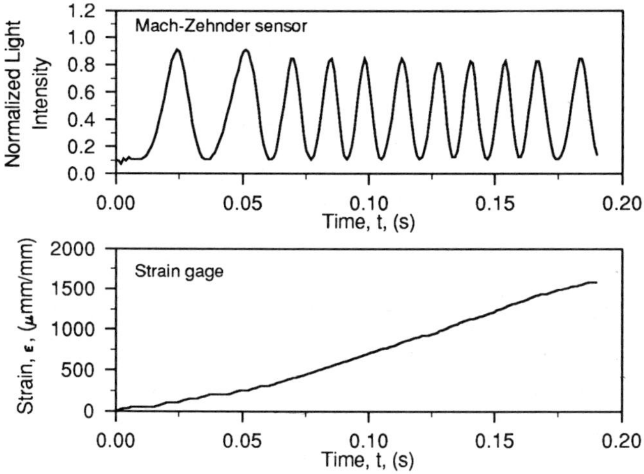

Figure 4. Plot of strain and light intensity variation as a function of time.

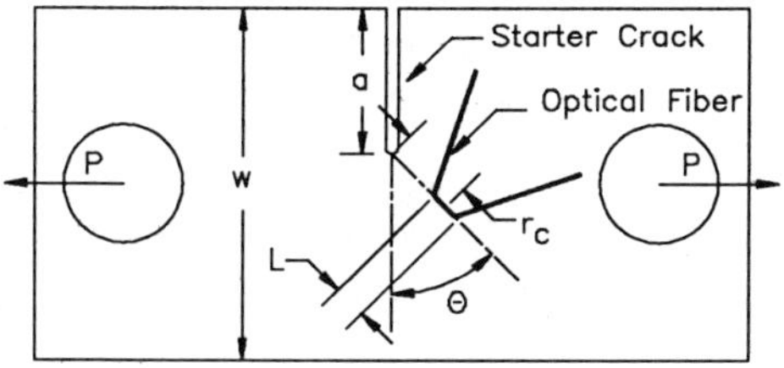

Figure 5. Single edge notched (SEN) specimen used in the static fracture experiment.

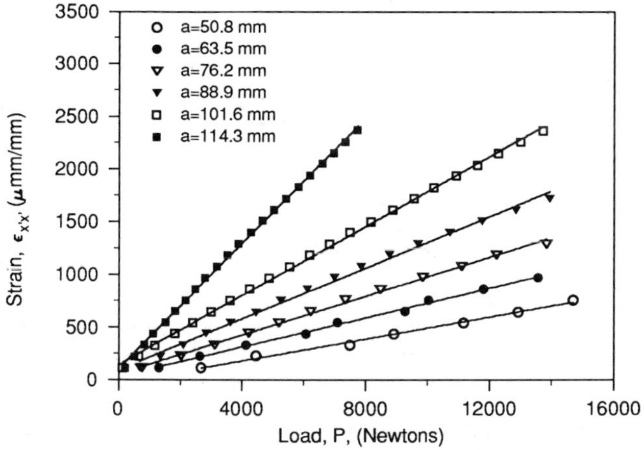

Figure 6. Plot of strain versus load for various crack lengths.

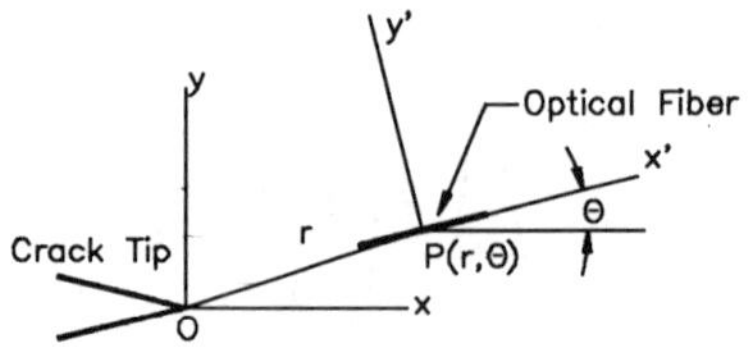

Figure 7. Definition of coordinate systems O_{xy} and $P_{x'y'}$.

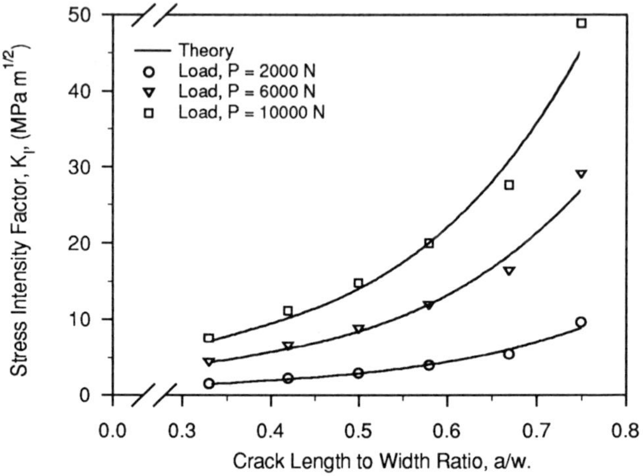

Figure 8. Stress intensity factor as a function of the crack length-to-width ratio.

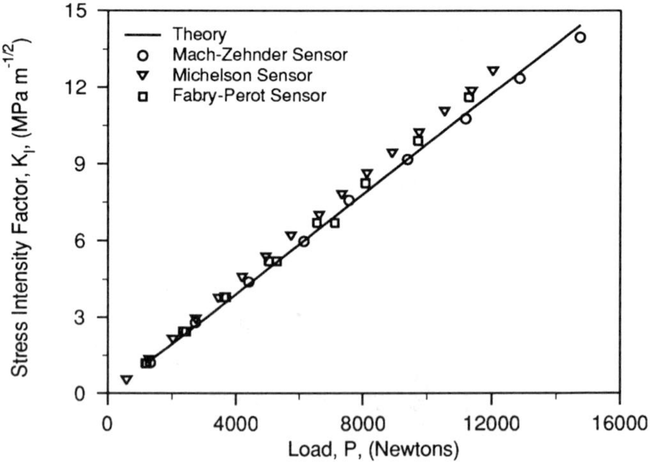

Figure 9. Plot of stress intensity factor versus load for the various sensors used.

where G is the shear modulus, ν is the Poisson's ratio, and B_0 is the first nonsingular stress field coefficient.

The contribution of the B_0 term can be dropped if $\cos(2\alpha)=-(1-\nu)/(1+\nu)$. For aluminum $\nu = (1/3)$, thus $\alpha = 60°$. The angle θ was also chosen equal to 60° because along this direction the singularity dominated zone is relatively large (Dally and Sanford). Thus, for $\alpha=\theta=60°$,

$$\varepsilon_{x'x'} = \frac{1}{2G} \frac{K_I}{\sqrt{2\pi r}} \left(\frac{3\sqrt{3}}{8}\right) \quad . \tag{8}$$

Rearranging Eqn.(8),

$$K_I = E\sqrt{\frac{8\pi r}{3}}\,\varepsilon_{x'x'} \quad . \tag{9}$$

Therefore, the stress intensity factor, K_I, can be determined from the strain $\varepsilon_{x'x'}$.

From the experimental data obtained, the value of K_I was evaluated, for various crack length-to-width ratios, as a function of increasing load. The results are shown in Fig.8. The experimental results were compared with the theoretical handbook values (Broek). For a single edge notched geometry

$$K_I = Y\sigma\sqrt{a} \ , \tag{10}$$

where

$$Y = 1.99 - 0.41\left(\frac{a}{w}\right) + 18.7\left(\frac{a}{w}\right)^2 - 38.48\left(\frac{a}{w}\right)^3 + 53.85\left(\frac{a}{w}\right)^4 , \tag{11}$$

and σ = far field stress. As seen from Fig.8, the experimental and the theoretical values agree very well. The experiment was repeated using the Michelson and the Fabry-Perot sensors. Figure 9 shows the stress intensity factor, K_I, as a function of applied load, for the three sensors, and the results are summarized in Table 1.

Embedded Sensors

The material chosen to embed the optical fiber sensors was plexiglass, which has a much lower Young's modulus than the glass fiber (Narendran *et al.*, 1991b). The optical fiber reinforces the host material. Therefore, the phase sensitivity predicted by Butter and Hocker will be the same as Sirkis and Haslach, and Eqn.(4) is valid.

Unlike the surface attached case, the fibers had to be bent, in some cases, to be embedded. The optical fibers used were very fragile without the buffer coating and it became necessary to leave this coating in place. Therefore, the calibration procedure was repeated with the buffer coating.

Calibration Procedure

A tension experiment, as shown in Fig.10, was utilized in the study of embedded fiber sensors. A small rectangular groove was milled into the specimen and an optical fiber was embedded as shown. The groove was filled with EnviroTex (polymer coating), the same material as plexiglass, and cured overnight. The embedded fiber formed the sensing arm of the Mach-Zehnder interferometer. The specimen was monotonically loaded and the corresponding optical fringes were recorded. Using the calibration constant, D, obtained from the surface-attached case (Narendran *et al.* ,1991a), the light intensity data was converted into axial strain, ε_x. Narendran *et al.* (1991b) give the plots of stress versus strain obtained for several embedding depths, d. The values of the elastic modulus of the host material, as determined from the stress-strain plots are summarized in Table 2. The deviation of the elastic modulus was up to 13% of that of the average value.

Embedded Fracture Experiments

Two fracture experiments are described here. These use inplane embedded optical-fiber sensors and transversely embedded optical-fiber sensors to determine stress intensity factor.

Inplane Embedded Optical Fibers

The optical fibers were embedded at a depth, d, from the surface as shown in Fig.11, and they formed the sensing arm of a Mach-Zehnder interferometer. The specimen was monotonically loaded and the load and the corresponding optical fringes were recorded simultaneously. As before, the light intensity data was converted to axial strain. Similar procedure was repeated with the fiber being embedded at different depths. The strain data obtained from the fiber-optic sensor was used to calculate the stress intensity factor using linear elastic fracture mechanics (Dally and Sanford). Figure 12 shows the stress intensity factor K_I as a function of load for several depths, d. The plots are linear but the slopes vary within 15% of the theoretical values. This is mostly due to the errors in the estimate of the effective gage length of the sensor.

Transversely Embedded Optical Fibers

In this case the fiber was embedded in the transverse direction, as shown in Fig.13. The specimen was loaded and the corresponding transverse strains were measured using the fiber-optic sensor. These strains were used to calculate the stress intensity factor, K_I, associated with the crack tip. Using linear elastic fracture mechanics Dally and Sanford have shown that,

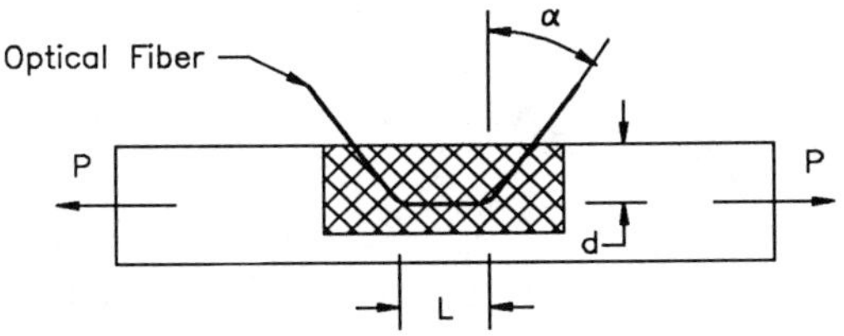

Figure 10. Tensile setup used for calibration of inplane embedded fiber-optic sensors.

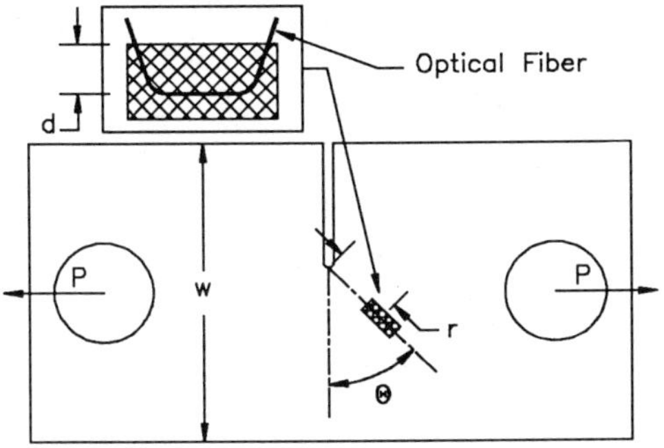

Figure 1:. Single edge notched specimen used in the inplane fracture experiments.

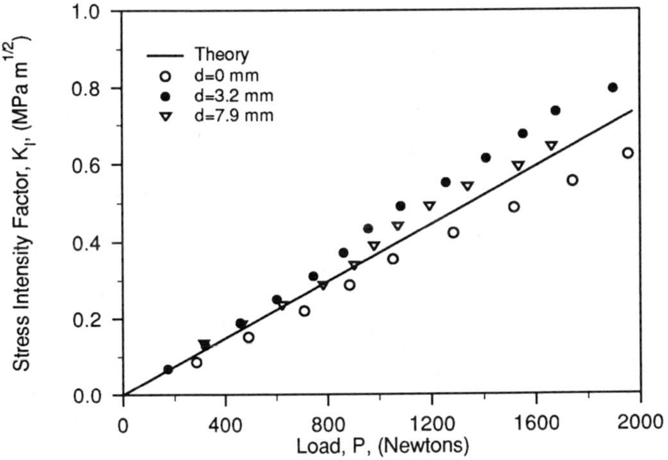

Figure 1 2. Plot of stress intensity factor versus load for various embedded depths.

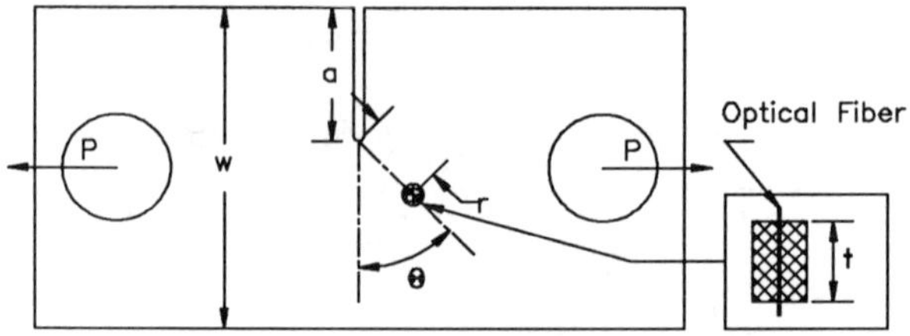

Figure 13. Single edge notched specimen used in the transverse fracture experiments.

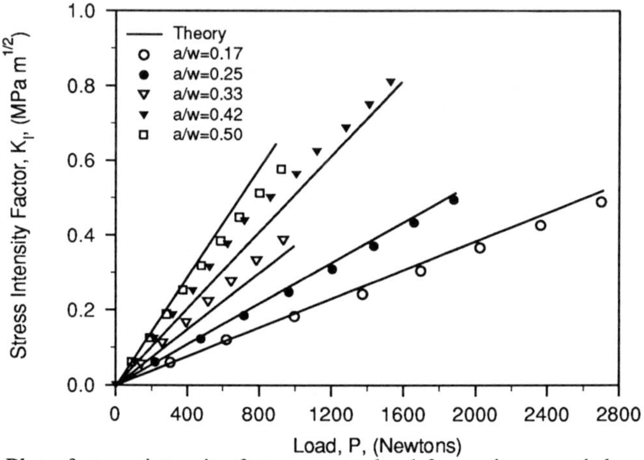

Figure 14. Plot of stress intensity factor versus load for various crack lengths.

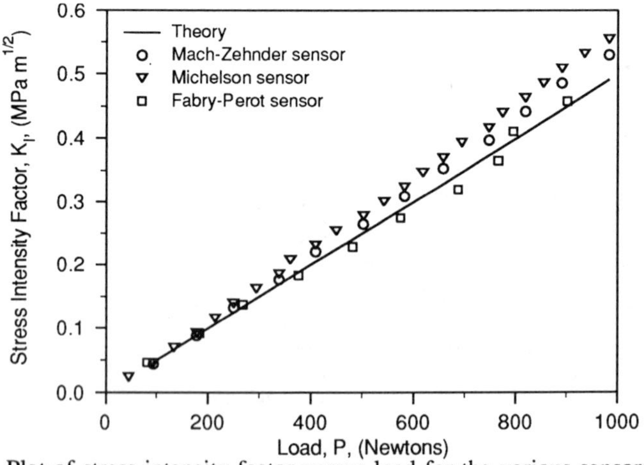

Figure 15. Plot of stress intensity factor versus load for the various sensors used.

$$K_I = \left[\frac{E\sqrt{2\pi r}}{2\nu \cos\left(\frac{\theta}{2}\right)} \right] \varepsilon_{zz} \quad . \tag{12}$$

where K_I is the stress intensity factor and ε_{zz} is the transverse strain.

The transverse strain data was used to determine the stress intensity factor, K_I. The procedure was repeated for several crack lengths. The variation of stress intensity factor with load for five different crack lengths is shown in Fig.14. As expected, these plots were all linear. The deviation of these values compared to the theory was within ±12%. The transverse embedded experiments were repeated using the Michelson and Fabry-Perot sensors and the corresponding plots are shown in Fig.15. The stress intensity factors determined by the Mach-Zehnder and the Michelson interferometer deviated more from the theoretical values than the Fabry-Perot sensor. Table 1 gives the deviation of the stress intensity factors from theory for the three interferometers.

DISCUSSION

Embedded and surface-attached fiber optic sensors have been used succesfully to measure opening mode stress intensity factor for fracture specimens (Narendran *et al.*, 1991a; Narendran *et al.*, 1991b; and Shukla *et al.* 1992). All the sensors studied have strain sensitivities of the same order of magnitude. However, they all have their own advantages and disadvantages, which are discussed below.

The Mach-Zehnder and Michelson sensors are the cheapest and easiest to construct and use in a laboratory environment. However, they are not as useful in field applications. The entire sensing fiber is sensitive to environmental effects such as temperature, strain, vibration, etc. This requires either a close coupling of the reference and sensing fibers, or a controlled environment.

The Fabry-Perot sensor is not as easy to construct. However, its unique construction has a number of advantages. Only one fiber is required which simplifies the setup. Both the reference and sensing beams propagate through the same fiber, which results in freedom from environmental effects. Also, the gage length can be controlled more accurately during sensor fabrication, as compared to the other two sensors. However, the presence of a thin hollow-core fiber makes the sensor relatively fragile, especially under bending and direct pressure, and makes surface attachment difficult (Shukla *et al.*, 1992). A possible alternative to this could be to use the Intrinsic Fabry-Perot sensor (Lee *et al.* and Liu *et al.*).

The Mach-Zehnder and Michelson sensors show greater deviations than the Fabry-Perot sensor for the transversely embedded configuration. These errors are primarily gage length dependent (Shukla *et al.*, 1992). Gage length dependent errors are also observed for the inplane embedded configurations of the Mach-Zehnder and Michelson interferometric sensor (Narendran *et al.*, 1991b).

ACKNOWLEDGEMENT

The support of the National Science Foundation under Grant No. MSS-9101514 is gratefully acknowledged.

TABLE 1. Experimental Errors for the Sensors used.

Sensor	Configuration	Deviation of K_I values from theory
Mach-Zehnder	Attached	6%
Michelson	"	3%
Fabry-Perot	"	5%
Mach-Zehnder	Embedded	11%
Michelson	"	17%
Fabry-Perot	"	5%

TABLE 2. Elastic Modulus Obtained at Various Depths. (Average = 2.81 GPa)

Normalized Depth, d/t	Tension Elastic Modulus (GPa)	Percent Deviation
0.00	3.13	11 %
0.13	2.55	9 %
0.26	2.56	9 %
0.38	3.02	7 %
0.51	2.45	13 %
0.76	2.88	2 %
1.00	3.10	10 %

REFERENCES

Broek, D., Elementary Enginnering Fracture Mechanics, (Martinus Nijhoff Publishers, 4th ed., (1986).

Bruinsma, A.J.A., P. van Zuylen, C.W. Lamberts, A. de Krijger, "Fiber optic strain measurement for structural integrity monitoring", *Proc. Soc. Exptl. Mech.*, Boston, 399-402, (1989).

Butter, C.D., and G.B. Hocker, "Fiber optics strain gauge," *Appl. Opt.* **17**, 2867-2869 (1978).

Dally, J.W., "Dynamic photoelastic studies of fracture," *Expt. Mech.* **28**, 349-361, (1979).

Dally, J.W., and R.J. Sanford, "Strain gage methods for measuring the opening mode stress intensity factor," *Proc. SEM Spring Conf.*, Las Vegas, 851 - 860, (1985).

Giallorenzi, T.G., J.A. Bucaro, A. Dandridge, G.H. Sigel, J.H. Cole, S.C. Rashleigh, R.G. Priest, "Optical fiber sensor technology", *IEEE J. Quant. Elect.* **QE-18**, 626-665 (1982).

Lee, C.E., H.F. Taylor, and A.M. Markus, "Optical-fiber Fabry-Pérot embedded sensor," *Opt. Lett.* **14**, 1225-1227, (1989).

Leonard, L., " 'Smart' composites: Embedded optical fibers monitor structural integrity", *Advanced Composites*, 47-50 (1989).

Liu, K., S.M. Ferguson, and R.M. Measures, "Damage detection in composites with an embedded interferometric optical fiber sensor," in *Fiber Optic Smart Structures and Skins III*, Proc. SPIE **1170**, 5-8, (1990).

Mathews, C.T., and J.S. Sirkis, "Experimental verification of phase-strain models for structurally embedded optical fiber sensors," *Proc. 1991 SEM Spring Conf.*, Milwaukee, 471-478, (1991).

Measures, R.M., N. Glossop, J. Lymer, M. Leblanc, J. West, S. Dubois, W. Tsaw, R.C. Tennyson, "Structural integrated fiber optic damage assesment systems for composite materials", *Appl. Opt.* **28**, 2626-2633 (1989).

Measures, R.M., "Advances towards fiber optic based smart structures," *Opt. Engg.* **31**(1), 34-47, (1992).

Murphy, K.A., B.D. Zimmerman, and R.O. Claus, "Embedded optical fiber sensors for internal material instruments", *Proc. Soc. Expt. Mech.*, 752-756, (1989).

Narendran, N., A. Shukla, and S. Letcher, "Application of fiber-optic sensor to a fracture mechanics problem," *Eng. Fracture Mech.*, **38**, 491-498, (1991).

Narendran, N., A. Shukla, and S. Letcher, "Determination of fracture parameters using embedded fiber optic sensors," *Expt. Mech.* **31**, 360-365, (1991).

Rosakis, A.J., C.C. Ma and L.B. Freund, "Analysis of the optical shadow spot method for a tensile crack in a power-law hardening material," *J. Appl. Mech.* **105**, 777-782, (1983).

Shukla, A., R.K. Agarwal, and H. Nigam, "Dynamic fracture studies on 7075-T6 aluminum and 4340 steel using strain gages and photoelastic coatings," *Eng. Fracture Mech.* **31**, 501-515, (1988).

Shukla, A., R. Singh, and N. Narendran, "Performance of Fiber Optic Sensors in Fracture Mechanics Applications," *1st Eur. Conf. Smart Struct. Matl.*, Glasgow, 285-288, (1992).

Sirkis, J., "Understanding Surface mounted interferometric optical fiber strain sensors," *Proc. Soc. Expt. Mech.*,Boston, 782-787, (1989).

Sirkis, J.S., and H.W. Haslach, Jr., "Complete phase-strain model for structurally embedded interferometric optical fiber sensors," *J. of Intell. Materials Systems and Structures*, **2**(1), 3-24, (1991).

Udd, E., *Fiber Optic Smart Structures and Skins III*, Proc. SPIE **1370**, (1990)

Valis, T., R.D. Turner, and R.M. Measures, "Distributed fiber optic sensing based on counter propagating waves", *Appl. Opt.* **28**, 1984-1990, (1989).

Wiencko, J.A., R.O. Claus, and R.E. Rogers, "Embedded optical fiber sensors for intelligent aerospace structures", *Proc. Sensors Expo.*, Detroit, Sept. (1987).

V - STRAIN AND TEMPERATURE

High Temperature Distributed Strain and Temperature Sensing using OTDR.

Keith H. Wanser[1], Michael Haselhuhn[2], and Michael Lafond[2]

Abstract

Operational characteristics of fused silica optical fibers at temperatures up to 1800°F are described with application to distributed strain and temperature measurements. In particular we report measurements of the temperature dependent Rayleigh/Brillouin backscatter in optical fibers. We find that operation of fused silica fibers at temperatures up to 1200°F is essentially reversible. At higher temperatures we observe irreversible effects associated with fiber structural relaxation and core dopant diffusion. The first tests on polyimide, aluminum, iron, and gold coated fibers at elevated temperatures using OTDR are also reported and survivability issues addressed. A high temperature fiber optic strain sensor and high temperature pipe loading test facility has been developed, and the first reported results from a pipe strain test at 1100°F using OTDR are described.

Introduction

There are a number of important applications for distributed fiber optic strain and temperature measurements at elevated temperatures[1-4]. Many utility plant applications require the sensors to be exposed to extreme temperatures. Some applications include; monitoring strain and deflection of high temperature (1200°F) steam pipes inside fossil fuel boilers, crack detection in high temperature steam pipes in high stress areas such as joints, and monitoring temperatures in various areas. Additional applications include distributed fire detection in buildings and tunnels, measurements of pressure and temperature in rocket fuel tanks and casings, and high temperature process monitoring.

The majority of prior work on optical fiber sensors has been at room temperature, or at most a few hundred °C. Some work at very high temperature (>1000°C) has been reported, including the use of Sapphire optical fiber[4] in very short lengths (~10 cm), but practical sensor packaging and survivability issues have

---

1 Associate Professor of Physics, Department of Physics, California State University Fullerton, Fullerton CA 92634

2 Department of Physics, California State University Fullerton, Fullerton CA 92634

not been adequately addressed. Often several meters of fiber are required to be subjected to a harsh, high temperature environment, especially in distributed sensing applications. In order to operate optical fiber sensors in environments over 1000°F, questions of fiber and sensor survivability and high temperature packaging must be addressed.

In a companion paper[5], the background theory and applications of distributed fiber optic sensors using optical time domain reflectometry (OTDR) were discussed. In the present work we address the application and operational properties of fused silica based optical fibers at temperatures up to 1800°F. First we discuss the background theory necessary to understand temperature dependent changes in optical fiber scattering properties using OTDR and its application to distributed temperature measurements using Rayleigh/Brillouin temperature dependent scattering. Then we present experimental results on the reversible and irreversible intrinsic fiber scattering properties and theoretical results on irreversible changes due to core diffusion. Next the high temperature properties of fiber coatings are evaluated. Results on fibers with coating materials of acrylate, polyimide, aluminum, iron, and gold are presented, and survivability and applications discussed. Finally, we present details of a high temperature pipe loading test facility that has been built and test results for packaged distributed fiber optic sensors with various coatings at elevated temperatures.

Temperature Dependent Scattering Properties of Optical Fibers

In order to understand how the basic factors affect the OTDR temperature dependent backscatter response, consider the fiber backscatter impulse response function given by[5]

$$P_B(t) = \frac{P_o}{2} \alpha_s S v \tau e^{-\alpha v t} \theta(t) \theta(\tau + \frac{2L}{v} - t) \quad , \tag{1}$$

where $P_B(t)$ is the returned Rayleigh backscattered power at the fiber input end at time t, α is the total power attenuation coefficient, including scattering, material absorption, and macro and micro bending loss, α_s is the scattering attenuation coefficient, S is the backscattering trapping factor, v is the average group velocity, L is the total length of the fiber, P_o is the power launched into the fiber, τ is the pulse width, and $\theta(t)$ is the Heaviside step function. Time t is converted to distance x (from the input end of the fiber) in the OTDR by the relation $t = 2\,x/v$. Equation 1 turns on at $t = 0$ and turns off at $t = \tau + 2\,L/v$.

The backscattering trapping factor S, which is the fraction of scattered power that is trapped in the backward propagating direction, is given for multimode fibers by[5]

$$S = \frac{3\,g\,NA^2}{8\,(g+1)\,n_{co}^2} \quad , \tag{2}$$

where NA is the fiber numerical aperture given by $NA = \sqrt{n_{co}^2 - n_{cl}^2}$, n_{co} is the fiber core refractive index (on axis), n_{cl} is the fiber cladding refractive index, and g is the index profile parameter (g = 2 for parabolic graded index fiber and $g = \infty$ for step index fibers). For a typical multimode fiber having an NA of 0.20 and core refractive index of 1.457, the trapping factor S is 4.71 x 10^{-3} for graded index and 7.07 x 10^{-3} for step index fiber. Thus for typical commercial multimode fibers, about one half percent of the Rayleigh scattered light intensity is captured (into guided mode power) by the optical fiber in the backward direction.

The impulse response function is fundamental to a quantitative understanding of the factors governing the OTDR. The factors of importance for high temperature operation of optical fibers are the total attenuation coefficient α, the scattering coefficient α_s, and the local trapping factor S.

The total scattering coefficient α_s consists of several contributions, which we can write as

$$\alpha_s = \alpha_{sr} + \alpha_{sb} + \alpha_{sra} + \alpha_{sf} \qquad (3.)$$

where α_{sr} is the temperature dependent Rayleigh scattering coefficient, α_{sb} the temperature dependent Brillouin scattering coefficient, α_{sra} is the temperature dependent Raman scattering coefficient, and α_{sf} is the static, "frozen in" Rayleigh scattering coefficient. The scattering coefficient α_{sf} is due to frozen in density and dopant concentration fluctuations[6] which consist of thermodynamic, and non-thermodynamic, processing dependent contributions.

The temperature dependent Rayleigh and Brillouin scattering contributions are comparable in size, and the Raman coefficient is about 100 to 1000 times smaller[7]. However, temperature dependent Raman scattering, although much weaker than temperature dependent Rayleigh/Brillouin scattering, has the advantage that one may obtain an absolute measurement of the temperature by measuring the ratio of the Stokes to anti-Stokes, frequency shifted light. This technique for distributed temperature measurements is called Raman ratio thermometry[7]. Raman ratio thermometry has good temperature sensitivity (of the order of a few °C) and commercial systems are available. However, due to the extremely low Raman scattering coefficient, the systems are very expensive compared to a comparable performance (i.e. spatial resolution and acquisition time) OTDR system. In addition, at high temperatures, the Raman ratio approaches one and the sensitivity of the method decreases.

For many distributed temperature sensing applications, a temperature resolution of 15-30°F is adequate. In some applications, such as distributed fire alarms for structures, simple high temperature alarm mode sensing is sufficient, and much more relaxed temperature resolution is adequate. In such cases, the temperature dependent Rayleigh/Brillouin scattering can be used for temperature monitoring. The spatial resolution of the distributed Rayleigh/Brillouin temperature measurement is just as high as the spatial resolution of the OTDR used to probe the system, and temperature differences from less than one meter sections of fiber are resolvable with commercial OTDRs. An example of how temperature is sensed using this method is shown in fig. 1. The upper trace in the figure shows an increased backscatter level from the section of the fiber that is being heated, corresponding to an increased scattering coefficient α_s. The magnitude of the front step height above the baseline from before the heated section to just inside the heated section is directly proportional to the absolute temperature. The return to the baseline signal level shown in the lower trace demonstrates a reversible temperature sensing mechanism for excursions of 1100°F, once the fiber has been initially annealed.

The reversible contribution to the scattering coefficient from temperature dependent Rayleigh scattering is given by[6,8,9]

$$\alpha_{sr} = \frac{8\pi^3}{3\lambda^4}\left[\frac{k_B T n^8 (p_{11}+p_{12})^2}{9B} + 4n^2\left(\frac{dn}{dT}\right)^2 \frac{k_B T^2}{\rho C_v}\right] \quad . \tag{4.}$$

The first term in eqn. 4 is from thermodynamic volume fluctuations and the second term is from thermodynamic temperature fluctuations. The reversible contribution to the scattering coefficient from temperature dependent Brillouin scattering is given by

$$\alpha_{sb} = \frac{8\pi^3 k_B T n^8 p_{12}^2}{3\lambda^4 \rho v_l^2} \quad . \tag{5.}$$

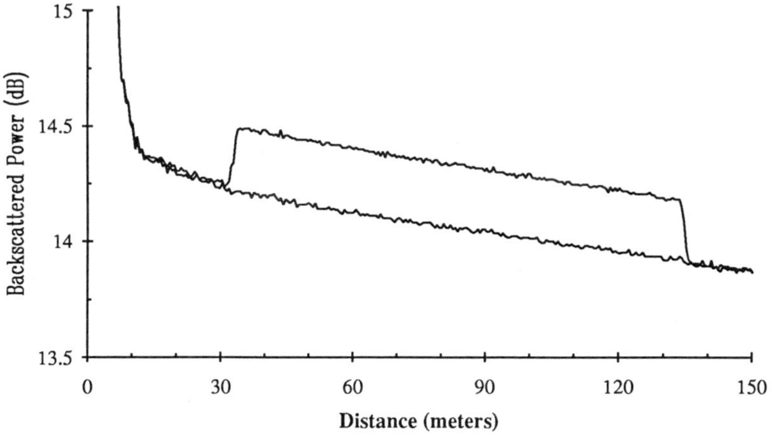

Figure 1. OTDR traces of 50/125 µm graded index fiber at 840 nm wavelength. 100 meter fiber segment in oven. Upper trace, 1100°F. Lower trace, cooling to 127°F.

The Brillouin scattering coefficient is due to scattering of light by thermally excited longitudinal acoustic waves, and eqn. 5 is the v-v polarization contribution. The transverse acoustic wave contribution is more than 15 times smaller. The parameters in eqns. 4 and 5 are, the refractive index n, absolute temperature T, bulk modulus B, wavelength λ, heat capacity per unit mass C_v, mass density ρ, Boltzmann's constant k_B, photo elastic constants p_{11} and p_{12}, and longitudinal speed of sound v_l.

Note that both eqns. 4 and 5 have an inverse fourth power wavelength dependence which is characteristic of scattering by objects that are much smaller than the wavelength. This fact can be used to advantage to enhance and calibrate the temperature dependent backscatter response by using a shorter wavelength or dual wavelength OTDR system. As an example, the temperature dependent scattering coefficient at a wavelength of 670 nm (a commercially available visible diode laser wavelength for OTDR) is a factor of 2.47 larger than the scattering coefficient at the more common 840 nm wavelength, thus enhancing the signal to noise ratio, and ultimately the temperature resolution obtainable with this sensing technique.

The second point concerns the linearity of the temperature response. For fused silica fibers, the second term in eqn. 4 is less than 1% of the size of the first term, and so both eqns. 4 and 5 are extremely linear in temperature for fused silica fibers. This linearity is shown in fig. 2. To get an idea of the size of the temperature dependent contribution to the attenuation, at a temperature of 81°F, the temperature dependent Rayleigh/Brillouin scattering contributes 0.135 dB/km compared to a total attenuation of ~2.3-2.4 dB/km at 850 nm (the Rayleigh and Brillouin contributions are of comparable magnitude). As an example of the agreement between theory and experiment, at a wavelength of 850 nm, eqns. 4 and 5 predict a total change in the

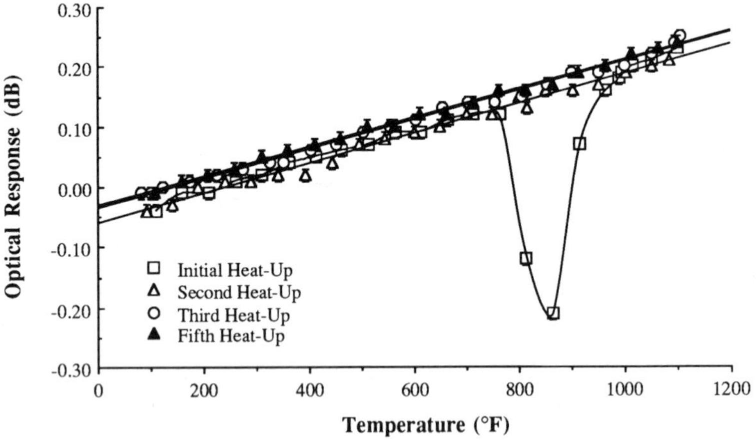

Figure 2. Temperature dependent response of 50/125 μm graded index optical fiber. Two point step size OTDR measurements 7 meters apart. Note the dip at ~800°F in the first heat-up associated with the fiber acrylate coating sublimation.

scattering coefficient of 0.257 dB/km for a 1025°F change in temperature, or a step change of 0.257 dB. This is in excellent agreement with the experimentally measured step size change in fig. 2, which is 0.263 dB over the same temperature interval. For a 0.01 dB OTDR measurement at 670 nm wavelength, this implies a temperature sensing resolution of 15.4°F.

Figure 2 shows an interesting dip in the first heat-up OTDR results. This phenomenon is associated with the sublimation of the acrylate coating. Figure 3 shows a series of OTDR traces below, at, and above the coating sublimation temperature. The large attenuation occurs as a result of microbend induced attenuation associated with shrinkage and sublimation of the acrylate fiber coating, which is quite noticeable at 800°F. The phenomenon does not appear to hurt the fiber. A similar phenomenon occurs for polyimide coated fiber at a higher temperature of 1052°F. Extended tests of polyimide coated fiber showed coating survivability at temperatures of 950°F. This fiber coating removal process suggests applications for fibers with special coatings as alarm mode temperature sensors, as the temperature at which the sublimation process induces significant attenuation is repeatable and controllable by the coating material used.

Temperature dependent backscatter data on gold coated fiber is shown is fig. 4. To our knowledge, this is the first reported temperature dependent backscatter data on gold coated fiber. As an indication of the reversible nature of the gold coated fiber at elevated temperatures, holding the fiber at 1100°F for 20 hours produced no observable change in the elevated or room temperature OTDR traces. This reversible data is significant, in that tests with aluminum and iron coated fibers

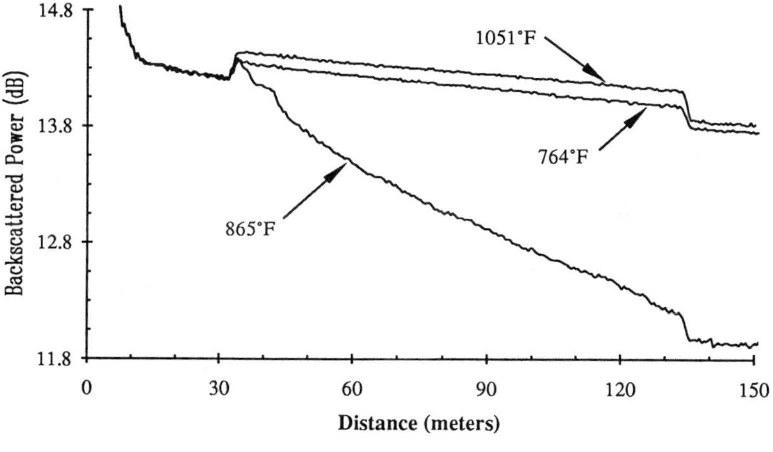

Figure 3. OTDR traces showing effect of fiber acrylate coating removal during first heat-up. Note the fiber attenuation after the sublimation process is complete (1051° F) is the same as before the sublimation (764°F), as indicated by the parallel traces in the heated region between 35 and 135 meters.

showed irreversible effects associated with annealing of excess coating induced attenuation, as shown in fig. 5.

Up to this point we have only discussed reversible temperature dependent scattering processes. An example of the OTDR backscattering response observed

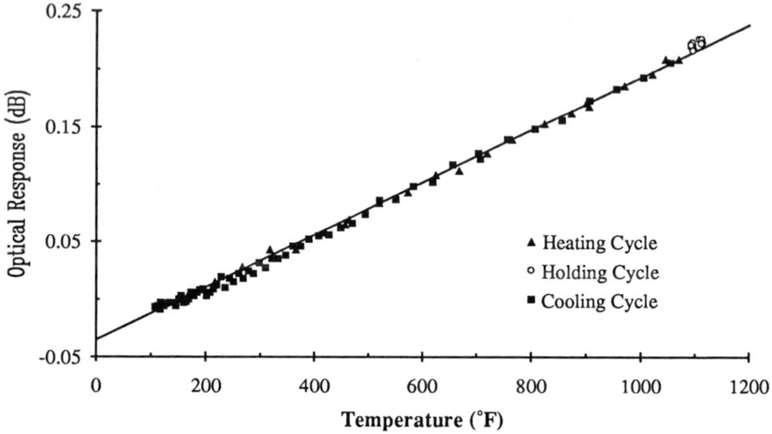

Figure 4. Temperature dependent response of 50/125/185 μm gold coated, graded index optical fiber showing reversible behavior. Two point step size OTDR measurements 2.5 meters apart. Holding cycle at 1100°F was 2 hours.

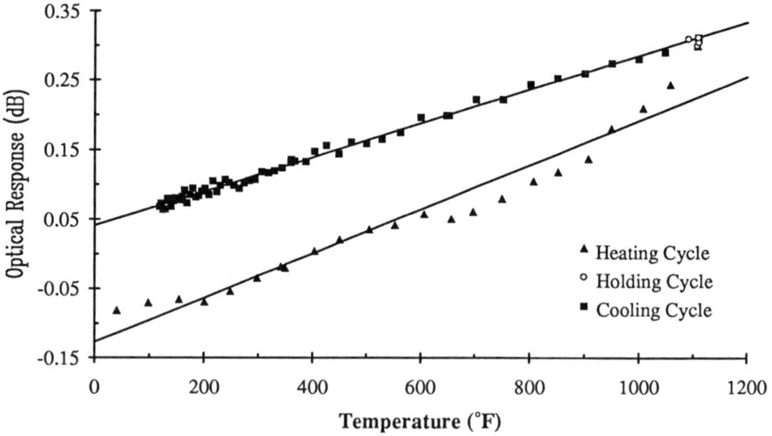

Figure 5. Temperature dependent response of 50/125/175 μm aluminum coated, graded index optical fiber showing irreversible behavior, initial heating and cooling cycle. OTDR step measurements 2.8 meters apart, 1 hour 1100°F holding time.

as a result of irreversible processes is shown in fig. 6. The data in fig. 6 show that the irreversible process has not induced fiber loss, as noted in the figure caption, but rather a lower backscattering. This lower backscattering results from two effects, a lower backscatter trapping factor and a lower scattering coefficient. In principle it is possible to separate out the two effects by using long (~1 km) lengths of fiber, and accurately measuring the slope of the heated region of fiber, as well as the step change associated with the level of the backscattered return power.

There are several temperature dependent irreversible processes which occur in the fibers at elevated temperatures that can affect the OTDR backscatter response. These processes are primarily structural relaxation of the amorphous material, and fiber dopant diffusion processes. The scattering from frozen in density fluctuations associated with the liquid state is responsible for the majority of the attenuation in silica optical fibers[6]. Thermodynamic considerations predict a value for the density fluctuations. It is possible to have density fluctuations in excess of thermodynamic values by rapid cooling of the optical fibers during their manufacture, which is process dependent. Relaxation of the density fluctuations towards smaller values predicted by thermodynamics can be achieved by annealing at high temperatures. The annealing point of bulk fused silica is approximately 1140°C (2084°F), as compared to a strain point of 1070°C, and a softening point of 1665°C. However, stress relief annealing and structural relaxation is expected to take place at somewhat lower temperatures in fused silica optical fibers, since the material has not been consolidated in the same manner as bulk fused silica.

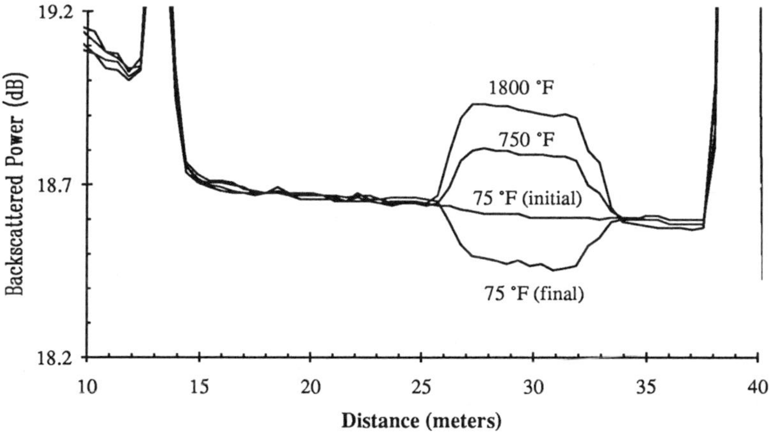

Figure 6. OTDR traces of 50/125 μm graded index fiber showing irreversible change after cycling fiber to 1800°F. Note (in the lower trace) the return to the same level of scattering after the heated up region as before, indicating no extra attenuation associated with the irreversible change.

The refractive index profile and waveguiding properties of optical fibers are controlled by suitable dopants in the core and cladding regions of the fiber. Commonly used dopants are GeO_2 (to raise the refractive index of the core region), Fluorine (to lower the refractive index of the cladding), and additional dopants such as Chlorine (to remove OH during fabrication), Boron, and Phosphorus. Each of these dopants makes a contribution to the refractive index profile and scattering coefficient of the fiber. The various dopants have differing diffusion coefficients at elevated temperature, and thus, depending on the fiber composition, different behaviors will occur in regard to altering the fiber numerical aperture (and thus trapping factor S), and scattering coefficient due to concentration fluctuations.

The (irreversible) diffusion of dopant atoms can be described by the diffusion equation

$$\frac{\partial C}{\partial t} = D\nabla^2 C \tag{6.}$$

where D is the diffusion coefficient of the dopant of interest (in SiO_2) and C is the mole concentration of the dopant. In a binary system such as GeO_2-SiO_2, the relation between the dopant concentration profile and the refractive index profile can be described by the equation[6]

$$n^2(r,t) = n_S^2 + C(r,t)(n_G^2 - n_S^2) \tag{7.}$$

where $n_S = 1.458$ and $n_G = 1.603$ are the refractive indices of pure fused SiO_2 and GeO_2 respectively, and r is the radial distance from the center of the fiber. Note from eqn. 7 that the refractive index profile mirrors the shape of the dopant concentration profile. The on axis ($r = 0$) solution to the diffusion eqn. 6 for an initial parabolic concentration profile (graded index fiber) is

$$C(0,t) = C_o\left\{1-(t/\tau_c)\left[1-e^{-(\tau_c/t)}\right]\right\} \tag{8.}$$

where C_o is the initial value of the on axis concentration and the core diffusion time τ_c is given by

$$\tau_c = \frac{a^2}{4D}\ , \tag{9.}$$

where $2a$ is the initial core diameter of the fiber. Note the unusual time dependence in the exponent in eqn. 8. Using eqn. 7, the fiber numerical aperture, which is proportional to the back scatter trapping factor, can be written as

$$NA(t)^2 = C(0,t)(n_G^2 - n_S^2) \tag{10.}$$

Thus we see from eqn. 2 that the back scattering trapping factor S is directly proportional to the on axis dopant concentration. In fig. 7 we plot exact solutions to the diffusion eqn. 6 for various times, showing the effect of core diffusion for an initially parabolic graded index profile fiber. Note that the on axis concentration, and thus the trapping factor, have decreased by 10% (corresponding to a step change in backscattering level of 0.458 dB) in as little as 0.1 core diffusion time. By way of comparison, the step change in fig. 6 is only ~0.17 dB, showing that as little as 0.04 τ_c is required to explain the results in the figure. Changes in the fiber backscattering trapping factor consist of NA changes from on axis index decrease, as well as slight profile exponent (g in eqn. 2) decreases due to core diffusion. Both effects decrease the trapping factor and are irreversible.

The diffusion coefficients of dopants in solids are strongly temperature dependent, and it is known that significant diffusion of germanium in fused silica occurs in a few hours at 1200°C[10]. Since dopant concentrations are considered proprietary, it could not be determined with certainty the degree to which dopants were responsible for the irreversible change. From estimates of core diffusion times based on published diffusion constants, it is believed that the diffusion of dopants, some with larger diffusion coefficients than germanium, plays a major role in the decreased level of backscattering shown in fig. 6.

High Temperature Pipe Strain Test Facility

In a companion paper[5], the construction details of a room temperature pipe loading test frame and test results obtained on optical fiber sensors attached to pipes in the frame were presented. For the high temperature work reported here, an insulated, computer controlled, removable oven was designed and fabricated around the existing pipe test fixture. This 8'x3'x3' oven was fabricated out of layered ceramic fiber insulation that can withstand temperatures of up to 2300°F. In order to

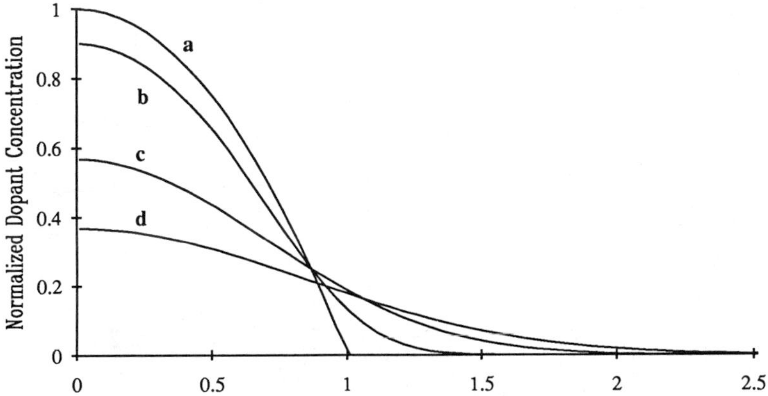

Figure 7. Optical fiber dopant concentration vs. radial distance for various times at elevated temperatures. The curves mirror the refractive index profile (eqn. 7). Curves are for the times, a.) t = 0, parabolic graded index fiber, b.) t = 0.1 τ_c, c.) t = 0.5 τ_c, d.) t = τ_c, where τ_c is the core diffusion time given by eqn. 9.

heat the oven, electric heating coils were placed on all four walls to reduce any thermal gradients that may exist within the oven. This elevated temperature testing facility, shown in figure 8, can achieve temperatures in excess of 1500°F and uses a 440 volt 30 amp outlet.

The high temperature pipe loading test facility simulates a 1/5 scale model of a selected header pipe in a fossil-fuel power plant boiler. The pipe tested in three point loading within the fixture was an 11 foot long, 6.625 inch diameter, high temperature, high carbon steel, schedule 80 pipe. A 25 ton actuator with load cell and a specially designed liquid cooled extension rod was below the pipe to measure the transverse force, and a Linear Variable Differential Transducer (LVDT) with thermal sinks above the pipe to measure and calibrate the transverse pipe deflection[5].

The testing fixture oven can be operated manually or by computer. With either method, the heating time can be controlled to take as long as the user desires to reach a given temperature, or hold at a specified temperature for testing. The computer controlled option allows for automated testing of the fiber optic sensor, including the measurement of temperature, applied load, vertical displacement, and optical response via a personal computer with data acquisition board and a GPIB controlled OTDR. Extensive computer software for this control and data acquisition function has been written.

Sensor Fabrication

The development of optical fiber strain and deflection sensors for monitoring pipelines and other structures at room temperature are described in a companion

Figure 8. Pipe loading test facility with computer controlled oven around pipe test fixture.

paper[5]. In order to test such structures at elevated temperatures, the optical fiber sensor had to be redesigned and manufactured. Extensive development work was performed on the design and manufacturing method of the sensor, which included real time monitoring of the optical fiber during sensor braid manufacturing using OTDR to ensure the functionality and sensitivity of the sensor. After several different combinations of wires, braids, and optical fibers, a sensor that had the type of sensitivity and durability that would be applicable to this type of elevated temperature structural monitoring was obtained. The final structure of the sensor included metal coated optical fiber(s) and sensitivity enhancing filler wire(s), surrounded by a specialized multi-wire stainless steel braid that could be mass produced in long continuous segments.

Elevated Temperature Strain Testing

The microbend sensors in these tests consisted of a metal coated 50/125 μm (core/cladding diameter), graded index multimode fiber. The fiber was surrounded by two 0.010 inch diameter and one 0.003 inch stainless steel wires packaged inside a custom designed stainless steel braid. During manufacturing, a small stainless steel tube was placed within the braid at one meter intervals, which served as an attachment point to the structure to be tested. The sensor was then held at these sites by special low profile attachment brackets shown in fig. 9, along with samples of the cabled optical fiber sensor. The attachment brackets were designed to be used with a sensor pretensioning tool, which allowed for adjustment of sensor response. The attachments were secured to the pipe at one meter gauge lengths, and the sensors installed utilizing these attachment sites. This design allowed for good sensitivity, as well as protection to the optical fibers, with little hysteresis.

Figure 9. Spools of stainless steel braided fiber optic microbend sensor, showing low profile pipe mounting brackets.

The first coating that was tested was an iron coated, graded index 50-125-238 optical fiber. The sensor consisted of two identical fibers and two 0.010" stainless steel wires encased in a stainless steel braid. The sensor was mounted on the testing fixture at top dead center of the pipe, as shown in fig. 10, and tested at room

Figure 10. Fiber optic strain sensor mounted on top of pipe in high temperature test facility. Note low profile sensor mounting bracket on the left side of the figure.

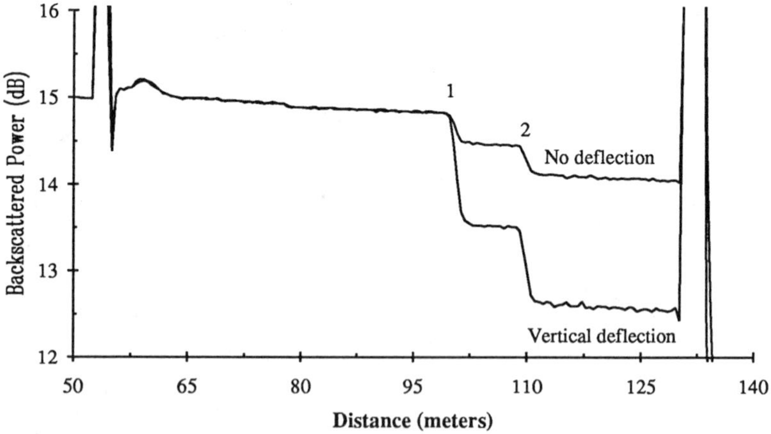

Figure 11. 75°F response of two ~1 m long iron coated fiber optic strain sensors tested in vertical deflection, mounted on top of the pipe shown in fig. 10. Vertical pipe deflection in the lower trace was 0.75 inches corresponding to a longitudinal strain of 0.145%.

temperature and elevated temperatures. Figure 11 shows two OTDR traces taken at room temperature with the pipe vertically displaced 0 and 0.75 inches respectively. Figure 12 shows the optical response of the iron coated fiber as it is tested between 0 and 0.75 inches. Figure 13 shows

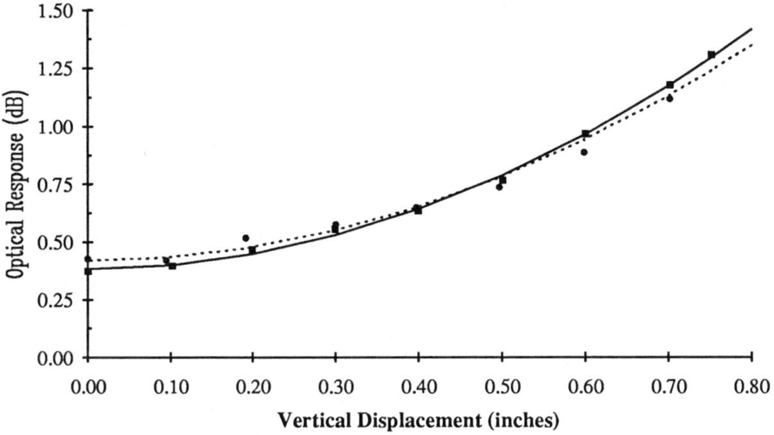

Figure 12. Optical Response of Iron coated fiber sensor 1 in fig. 11, 75°F. Curves are pure quadratic least square fits to data (i.e. $y = m x^2 + b$).

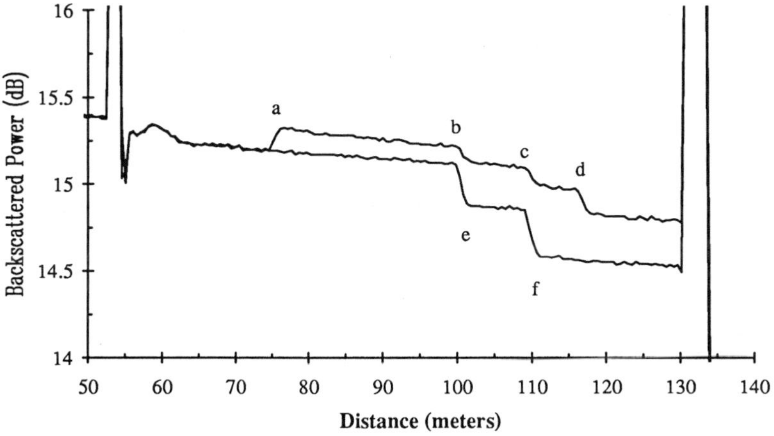

Figure 13. OTDR response of ~1 meter long iron coated fiber sensors. Upper curve, ~650°F, a.) temperature step going into oven, b.) location of microbend sensor 1, c.) location of microbend sensor 2, d.) temperature step associated with going out of oven to room temperature. Lower curve 75°F, e.) baseline prebias of sensor 1, f.) baseline prebias of sensor 2.

the combined effects of distributed temperature sensing and distributed strain sensing, as described in the figure caption. Note relaxation of the sensor prebias at high temperature for this sensor design. Also note temperature step heights a and d in the figure are the same. The iron coated fiber in the tests shown was very brittle and difficult to handle without breaking. The sensor repeatedly broke. This problem was not intrinsic to the iron coating, but rather was due to the prototype nature of the fiber drawing and coating process that was used. Due to difficulty in obtaining fibers made with iron, and recent availability of gold coated fiber in reasonable lengths, the iron coated sensors were not pursued further.

The second metal coating that was tested utilized a gold coated fiber in order to withstand high temperature structural monitoring (gold melting point 1947°F). Gold has excellent corrosion resistance properties, as well as good mechanical properties for an optical fiber coating. A gold coated fiber optic sensor was manufactured, as described earlier, and was installed onto the pipe testing facility as shown in fig. 10. The fixture was then heated, and when the desired temperature was achieved, the pipe was vertically deflected. The first results of high temperature strain measurements using this sensor interrogated by OTDR are presented in fig. 14. The figure shows the response of the same sensor at room temperature and at 1100°F. Although there are some temperature dependent sensitivity effects, the sensitivity of the sensor actually increased at high temperature, due a thermal expansion change in its initial prebias point. These preliminary results are encouraging and show promise for high temperature distributed sensing using this approach. It is expected that additional refinement of sensor packaging materials and attachment techniques will further improve the sensor response and temperature stability characteristics.

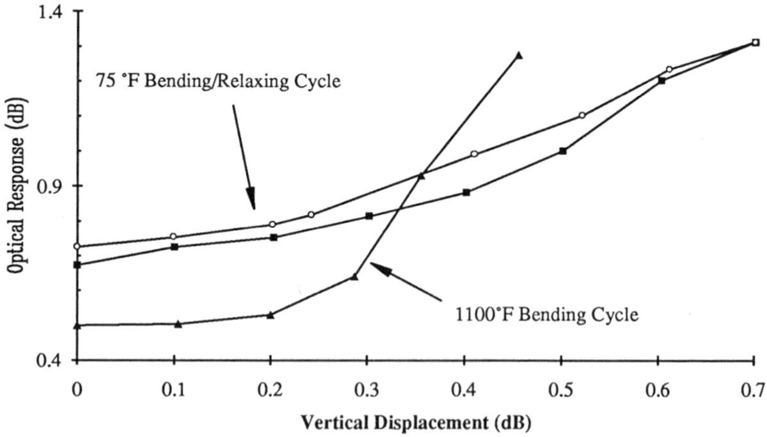

Figure 14. Response of ~1 meter long gold coated optical fiber sensor attached to pipe as shown in fig. 10. Vertical displacement of 0.25 inch is longitudinal strain of 0.048%

Acknowledgments:

Support for much of the work reported here was provided by G2 Systems Corporation, the California Department of Commerce, and Southern California Edison. The contributions of Pacific Braid and Cable to prototype sensor fabrication are gratefully acknowledged. We thank Mr. Jim Keich of the CSUF Civil Engineering Department for technical assistance with various aspects of this work.

References

1.) Weiss, J., Esselman, W., and Lee, R., "Assess fiberoptic sensors for key powerplant measurements", *Power*, pp. 55-58, Oct. (1990).

2.) Griffiths, R. W., "Fiber Optic Monitoring in the Electric Utility Power Plant Environment", *EPRI Conference Proceedings, Optical Sensing in Utility Applications*, San Francisco, May (1991).

3.) Schad, H., "Fiberoptical Distributed Temperature Sensing of Temperature Profiles of a Transformer Model Coil", *Proc. SPIE*, **985**, *Fiber Optic and Laser Sensors VI*, pp. 64-68 (1988).

4.) Murphy, K. A., Koob, C. E., Plante, A. J., and Claus, R. O., "High Temperature Sensing Applications of Silica and Sapphire Optical Fibers, *Proc. SPIE*, **1370**, *Fiber Optic Smart Structures and Skins III*, pp. 169-178 (1990).

5.) Wanser, K. H., Haselhuhn, M., Lafond, M., and Williams, J., "Distributed Fiber Optic Sensors for Civil Structures using OTDR", *Applications of Fiber Optic Sensors in Engineering Mechanics*, American Society of Civil Engineers, New York, (1993).

6.) Olshansky, R., "Propagation in glass optical waveguides", *Rev. Mod. Phys.*, **51**, pp. 341-367, April (1979).

7.) Dakin, J. P., et al, "Distributed optical fibre Raman temperature sensor using a semiconductor light source and detector", in *The Distributed Fiber Optic Sensing Handbook*, Dakin, J. P., ed., IFS Publications, UK pp. 143-145 (1990). Also pp. 16, 119-167 for papers on several methods of distributed temperature sensing

8.) Landau, L. D., and Lifshitz, E. M., *Electrodynamics of Continuous Media*, p. 389 Pergamon, New York, (1975).

9.) Crosignani, B., Di Porto, Paolo, and Bertolotti, M., *Statistical Properties of Scattered Light*, p. 51, Academic, New York, (1975).

10.) Shiraishi, K., Aizawa, Y., and Kawakami, S., "Beam Expanding Fiber Using Thermal Diffusion of the Dopant", *J. Lightwave Technol.*, **LT 8**, pp. 1151-1161 (1990).

Distributed Fiber Optic Temperature Sensing

A. A. Boiarski[1]

Abstract

A special fiber optic sensor can measure local temperatures within structures, material processes, and mechanical systems using a distributed sensing format. Several one-kilometer lengths of prototype fiber were produced by modifying the cladding material coated on standard core glasses. Changes in Rayleigh back-scattered light from the fiber cladding indicated changes in local temperature at points along the entire fiber length. A demonstration system was assembled consisting of twenty meters of fiber and a special high-resolution optical time domain reflectometer. Data obtained under controlled conditions showed that ten-centimeter long hot spots were detected to within ±5°C accuracy over a 0 - 150°C temperature range. In addition, field test results located hot spots in a switchgear circuit breaker at an operating power plant.

Introduction

Measurement of temperature can provide important information concerning safe and efficient operation of structures and equipment. Further, monitoring temperature of materials during fabrication can lead to *in situ* process modifications that reduce waste and improve performance. Although beneficial, temperature measurement in structures and during materials processing has some major constraints. For example, both internal and external measurements must be made at multiple points and often within large volumes. Therefore, many sensors are normally required. The spatial resolution of each measurement should be small

----

[1]Senior Research Scientist, Battelle, 505 King Avenue, Columbus, Ohio 43201-2693

(e.g., 10 cm or less) and the sensor itself must also have a compact size, rugged construction, and low overall cost per measurement site.

To provide continuous, high-spatial resolution monitoring of temperature at many locations within a structure or material process, a high- resolution distributed fiber optic temperature sensor (DFOTS) system was developed. The key DFOTS advantage is the fact that one fiber can be used to obtain measurements at many points along its length. Further, optical fibers are small and relatively inexpensive, especially for the distributed sensing format. The overall temperature measurement accuracy goal was $\pm$5°C with a range of 20-150°C. The sensor length goal was 20 meters and the length resolution target was 10 cm. Measurement length resolution was a particularity stringent goal in that no other DFOTS method has achieved this resolution limit.

In the text of this report, distributed fiber optic sensing is first described in general, then temperature sensing is discussed in some detail. Alternate DFOTS approaches are described and the preferred clad-scattering approach is compared to the other possible methods. Both laboratory and field test data are also provided to demonstrate the DFOTS system performance relative to the above goals.

Distributed Fiber Optic Sensing

The approach used for distributed fiber optic temperature sensing is based on standard Optical Time Domain Reflectometry (OTDR) technology. The OTDR was developed for characterizing fiber optic communication systems. Using the OTDR approach involves transmitting a light pulse along an optical fiber where it is gradually attenuated by a combination of absorption and scattering effects. The scattered light returns in a direction opposite to the forward-propagating pulse (backscattered) and is detected. The backscattering-based OTDR method permits measurement of local attenuation and scattering intensity for optical fibers without cutting the fibers; thus a measurement can be made from either fiber end.

The physical distance of a particular change in an optical fiber's backscattered signal can be calculated by measuring the elapsed time of the returned pulse. If the time required to propagate back and forth is τ, then the location, L, of the change is given by

$$L = c\tau/2n \ , \qquad (1)$$

where

c = velocity of light in a vacuum (3×10^8 m/s),

and

n = refractive index of the fiber.

To use the OTDR approach as the basis for a distributed fiber optic sensor, the fiber must be modified along its length. The modification causes a change in the loss and/or backscattering characteristics of the fiber when a particular parameter (e.g., temperature) changes. The change in some parameter is monitored by measuring the amount of change in the local backscattered light characteristics (local change in backscattered signal). Furthermore, the location (in time) of the OTDR signal change can be used to locate (in space) the parameter change.

Previous Distributed Temperature Sensing Approaches

By far, the most active area of distributed fiber optic sensing has been that of temperature measurement. In fact, several review articles on the subject appear in the literature (Twerdochlib et al, 1989 and Rogers, 1986). DFOTS approaches which have been experimentally demonstrated are discussed below.

The amount of backscattered light, $P_{bs}(\ell)$, from a given location, ℓ, along the fiber and within the scattering element, dx, can be written as

$$P_{bs}(\ell) = 1/2\ P_o\ \Delta t\ \upsilon_g\ C_s\ NA^2 \exp\left(\int_o^{\ell} -2\alpha\ dx\right) \qquad (2)$$

where P_o is the launched power; Δt, the source pulse width; and υ_g, the pulse group velocity. In Equation 2, C_s and α are the scattering constant and the total loss coefficient, respectively. The term NA in Equation 2 is the fiber's local numerical aperture (i.e., light capturing efficiency), which is dependent on the index of refraction of the core and clad materials.

In order to use the backscattered light pulse intensity in a DFOTS, it is clear (from Equation 2) that α, NA, or C_s must be dominant functions of temperature. Now C_s and α can be written as follows:

$$C_s \simeq (\alpha_R)_{co} + (\alpha_s)_{co} + P_c/P_t\ (\alpha_s)_{c\ell} \qquad (3)$$

$$\alpha \simeq \alpha_{co} + P_c/P_t\ (\alpha_{c\ell}) \qquad . \qquad (4)$$

In Equations (3) and (4), α_R is the Raman scattering coefficient and α_s is the Rayleigh scattering coefficient. The factor, P_c/P_t, is the fraction of propagating power that exists in the cladding due to evanescent wave effects. Subscripts co and $c\ell$ are associated with the core and cladding, respectively.

Equations 2 through 4 reveal that there are many possible parameters which could be exploited to provide a distributed temperature sensor. The various techniques used by researchers to form a DFOTS system can be discussed in terms of the variables in these equations.

Researchers at Southampton University produced a distributed temperature sensor based on temperature-dependent attenuation of the core glass (i.e., $\alpha_{co} = \alpha_{co}(T)$). For example, the fiber core can be doped with neodymium and a 904 nm pulsed laser source used to obtain DFOTS data (Farriers and Fermann, 1987). At this wavelength, the temperature-dependent core absorption, $\alpha_{co}(T)$, is linear over a 50°C interval. Holmium doped fibers have also been investigated (Farriers et al, 1987), but these fibers are only sensitive at temperatures less than 40°C.

The first demonstration of a distributed fiber optic sensor used a special fiber having a liquid core (Hartog, 1983). In this case, increasing the fiber temperature causes an increase in the scattering coefficient of the liquid core, $(\alpha_s)_{co}$, by 0.02 dB/°C. Also, the liquid refractive index decreases with increasing temperature resulting in a decreased NA. The NA and $(\alpha_s)_{co}$ effects tend to cancel one another. A mode filter, made up of a launch fiber of lower NA, eliminates the sensing fiber NA effect.

Scattering in optical fibers is caused principally by the Rayleigh effect which results from glass inhomogeneities formed in the production process. This type of scattering is at the same wavelength (elastic) as the incident light and it is largely independent of temperature changes (except for liquid core fibers described above). There is, however, a small (100-1000 times less) contribution to the scattered power from the Raman effect which originates from molecular and crystalline effects within the core glass. A key attribute of the Raman effect is that it results in scattering at a different wavelength than the incident light (i.e., inelastic).

Initial Raman work, carried out by Plessey at the CEGB (UK), led to a practical Raman DFOTS device incorporating solid state components (Dakin et al, 1985). The Raman Stokes to anti-Stokes scattering intensity ratio was used to determine temperature along the sensor length independent of attenuation losses. York Technology (Hartog et al, 1985) measured the anti-Stokes backscattered light from the fiber and utilized a different referencing scheme to cancel attenuation effects. Efforts have recently been made to lower the 7.5 m spatial resolution of the Raman DFOTS to 1 meter (Ogawa et al, 1989) using narrow pulsed laser sources. However, Raman scattering produces such a weak signal that it will always have limited resolution capability. A key advantage to the system is that long lengths of ordinary communications grade fiber can be used as the distributed temperature sensor.

Clad Loss and Scattering Change Approach

Since evanescent wave effects cause a small fraction of the guided light to reside in the cladding, a low-loss material is normally chosen to minimize overall fiber scattering and absorption losses. However, if scattering centers are purposely added to the cladding material, fiber loss and backscattering changes will occur. The clad contribution to overall loss and backscattering is shown in Equations 3 and 4 where P_c/P_t is the fraction of power traveling in the clad material. According to a model for weakly guiding fibers (Gloge, 1971), this fractional power is:

$$P_c/P_t = 2\ \lambda/[3\pi a\ (2n_{co}\ \Delta n(T))^{1/2}]\ , \qquad (5)$$

where
λ = wavelength of incident light source
a = radius of fiber core.

As the core or clad index changes with material temperature, this change alters the index-difference term $\Delta n(T)$. More or less power will, therefore, be present in the cladding and more or less fiber transmission loss and backscattering change will occur.

The clad-scattering approach to provide a DFOTS system incorporates a special cladding that scatters light. Some of this clad scattering is coupled back into the core and this scattering dominates the backscattered light intensity collected by the OTDR. The amount of clad scattering is, itself, dominated by the fraction of the propagating light that travels in the cladding due to evanescent wave effects. This fraction of clad light intensity is a function of temperature because the cladding's refractive index is close to that of the core. Also, the clad index varies with temperature. Thus, the clad power fraction, hence the total clad scattering, and scattering coefficient, C_s, also vary with temperature.

Besides scattering light back into the core, scattering in the cladding material attenuates the forward propagating pulse. Therefore, the fiber loss coefficient, α, is also a function of temperature. A combined loss and scattering approach is used to describe changes in backscattered light intensity since both of these effects are present; namely,

$$P_{bs}(\ell,T) = C_0\, C_s(T)\, \exp\left(\int_o^{\ell} -2\alpha(T)\, dx\right) , \qquad (6)$$

where C_0 is a constant and $C_s(T)$ and $\alpha(T)$ are changes in the temperature-induced backscattering and attenuation coefficients, respectively. These two coefficients are given by Equations 3, 4, and 5 as:

$$C_s(T) = C_1(\alpha_s)_{c\ell} / [\Delta n(T)]^{1/2} \qquad (7)$$

and

$$\alpha(T) = \alpha_{co} + C_2\alpha_{c\ell}/[\Delta n(T)]^{1/2} \quad . \qquad (8)$$

where C_1 and C_2 and constants. The index-difference term $\Delta n(T)$ can be written as:

$$\Delta n(T) = n_{co} - n_{c\ell} \quad . \qquad (9)$$

Assuming a constant core index, n_{co}, and a linear change in clad index, $n_{c\ell}$, with temperature, Equation 9 becomes:

$$\Delta n(T) = \Delta n(25°C) + dn_{c\ell}/dT\ (T - 25) \qquad (10)$$

where,

$\Delta n(25°C)$ = Constant index difference at 25°C

and

$dn_{c\ell}/dT$ = Constant clad index temperature change.

In Equations 7 and 8, the terms $(\alpha_s)_{c\ell}$, α_{co}, and $\alpha_{c\ell}$ are assumed to be constant. Therefore, the entire change in backscattered light intensity is controlled by the rate of change of the clad index ($dn_{c\ell}/dT$) with temperature. For most polymer materials $dn_{c\ell}/dT < 0$. Therefore, as temperature increases, the index of polymer materials usually decreases. From Equation 10, $\Delta n(T)$ will, therefore, increase. Given this increase, Equation 5 indicates that P_c/P_t will decrease (i.e., less light in clad) so C_s and α decrease. Therefore, the OTDR's backscattered light intensity signal decreases in a hot zone. Further, the slope of the backscattered light signal versus distance plot (i.e., the loss coefficient) also decreases (hot fiber becomes less lossy). Following the hot zone, the local backscattered intensity and OTDR slope return to values indicative of local conditions. There is no permanent intensity loss across the hot zone. In fact, the signal can actually be greater after a hot zone than it would normally be if the hot zone was not present. This occurs because the lower loss coefficient in the hot zone causes less transmission loss through that zone. The fiber temperature profile is determined by measuring the change in backscattered light intensity along the fiber length and use of Equation 6.

Comparison of DFOTS Approaches

Of the approaches described above, only Rayleigh scattering from a special cladding, Raman scattering, and the doped core method have commercial viability. All three approaches have similar temperature ranges and measurement accuracies. The key differences concern the sensor length and length resolution.

The Raman approach has the longest sensor length (1-2 km) and the highest accuracy (±1°C). However, because of the low backscattered intensity, the length resolution is greater than 1 meter. The special clad-scattering method provides a medium sensor length (20-100 m), good accuracy (±5°C), and the best length resolution (≤10 cm). The doped core method results in an intermediate length sensor with reasonable accuracy, but the length resolution is greater than 1 meter.

The Clad-scattering method meets the length resolution requirements of most smart structure and smart materials processing applications. On the other hand, the Raman approach would be most applicable to monitoring large structures. Raman system cost is approximately twice that of the other approaches. This is mainly due to the special OTDR that must be provided for the Raman method. Further, the clad-scattering and Raman approach are not based on increased attenuation. Therefore, the dynamic range of these methods will be greater than the doped core approach, especially for multiple hot spots along the fiber length.

Prototype DFOTS System

A prototype DFOTS system consisted of a high-resolution OTDR, portable computer, and a 20-meter fiber sensor with connector and end termination. Several sensor types were produced using borosilica and silica glass with fiber core diameters of 100 μm.

Special liquid cladding materials were formulated for each glass type by mixing various additives into a base material of proper refractive index. These materials were sent to several commercial fiber-pulling facilities where the liquids were coated onto the core following fiber drawing. A dip coating process followed by ultra-violet light curing provided nominal clad diameters of 140 μm. Lengths of up to 1 km were produced at costs of less than $1/meter. The sensing fibers were terminated on one end with a connector and on the other end with a tube filled with index matching liquid and sealed with an appropriate epoxy.

A special OTDR (Opto-Electronics Model TDR30) was used to obtain the desired length resolution. The TDR30 instrument used a 904 nm laser. Laser pulse widths of 100 ns and gated photon counting detection provided a length resolution of 2, 8, or 30 cm. Signal resolution was 0.01 dB and the dynamic range was 32 dB. The average measurement time was 45 seconds when 2048 laser pulses were averaged. A GPIB interface was used to transfer OTDR information to a portable computer for off-line processing and display. Appropriate software code was written to convert OTDR intensity output to temperature profile results.

Prototype System Tests

The prototype sensor system was tested using a small electric oven with appropriate entry and exit ports and a thermocouple for oven temperature monitoring. A small length of copper tubing wrapped with heater tape and fitted with a separate thermocouple was also placed in the oven to produce hot spots within the oven hot zone. Various lengths of sensor fiber were located in the oven and heater tube to generate a variety of temperature profiles.

Typical backscattered light intensity data are shown in Figure 1 for a 4-meter long hot zone (80°C) located 16 meters from the input end of a silica core fiber sensor. A reference OTDR plot (i.e., room

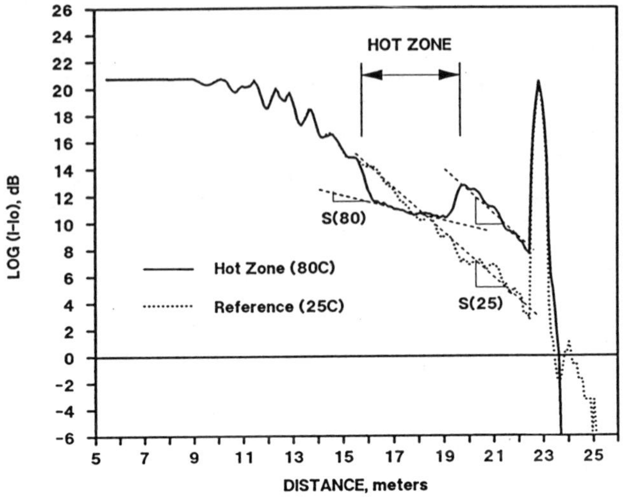

Figure 1. OTDR Data for 25°C and a 4 m Hot Zone (80°C)

temperature conditions of 25°C along entire fiber length) is also shown in this figure. Note in Figure 1 that the room temperature fiber loss coefficient (slope of OTDR plot in 25°C region) was S(25) = α = 1.3 dB/m for the 904 nm source. Other fiber sensor types had 904 nm loss coefficients from 0.6 to 1.6 dB/m. At 850 nm wavelengths, the silica-core fiber loss was a factor of approximately 2 less (0.3 to 0.8 dB/m).

Figure 1 shows that in the 4-meter hot zone two effects occur; namely:

- Decrease in backscattered light intensity (decrease in C_s) at beginning of hot zone and corresponding increase at end of heated fiber
- Decrease in fiber loss coefficient, α, resulting in a decrease in OTDR slope.

Both of these effects were predicted in the model outlined above and both are due to changes in clad scattering. Therefore, the hot-zone data validate the basic clad-scattering model which indicates that the dominant sensing mechanism is the change in refractive index of the clad with temperature.

Figure 2 shows loss coefficient data as a function of temperature for various sensing fibers. Note that for the silica core DFOTS with large clad index change (i.e., large dn/dT), the change in α with temperature

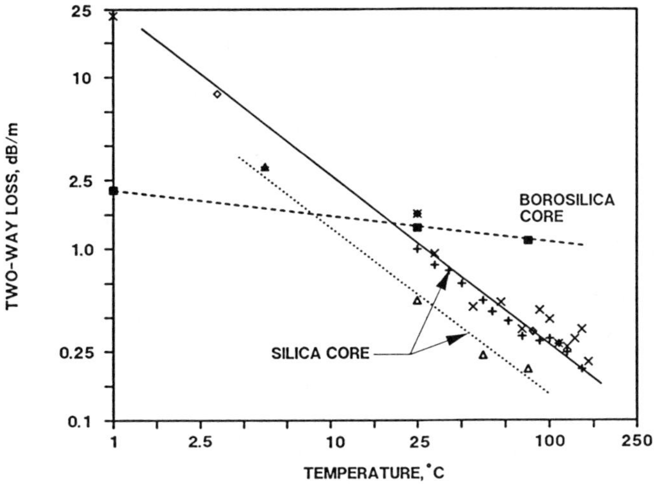

Figure 2. Temperature Effect on Two-way Fiber Loss

is also large compared to the borosilica fiber case. The change in backscattering constant, C_s (assuming C_s = 0 for room temperature conditions) is shown in Figure 3 for various fiber temperatures. Note that the backscattering constant decreases (See Equation 7) as temperature increases. The various sensing fiber types experience different changes in scattering coefficient. Also note that the greatest change in C_s occurs from 0 to 60°C. Above 60°C , the change is less dramatic.

Figures 2 and 3 indicate that both loss and scattering coefficient measurements can be used to measure temperature along the fiber length. For a hot spot, the process of converting OTDR data to temperature is a simple matter of first subtracting the room temperature reference OTDR output from the signal OTDR output. Such a difference plot for a 0.5 meter, 100°C hot spot in a borosilica core sensor is shown in Figure 4. In this case, the scattering coefficient change dominates. A simple algorithm, determined from a curve fit to the data in Figure 3, can then be used to convert the difference plot in Figure 4 into a temperature plot. The results of such a conversion are shown in Figure 5 where temperature of the hot spot, measured by the DFOTS system, is shown to agree to within $\pm$5°C of the thermocouple value.

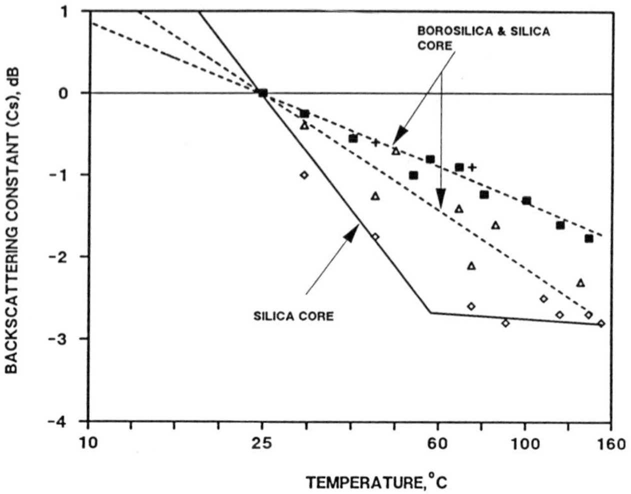

Figure 3. Temperature Effect on Backscatter Coefficient

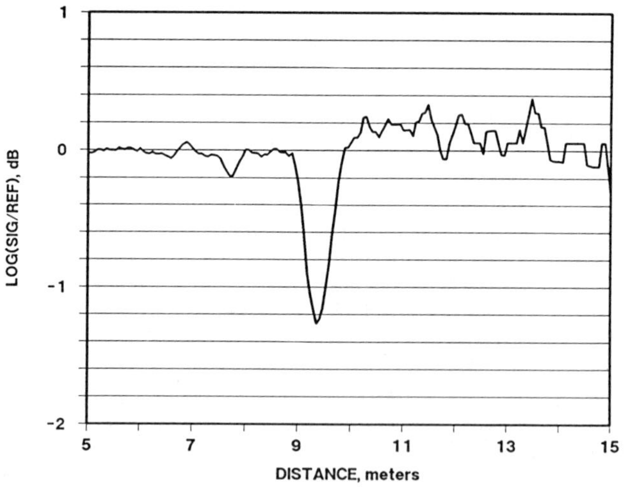

Figure 4. Difference Plot for 0.5 m Hot Spot (100°C)

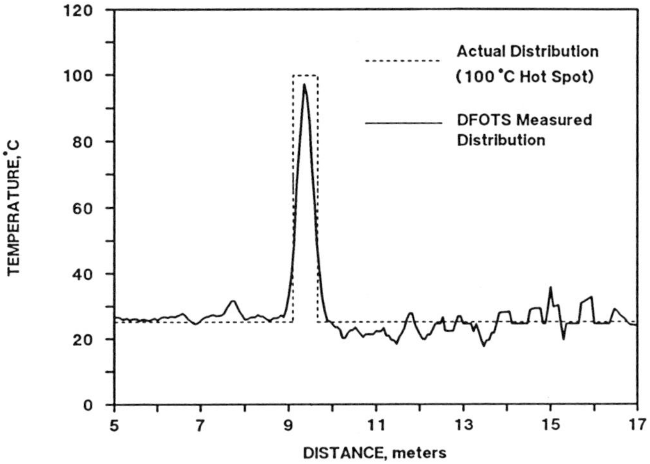

Figure 5. DFOTS Measured Temperature for 0.5 m Hot Spot

For a 0.5 m hot spot within a 6 meter hot zone (50°C) the difference plot is somewhat more complex (See Figure 6) because both C_s and α effects are equally important. In this case, a curve fit of data in both Figures 2 and 3 must be used to extract the temperature information using an iterative procedure.

Field Test Data

After 35 years of nominal operation, a fire occurred in one of the switchgear breakers at Ohio Edison's R. E. Burger plant. The root cause was not determined, but signs of overheating in the finger cluster region of the 2000 amp breaker were noted after the fire. Standard infrared imaging techniques could not be used because the finger cluster was located within the cell and could not be accessed with this line-of-site equipment.

The DFOTS fiber was installed on a typical switchgear circuit-breaker cell to validate that the finger cluster region was a location of high heating. Further, temperature measurements of the cell interior could also assess the effect of newly installed cooling fans. One phase of a typical breaker was instrumented with eight meters of the borosilica core DFOTS fiber.

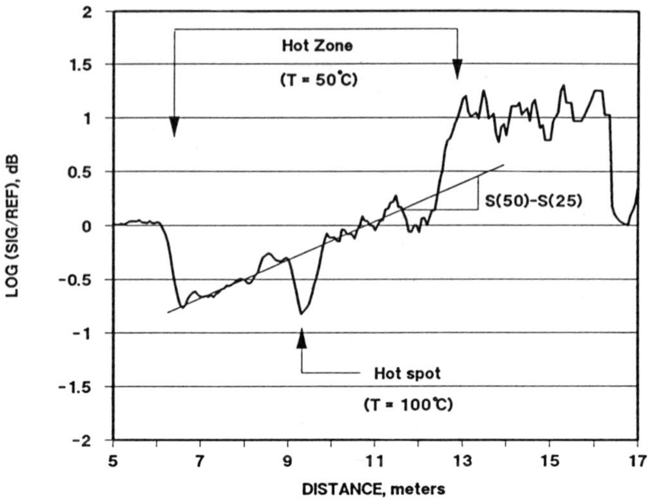

Figure 6. Difference Plot for 0.5 m hot spot (100°C) in 6 m Hot zone (50°C)

The finger cluster region was covered in a serpentine fashion with sensing fiber attached to the fingers with epoxy at various points along the fiber length. A Luxtron fluoroptic™ thermocouple (Model 750) was epoxied to one point on the forward edge of the fingers. This fluoroptic™ sensor was located near a point on the DFOTS fiber to provide a one-point comparison of the two measurement approaches.

OTDR data were obtained at various intervals over a nine-hour period after switch activation. During this time period, the load was increased periodically resulting in step changes in circuit current as auxiliary equipment came on line. Figure 7 shows a plot of DFOTS-measured temperature as a function of time after switching for a point near the Luxtron sensor and also one at the back of the finger cluster. As current increased, both sensor types indicated that the finger-cluster temperature increased. When the cooling fans were activated, both approaches also indicated a 15-20°C drop in conductor temperature. The measured DFOTS temperature was within approximately ±5°C of the value indicated by the nearby Luxtron point sensor. Data from the DFOTS fiber indicate that the highest temperature occurred near the rearward portion

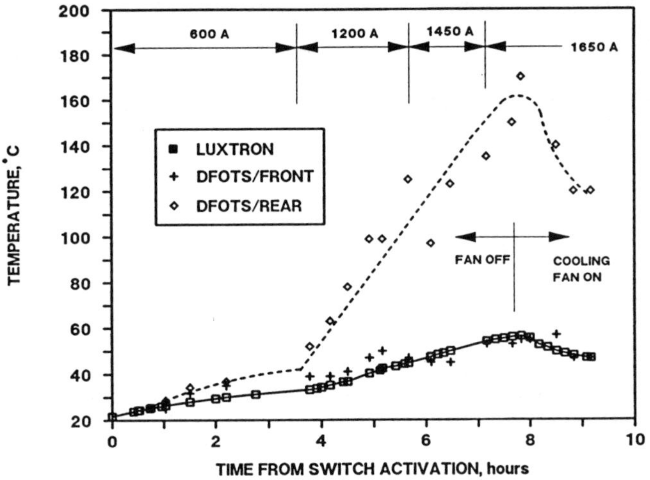

Figure 7. Temperature Results in Finger Cluster Region for DFOTS and Luxtron Fluoroptic Thermometer

of the finger cluster at the point where it was attached to the horizontal conductor bar. Temporal variation of temperature in this rearward region followed the forward finger cluster trend but peak temperature values were nearly 100 °C higher.

Conclusions

A distributed fiber optic temperature sensor based on Rayleigh scattering in the cladding is feasible and practical. The system can provide high-length-resolution temperature profile data in various structures, equipment, and material processes. The current sensor system can measure temperatures from 0 to 150°C with an accuracy of ± 5°C. Fiber lengths from 20 to 100 meters are attainable, depending on fiber type and laser wavelength. The measurement length resolution was less than 10 cm.

Experimental results for changes in backscattered light intensity agreed reasonably well with trends predicted by a clad-scattering model. Changes in both backscattering and loss coefficients for the sensing fiber were controlled by the cladding's refractive index change with temperature.

The DFOTS system successfully located hot spots in the finger-cluster region of a switchgear at the R. E. Burger plant. These field tests demonstrate the value of using a DFOTS approach for diagnosing overheating conditions within electrical equipment.

Acknowledgments

The work reported herein was co-funded by the Electric Power Research Institute and The Consolidated Edison Company of New York, Inc.

References

Dakin, J.P. et al, Elect. Lett., 21 (13), 569-70 (1985)

Farriers, M.C. and Fermann, M.E., Proc. of SPIE Conf. on Fiber Optic Sensors, The Hague, Netherlands, Vol 798, 115-7 (1987)

Farriers, M.C. et al, OFC/IOOC '87, 6th Inter. Conf. of Integ. Opt. and Fiber Comm., Reno, NV, Paper W15 (1987)

Gloge, D., Appl. Optics 10 (10), 2252-8 (1971)

Hartog, A.H., J. of Lightwave Tech., LT-1 (3), 498-509 (1983)

Hartog, A.H. et al, Elec. Lett, 21 (32) 1061-2 (1985)

Ogawa, K. et al, Springer Proceedings in Physia, V 44, Optical Fiber Sensors, Springer-Verlag Berlin (1989)

Rogers, A.J., J. Appl. Phys. 19, 2237-55 (1986)

Twerdochlib, F.T. et al, EPRI Report 2487-1 (1989)

New Methods for Distributed Optical-Fibre Measurement of Strain and Temperature in Large Structures

A J Rogers[1]

Abstract

Distributed optical-fibre sensing (DOFS) provides the means by which both the spatial distribution and the temporal behaviour of a measurand field can be measured simultaneously. Thus the distribution of strain and temperature (for example) in large critical structures may be monitored continuously in order that remedial action can be taken when required, or in order to understand better the behaviour of the structure under extreme or otherwise anomalous conditions. DOFS methods hitherto have concentrated on backscatter techniques. These have important limitations, especially in respect of spatial resolution. New methods which rely on forward-scatter are now being explored. Two such methods, based on the optical Kerr effect, are described in this paper. Spatial resolutions of ~0.1m and measurement accuracies of better than 1% should be achievable.

1. Introduction

Distributed optical-fibre sensing (DOFS) comprises a set of techniques for measurement of the spatial and temporal behaviour of physical fields such as temperature, strain, pressure, displacement, electric field, magnetic field, radiation level, etc. Various methods for realising DOFS have been explored [1,2], each, as always, with its own advantages and disadvantages for particular application areas.

Research in DOFS has been stimulated by a number of important specific applications. For example, the requirement to measure the distribution of temperature in equipments vulnerable to 'hot-spots' or to anomalous temperature gradients, such as power transformers or multilayered material structures; the requirement to measure strain distributions in critical structures such as pressure vessels, boilers, dams, bridges, tall buildings; and the requirement to locate, spatially, anomalously-large electric and/or magnetic disturbances in malfunctioning electrical and electronic equipment. More recently, the growing interest in structures which, effectively, have the ability to self-monitor, and perhaps also to self-adjust ('smart' structures), has provided added impetus to

[1]Professor of Electronics, King's College London, Strand, London WC2R 2LS, UK

DOFS research [3].

The optical fibre possesses singular advantages for distributed sensing : it is dielectric, insulating, flexible and sensitive (in its optical propagation properties) to a wide range of external physical fields. Its flexible, 'one-dimensional' nature makes it very suitable for the variety of geometrical configurations needed for distributed sensing in myriad industrial applications.

Additionally, some emphasis should be placed on the potential importance of DOFS in research and in manufacturing diagnostics. Here the importance lies largely with its ability to probe the behaviour of systems more satisfactorily as a result of access to more detailed spatial and temporal information, and to detect and locate departures from prescribed behaviour with much greater facility.

2. Forward - Scatter DOFS

Most DOFS explorations to date have relied on backscatter arrangements [1-4]. In these, a pulse of light is launched into an optical fibre and is continuously Rayleigh backscattered as it propagates (fig 1). Time resolution of the emerging backscattered light then reveals the spatial distribution of any external field which is capable of modulating, in a deterministic way, some characterizing property of the light, this latter being demodulated by the detector. Such a one-dimensional 'lidar' arrangement has the clear advantage of simplicity, but it has a number of disadvantages.

The primary disadvantage is that Rayleigh backscatter is at a low level, of order 10^{-5} of the forward travelling energy, per metre of fibre. Moreover, since the fibre attenuation (in a propagation 'window') will depend almost entirely on Rayleigh scatter, there will be an optimum length, for a given scatter coefficient, with regard to both sensitivity and resolution of the DOFS system.

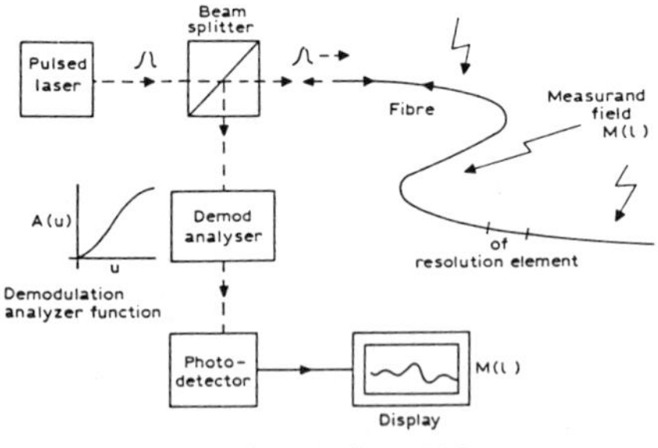

Fig 1. OTDR for Distributed Measurement

Suppose, however, that there is no necessity to rely on backscattered light. Consider the forward-scatter arrangement shown in fig 2. In this arrangement a continuous wave (CW) source is launched, at one end (F) of the fibre, into one of two possible modes T and R, say. These might be two propagation modes, two polarization modes, or even two separate cores within the same cladding. The important conditions which must obtain are that the two modes must be separately identifiable at the output from the fibre, and that light coupling between them must both be possible and dependent, in some way, on the external field to be measured (measurand). The mode into which the CW is launched will be labelled the 'filled' (in this case T) mode whilst the other is the 'free' (in this case R) mode.

An optical pulse is now launched into the other end (B) of the fibre in such a way as to allow it, whilst propagating in the opposite direction to the CW, to couple the latter from the 'filled' to the 'free' mode. If, as was required above, the coupling mechanism is modulated by the measurand in a deterministic way, then, by time-resolving the CW power level emerging from end B in the 'free' mode, one obtains the spatial distribution of the measurand field.

The important advantage which this method offers over the backscatter technique is that of increased received power level. There is no longer an optimum attenuation coefficient for a given fibre length, since one is not relying on backscatter (although the mode-coupling requirement might impose other limitations on the minimum attenuation allowable). The increased power level implies improved signal-to-noise ratio and thus an enhanced 'sensitivity x resolution' product. It must be remembered, however, that this is at the expense of requiring a more sophisticated optical source than is required for backscatter systems, one capable of accessing the non-linear optical regime via a large peak power density in the optical fibre ($\sim 10^9$w.m^{-2})

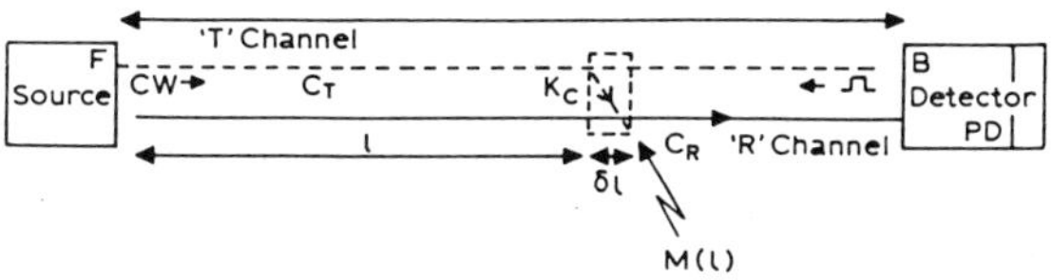

Fig 2. Schematic for Forward-scatter DOFS

Analysis shows that such forward-scatter arrangements should improve the overall performance of the system by at least two orders of magnitude, allowing spatial resolutions down to 0.1m and measurement accuracies better than 1%. The first forward-scatter (F/S) method for DOFS to be studied relied on a Raman interaction between a counterpropagating pulse and a CW probe in a low-birefringence fibre [5]. In this, the CW probe received Raman gain from the pulse, and this gain depended on the relative polarization states of the probe and the pulse, these, in turn, being dependent upon the strain distribution along the fibre. As a demonstration of the basic principle of F/S DOFS this was satisfactory, but it was not developed further since the signal processing required to determine the strain distribution was too complex for practical implementation. However, polarization methods are to be preferred since they are both very sensitive and independent (to first order) of fibre attenuation.

High birefringence (hi-bi) fibres are now quite readily available. Such fibres possess a natural linear birefringence which allows good control over the propagation state of the propagating light, over long distances. Linear birefringence relies on an intrinsic cross-sectional asymmetry which leads to different refractive indices for two particular directions of linear optical polarization.

The first attempt to use such fibres in DOFS involved a backscatter method [6,7] which was a variation on POTDR [4]. In this method the propagating pulse, equally disposed in each of the two hi-bi eigenmodes, accumulated a linearly increasing phase difference between its two modal components as it propagated, and this was doubled for backscatter from any given point. By mixing the two backscattered eigenmode signals on emergence, a frequency was generated, and the time dependence of this frequency mapped the distribution of birefringence along the fibre, corresponding to the pulse's passage along it (fig 3). Any external field capable of modifying that birefringence was mapped correspondingly. Whilst this method worked encouragingly well, it suffered still from that fact that it was a backscatter method, and thoughts turned towards attempts to embody the same, or a similar, idea in a forward-scatter arrangement.

The first thoughts along these lines involved the use of 'resonant' coupling. If a pulse propagating in a hi-bi fibre could be arranged to impose a suitable periodic perturbation with a spatial period equal to the hi-bi beat length (distance over which the differing polarization mode group velocities impose a phase difference of 2π between mode components) then this would lead to resonant coupling, from one mode to the other, of a counterpropagating CW probe. The birefringence could then be mapped via the coupling efficiency (level of coupled light) as a function of time: a measurement of the coupled power as a function of position allows a measurement of the distribution of birefringence along the fibre, and hence also of any distributed field, such as strain or temperature, which modifies it in a deterministic way.

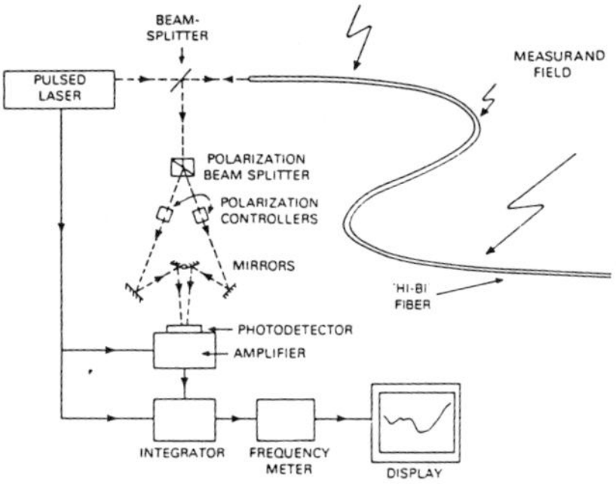

Fig 3. Frequency-derived Backscatter DOFS

3. Kerr-effect Methods for Forward-Scatter DOFS

When an electric field is applied to an isotropic material (such as fused silica), linear birefringence is induced in the material, with the slow axis in the direction of the field and the fast axis orthogonal to it. The birefringence, the difference between the two linear refractive indices, is given by :

$$\Delta n_k = \lambda_0 b E^2 \qquad - (1)$$

where λ_0 is the vacuum wavelength, b the Kerr constant and E the electric field.

In the optical Kerr effect the electric field is that associated with an optical wave, so that the birefringence induced by an intense wave in a material may be probed via the effect it has on another wave, either co-propagating or counterpropagating.

Two F/S DOFS arrangements, which use the optical Kerr effect, have been explored:-

(i) Frequency-Derived Forward-Scatter DOFS

Frequency-derived backscatter DOFS has already been described in Section (2) and has also been reported in the literature [e.g. 6,7]. The forward-scatter version of this idea utilises the optical-Kerr effect.

Suppose that an intense optical pulse is launched into a hi-bi fibre, with equal power in each of the two eigenmodes. As the pulse propagates, the two components will come into phase once per beat length and into antiphase at the intervening half beat lengths (fig. 4). The electric field of the resultant wave thus will maximise in a direction lying at $\pm 45^0$ to the intrinsic axes, every alternate half beat length. The Kerr-effect-induced birefringence resulting from this field consequently will cause the resultant birefringence axes to rock around the intrinsic direction with a spatial period equal to the beat length.

Consider now a CW, with the same wavelength as the pulse, launched, again equally into the two eigenmodes, in the opposite direction to the pulse. When it encounters the pulse, the Kerr-induced rocking will produce optimal phase-matched coupling (since the rocking occurs naturally at the 'resonant' frequency), as a result of which probe power will be coupled from one mode to the other. When pump and probe are at the same wavelength there will be an exact correspondence in beat length which will maintain the coupled power constant as the pulse propagates down a fibre with uniformly distributed birefringence. However, if pump and probe are at different wavelengths the differing beat lengths will lead to the condition that optimum coupling will occur only when the rocking axes can couple in-phase probe components, i.e. with an effective beat-length given by L_e where

$$\frac{1}{L_e} = \frac{1}{L_1} - \frac{1}{L_2} \; ; \; L_e = \frac{\lambda_1 \lambda_2}{\Delta\lambda \, \Delta n_I} \qquad (\Delta\lambda = \lambda_1 - \lambda_2)$$

From this we may conclude that the optimum coupling occurs at a 'beat' frequency given by :

$$f_D = c \, \Delta n_I \cdot \frac{\Delta\lambda}{\lambda_1 \lambda_2} \qquad (2)$$

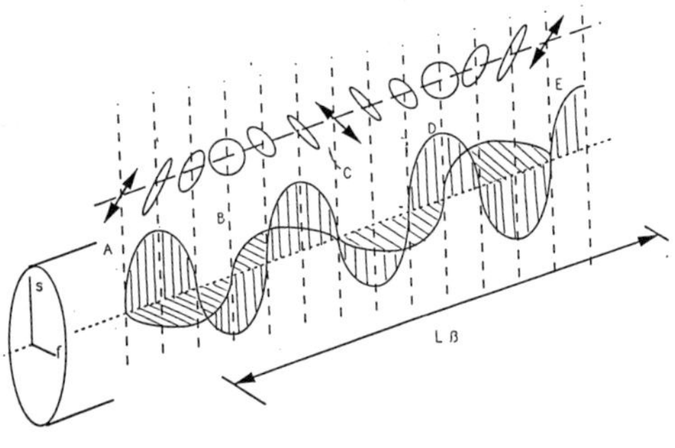

Fig 4. Evolution of Polarization State in Hi-Bi Fibre

where λ_1 is the wavelength of the pulse light, λ_2 that of the probe light, and L_1 , L_2 the respective beat lengths; c is the velocity of light in the fibre and Δn_I is the intrinsic birefringence of the fibre.

The experimental setup is illustrated in fig (5). The fibre used was 24m long with an elliptical core (supplied by Andrew Corporation) and was single mode at 633nm, where it exhibited a beatlength of 4 mm. The pump pulses were generated by a dye laser pumped by a Q-switched, frequency-doubled Nd: YAG laser. Pulses of 8 ns (FWHM) maximum duration were generated with a repetition rate of 50 Hz. To allow useful differences between the pump and probe wavelengths, the duration of the pump pulses was required to be shorter than the pulse width actually available from the dye laser. Accordingly a fast optical shutter was used to provide a sufficiently short pulse. The pump beam was chopped by a fast Pockels cell to provide pulses of 2 ns (FWHM) and was circularly polarized by the quarter wave plate so as to excite equally the two eigenmodes of the fibre. The probe was provided by a cw Argon-pumped dye laser which was counter-propagating with respect to the pump, and linearly polarized at a small angle to the fibre's birefringence axes, with a launched power of ~ 300 mW. In principle the strongest probe signal should be provided when the eigenmodes are equally excited by a launched polarization set at 45° to the fibre's principal axes. The signals would be then superimposed on a high mean received power. However, an avalanche photodiode detector was used, and this rendered high mean signal levels undesirable. After reflection from a beamsplitter at near-normal incidence, the probe beam was analysed with a Glan Thompson analyser. A monochromator was placed before the avalanche photodiode detector in order to block the large pulse predominantly produced by the front face reflection from the fibre. This pulse otherwise would have saturated the input amplifier. The pump and probe beams were set at various wavelength differences up to 2.4 nm apart, generating derived signal frequencies up to ~ 200 MHz. The signals were recorded and analysed using a computer-interfaced digital storage oscilloscope which was used to average the results from 256 pump pulses. The averaging process occupied approximately 10 seconds.

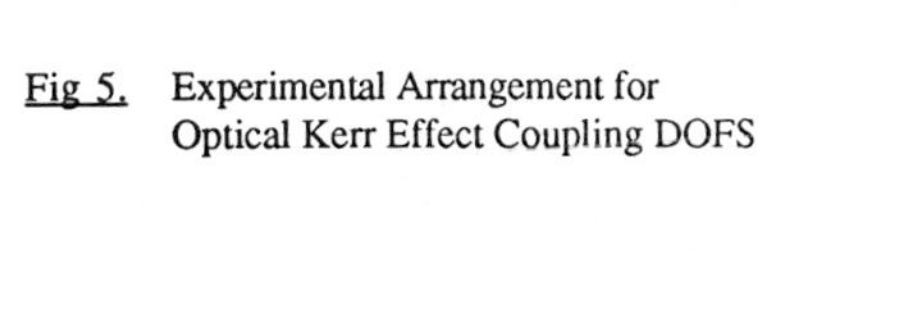

Fig 5. Experimental Arrangement for Optical Kerr Effect Coupling DOFS

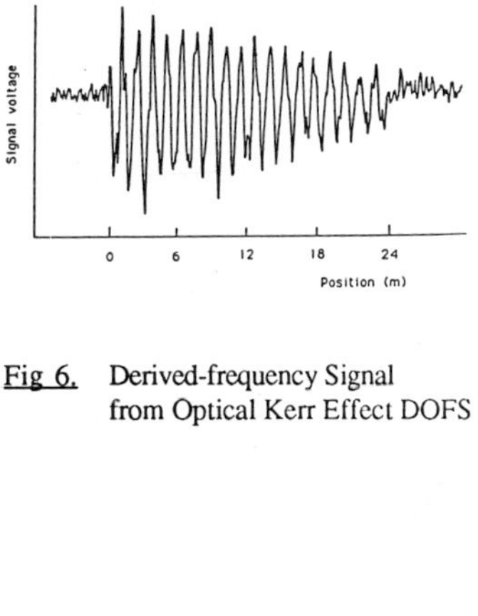

Fig 6. Derived-frequency Signal from Optical Kerr Effect DOFS

The variation of the derived frequency with wavelength offset was measured by the following method: the pump and probe were initially set at the same wavelength of 646 nm. Then, by tuning the pump wavelength with respect to the probe, the required wavelength shift was set. An example of a derived frequency signal averaged over 256 pump pulses is shown in fig (6). The wavelength shift between the pump and probe in this measurement is 1.5 nm and the corresponding derived frequency is 71MHz. The measured signal to noise ratio was 19.3 dB.

Using the novel technique presented here, the spatial variation of the birefringence of a polarization maintaining fibre can be measured remotely in a short time, and, since the signal is in the form of a frequency, it is immune from the common error sources present in intensity - coded systems. Any change in the applied stress or ambient temperature will tend to change the local birefringence and will therefore be indicated by a shift in the frequency due to the external perturbation. Thus this technique may be applied to implement a distributed optical-fibre sensor. In this application the tuneability of the derived frequency could provide some extra flexibility in the system design.

(ii) Polarization Dependent Kerr Effect Forward-Scatter DOFS

In this second method the emphasis presently is on the spatial location of a perturbation rather than the measurement of its magnitude, although suitable processing is capable, in principle, of revealing the latter. In its present form it is capable of providing good spatial resolution and rapid response for application to, for example, intrusion monitoring or vehicle location.

The optical arrangement employs a length of polarization maintaining (hi-bi) fibre carrying two counter-propagating beams. A CW probe beam is launched from one end of the fibre so as to excite equally the two eigenmodes, and the polarization state of this beam is detected at the far end of the fibre by means of a beamsplitter and analyser oriented at forty-five degrees to the birefringence axes. An intense, pulsed, pump beam is launched on one of the birefringence axes. This arrangement is similar to the well-known Kerr-effect shutter system [8]

As in the Kerr shutter, the pump pulse causes a phase shift between the eigenmodes of the probe beam, leading to a change in the output polarization state of the probe. This is detected as a sharp change in the probe intensity passed by the analyser when the pump is initially launched into the fibre.

If, now, a force acts at an angle to the axes along a section of fibre, coupling of the pump light to the other axis will occur, and the Kerr effect on the probe will thus be modified. The probe light itself will also experience mode coupling, which will further modify the output polarization state. The actual change which occurs will depend, inter alia, on the states of polarization of the beams as they enter the perturbed region and thus, unless the birefringence perturbation is very small compared with the intrinsic birefringence, there will exist a mutual dependence of effects from different measurement locations, which only fairly complex signal processing would be able to resolve. However, it is clear that for any change in the direction of birefringence axes consequent upon the perturbation by a measurand, there will, in general, result a change in optical Kerr effect. A differentiated signal thus will, at least, indicate differential features of the measurand distribution, even though a fully quantified spatial distribution is more elusive.

The fibre used in the experiments [9] was a mono-mode high-birefringence fibre, manufactured by Andrew Corporation, with a diameter of 67μm, attenuation of 35dB/km at 633nm and core to cladding refractive index difference Δn = 0.032. The length of the fibre was about 100m. The arrangement of the experiment is shown in Fig (7). Pump pulses (617nm) of 8ns (FWHM) duration were generated in a dye laser with a repetition rate of 50Hz. These pump pulses were launched on to one of the birefringence axes of the fibre with the help of a halfwave plate and with a peak power of 3W measured at the output end of the fibre. The linearly polarized probe beam of wavelength 633 nm, from a He-Ne laser, was launched into the fibre at 45° to the birefringence axes. On emergence, the probe beam was directed by a beam splitter to the detector, via the polarization analyser; its average power at the detector was about 25μW. The He-Ne laser and the detector were protected from the pump light by use of band-pass filters at 633 nm. Force was exerted by pressing metal rods on the fibre. The received signals were recorded, averaged and differentiated using the functions of the digital storage oscilloscope.

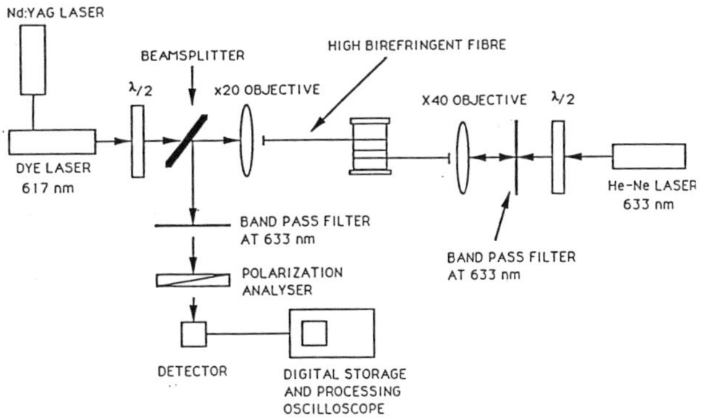

Fig 7. Experimental Arrangement for Polarization-state Dependent Kerr Effect DOFS

In the absence of any measurand-induced perturbation the Kerr effect of the pulse is to modify the local value of birefringence as it propagates, a modification which is sensed by the probe beam as a phase shift between the eigenaxes, and which is, in principle, constant for the duration of the pulse's passage through the fibre. The effect of this phase shift on the optical signal passing through the analyser is shown in fig (8). In practice, the slow fall in the value of the phase shift is due to the attenuation of the pump pulse with distance along the fibre. Fig (8) also shows the effect of differentiating this signal with respect to distance.

Fig (9) shows the fluctuating analyser signal when the fibre was perturbed at two points, and its differential with respect to distance. The points at which the weights were applied are clearly evident. Such a system, even as it stands, could be used as an intruder alarm or as an indication of anomalous disturbance of almost any kind.

Thus is demonstrated the use of the optical Kerr effect to determine the locations of discrete mode coupling points space along a polarization maintaining fibre. Differentiation of the received signal with respect to time provides a simple way to reduce confusing interactions when multiple coupling points are present. Further work on this method is continuing, with a view to developing practical real time processing and theoretical analysis of the potentialities of the technique.

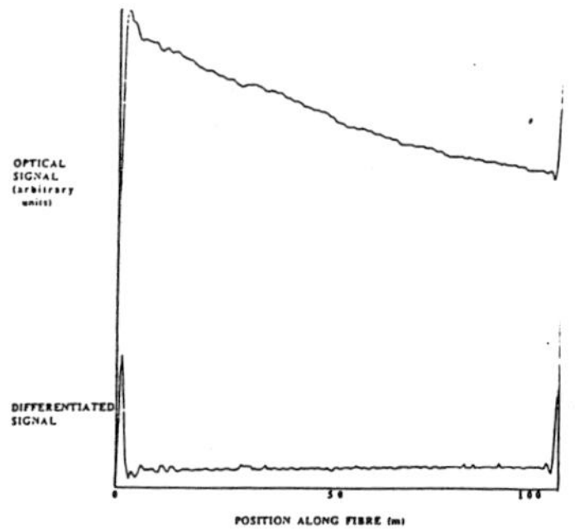

Fig 8. Unperturbed Signals

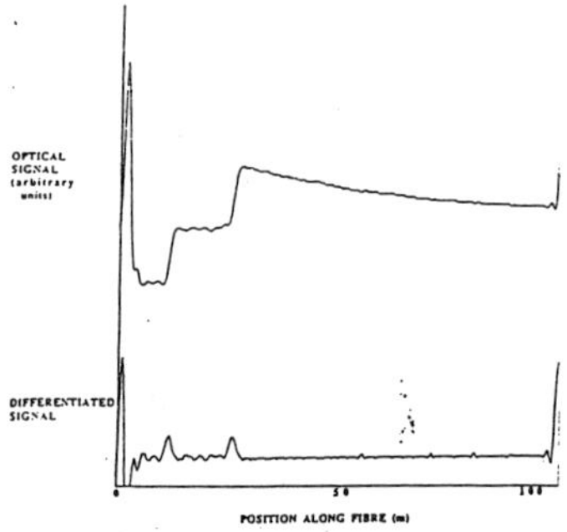

Fig 9. Signals with Two Points of Stress Perturbation

4 Conclusions

Forward-Scatter DOFS offers an important improvement in performance over backscatter methods. The enhanced SNRs imply spatial resolutions of order 0.1 m and measurement accuracies better than 1%. The primary disadvantage is that higher peak-power sources are required in order to access the optical non-linear regime.

Two forward-scatter DOFS systems utilising the optical Kerr effect have been explored. Both systems are polarimetric arrangements which take advantage of the properties of high-birefringence optical fibre. These systems are both capable of further development to the point where they could provide the means for practical strain and temperature distribution measurement on large critical structures. This class of DOFS measurement holds promise for practical systems in a variety of industrial applications including 'smart' or self-adjusting structures.

5 Acknowledgements

The work described above has benefited from contributions from a number of colleagues, most notably Dr. V. A. Handerek, Mr S U. Ahmed, Mr I. Cokgor and Mr F Parvaneh. The work has been supported by the UK's SERC, the Royal Society and the UK Optical Sensor Collaborative Association (OSCA).

REFERENCES

1. Rogers A. J. 'Distributed Optical-fibre Sensors for the Measurement of Pressure, Strain and Temperature' Phys. Rep. Vol 169, No. 2, 1988, pp 99-143.

2. Culshaw B. 'Distributed and Multiplexed Fibre-optic Sensor Systems' Proc. NATO Adv. Study Inst (Erice, Italy) 1986 (Nijhoff, The Hague).

3. Handerek V. A. and Rogers A. J. 'Sensor System Architectures for Spatially-resolved Dynamic Strain Measurement Using Optical Fibres', Proc. 1st European Conference on Smart Structures and Materials, Univ. of Strathclyde, May 1992.

4. Rogers A. J. 'Point and Distributed Polarimetric Optical-fibre Sensors' Int J. Opt. Sensors, Vol 1, No 6, 1986, pp 457-472.

5. Farries M. C. and Rogers A. J. 'Distributed Sensing Using Stimulated Raman Action in a Monomode Optical Fibre', Proc. 2nd Int. Conf on Optical-fibre Sensors, Stuttgart, 1984, pp 121-132.

6. Handerek V. A., Parvaneh F. and Rogers A. J. 'Frequency-derived Distributed Optical-fibre Sensing : Signal Frequency Downshifting', Elect. Lett, Feb. 1991, Vol 27, No 5, pp 394-396.

7. Rogers A. J. and Handerek V. A., 'Frequency-derived Distributed Optical-fibre Sensing: Rayleigh Backscatter Analysis', Appl., Opt. Vol 31, No. 21., 1992, pp 4091-4095

8. Dziedzic S. M., Stolen R. and Askhin A., 'Optical Kerr Effect in Long Fibres', Appl. Opt. Vol. 20, 1981, pp 403-406.

10. Handerek V. A., Rogers, A. J. and Cokgor I., 'Detection of Localized Polarization Mode Coupling using the Optical Kerr Effect', Proc. 8th Int. Conf. on Optical fibre Sensors (OFS 8), Monterey, California, January 1992, pp 250-253.

Strain and Temperature Measurement Using Optical Fiber

Faramarz Farahi[1]

Abstract

A structure which possesses microsensors for determining the state of the structure and the environment in which it is operating is called *sensing structure*. These sensors which are an integrated part of the sensing structure are used for strain and temperature measurement, damage identification and condition monitoring during fabrication process. Strain and temperature measurement are also playing an increasingly important role in many other industrial applications. In this work a summary of various methods for measuring strain and temperature using optical fibers is presented. Interferometric and polarimetric techniques for the recovery of strain and temperature induced modulations of the fiber guided beam are studied and sensors based on non-linear optical phenomena in optical fibers are introduced.

Introduction

Optical sensing techniques have long been associated with precise and non-intrusive measurements. In conventional optical instruments considerable mechanical stability is required in order to maintain alignment, thus restricting applications to benign environments. In the past decade there has been considerable development in the field of fiber optic technology. The availability of high quality optical fibers and fiber components has facilitated the realization of fiber optic based sensors, which do not share the extent of problems associated with conventional optical systems. Their inherent high resolution, light weight, and immunity to electromagnetic interference give optical fiber sensors an advantage in diverse industrial applications.

---

[1] Associate Professor of Physics, University of North Carolina at Charlotte, Charlotte, N.C. 28223, U.S.A.

A beam of light is fully characterized by its wavelength, phase, intensity, polarization state ellipticity, polarization azimuth, degree of polarization and degree of coherence. In an optical sensor any one, or a combination of these parameters, may be modulated in response to external fields such as temperature and strain. To date, no sensor based on the modulation of degree of polarization state or degree of coherence has been demonstrated, although these two properties of light have been used for enhancing the measurement range and increasing signal to noise ratio (Dakin and Culshaw, 1988). Phase modulation is usually related to the optical path length modulation and is conventionally detected by interferometric techniques. Sensors based on this principle have very high resolution. The operational principle of sensors in which the external field induces change in the polarization state ellipticity or azimuth, is the change in the linear or the circular birefringence of the medium respectively. The optical fibers which preserve phase and polarization information of guided beam are typically monomode fibers. Sensors using multimode optical fibers are based on the intensity modulation of light. Multimode sensors are more robust and simpler to employ in practical conditions, but they do not have the inherent resolution attainable in monomode sensors. In many applications such high resolutions are not necessarily required and those obtained using multimode sensors generally suffice.

Although the scope of this work is monomode fiber optic sensors, multimode fiber sensors for strain measurement will be briefly introduced.

Multimode Fiber Optic Sensors

A complete spectrum of environmental parameters has been monitored with multimode fiber sensors (Jones, 1985). Multimode fiber sensor are mostly configured as extrinsic sensors, but intrinsic sensors using microbending technique have also been widely used for pressure and strain measurement (Fields *et al.*, 1980). These devices are made when a length of multimode fiber is sandwiched between two corrugated plates and application of pressure or strain bends the fiber and causes coupling to lossy modes thereby lowering the amount of light transmitted by the device. The shot noise limited resolution for a pressure transducer using microbending technique was calculated to be 100 μP in a 1 Hz bandwidth (Fields *et al.*, 1980). The microbending device uses the fiber itself in the transduction process, rather than just as a convenient means to transmit the light to and from a bulk optic device.

Another interesting multimode fiber optic sensor is based on the principle of frustrated total internal reflection (Palmer *et al.*, 1983, Spillman and McMahon, 1980). In one embodiment of this technique, two fibers are cleaved at an angle to the normal to the fiber axis so that light coming to the end of one of the fibers would normally be reflected sideways out of the fiber and lost. However, bringing the second fiber close to the first permits light to be coupled across into the second fiber by the principle of frustrated total internal reflection. One of the fibers moves under the influence of pressure or strain thereby varying the amount of light coupled across.

Monomode Fiber Optic Sensors

In a single mode fiber the coherence of the beam is maintained giving rise to the realization of phase and polarization type sensors. Single mode fiber optic interferometric sensors form the topic of next section. Interferometric detection techniques for the recovery of measurand induced phase modulation is discussed, where the external fields are strain and temperature. Due to the periodic nature of the interferometer's transfer function, the operating range in such sensors is restricted to 2π radians phase change. Through the use of highly birefringent optical fibers it is possible to recover the measurand induced polarization state modulation of the guided mode in the fiber, realizing fiber optic polarimetric sensors. The recovery of polarization modulation and combined interferometric-polarimetric techniques yielding both the phase and polarization state information. Such technique enables sensor systems with greatly increased ranges to be implemented, and in some cases simultaneous measurement of temperature and strain becomes available. Finally strain and temperature sensors based on formation of holographic gratings in optical fibers will be briefly discussed.

Interferometric Strain and Temperature Sensors

Many optical fiber interferometric temperature and strain sensors possessing very high resolutions have been reported (Butter and Hocker, 1978, Hocker, 1979 and Eickhoff, 1981). The simplest fiber optic interferometer is the Michelson which is shown in the schematic diagram of Figure 1 (Corke *et al.*, 1983). Light from a laser is amplitude divided by the fiber coupler to produce reference and signal beams which propagate in the arms of the interferometer. After traversing the interferometer arms, the beams are recombined at the coupler where coherent mixing takes place with the output being monitored by a photo detector.

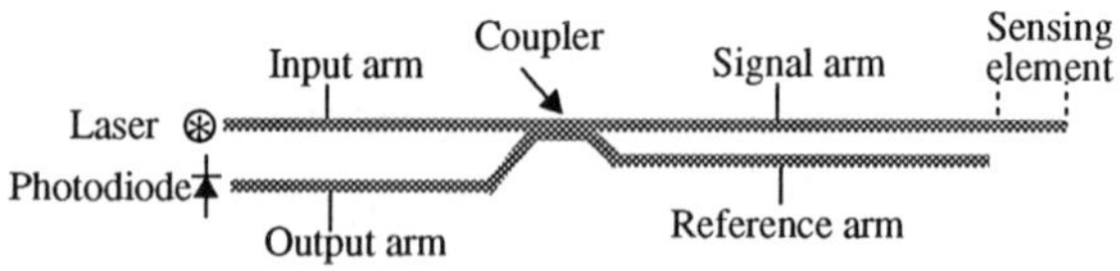

Figure 1: Fiber optic Michelson Interferometer

At a given source wavelength λ, phase retardation ϕ is given by;

$$\phi = \frac{2\pi Ln}{\lambda} = \beta L \qquad \text{(EQ 1)}$$

where L is the length, n is the refractive index of the fiber, β is the propagation constant, and c the velocity of light. The principle behind interferometric sensing is that a measurand induced change in n (hence in β) and L may be recovered by corresponding change in ϕ. The external field, x, may affect either L or n, and the change in ϕ may be described as

$$\frac{1}{L}\frac{\partial\phi}{\partial x} = \frac{\beta}{L}\frac{\partial L}{\partial x} + \frac{\partial\beta}{\partial x} \qquad \text{(EQ 2)}$$

The optical detectors available are not capable of operating at optical frequencies, and hence it is not possible to recover the optical phase directly. Instead, the phase may be obtained indirectly by interfering the phase modulated beam with mutually coherent light in a reference beam.

The intensity of the interference signal between two mutually coherent beam of an interferometer is given by

$$I = I_0 [1 + V\cos\phi] \qquad \text{(EQ 3)}$$

where I_0 is an intensity constant, and $V \le 1$ is the visibility of the fringes at the output of the interferometer. Equation (3) shows that the measurand induced phase change given by Eq.(2) produces intensity modulation at the output of interferometer. This is the mechanism in which the effect of external fields are measured interferometrically.

Since the reference and the signal arms are both sensitive to the external fields, the signal and reference arms are usually closely coupled together along their entire length except the sensing region (Corke *et al.*, 1983, Valis *et al.*, 1991). This provides a high degree of common mode rejection and substantially reduces the effect of environmentally induced perturbations.

A fiber optic Mach-Zehnder interferometer is shown in Fig. 2a. Although the system requires two fiber couplers and operates in transmission mode, it has the same transfer function as Michelson interferometer. The low finesse remote fiber Fabry-Perot interferometer (Fig. 2b) has also been used for strain and temperature measurement (Farahi *et al.*, 1987, Valis *et al.*, 1990). This interferometer is considered to have a transfer function similar to Michelson and Mach-Zehnder, and it offers the same sensitivity as the Michelson interferometer. The advantages of Fabry-Perot over other interferometers are its small size which allows localized measurements to be made and its remote operation.

When the sensing length of a fiber optic interferometer is subject to an external field the phase of the output signal is modified in the following way;

$$\delta\phi_T = LA\,(T, \lambda)\,\delta T$$
$$\delta\phi_\varepsilon = LB\,(\varepsilon, \lambda)\,\delta\varepsilon \qquad \text{(EQ 4)}$$

where A and B are the thermal and strain sensitivity coefficients of the fiber and the external effects are considered to be temperature and strain.

Although from Eq.(4) it seems that there is a simple and linear relation between the interferometer's phase change and applied temperature or axial strain, there are some inherent problems associated with the interferometric techniques. Firstly, the transfer function of an interferometer is periodic, so that unambiguous measurement range generally corresponds to a phase change of 2π rad. Secondly, interferometric techniques are susceptible to cross sensitivities; for example, if the sensing element simultaneously experiences both axial strain and a temperature change, the resulting interferometric phase change is caused by a combination of both effects, and neither can separately be identified, i.e.

$$\Delta\phi = \delta\phi_T + \delta\phi_\varepsilon = L\,[A\,(T, \lambda)\,\delta T + B\,(\varepsilon, \lambda)\,\delta\varepsilon] \qquad \text{(EQ 5)}$$

Some of the solutions to these problems are closely related, such as using combined interferometric and polarimetric techniques (Leilabady *et al.*, 1986) or using two wavelength technique (Meltz *et al.*, 1986) which can be used either to extend the measurement range or to simultaneously measure temperature and strain. These method along with a compensated polarimetric technique using polarization maintaining fiber (Dakin and Wade, 1984, and Farahi *et al.*, 1988) will be described in the following sections.

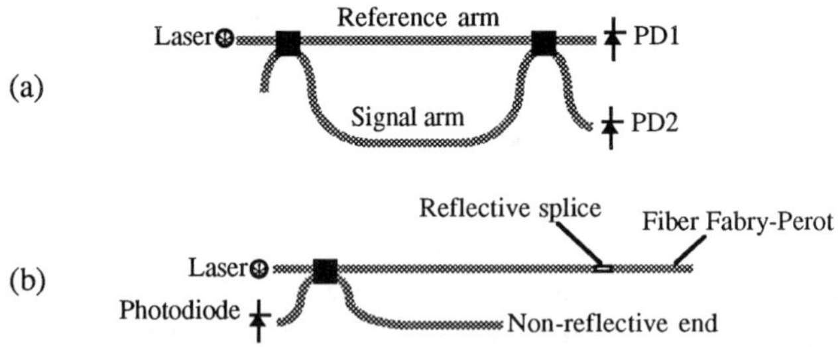

Figure 2: Fiber optic Mach-Zehnder (a), and remote Fabry-Perot (b) interferomters.

Polarimetric Strain and Temperature Sensors

In this type of sensors the degree of linear birefringence of the sensing fiber will be modified under the influence of temperature change or applied strain (Fig. 3).

The azimuth of the linearly polarized input beam makes an angle of $\pi/4$ with the fast and slow axes of highly birefringent fiber. The output beam after passing the analyzer has a transfer function similar to that of the two beam interferometer (Eq.3), with the phase retardation is now given by

$$\Delta\phi_{fs} = L[(A_f - A_s)\,\delta T + (B_f - B_s)\,\delta\varepsilon] \quad \text{(EQ 6)}$$

where A_f, B_f, A_s, and B_s are thermal and strain coefficients for the fast and slow modes respectively. It is clear from Eq. (6), that these devices are inherently less sensitive than their interferometric counterparts, the factor by which their unambiguous measurement range is increased. The application of polarimetric sensors for damage and strain detection in composite materials has been shown by Waite *et al.* (1988), and an arrangement which permits remote operation of polarimetric sensors was suggested by Corke *et al.* (1984) for temperature measurements. The resolution of a polarimetric sensor designed to sense strain, for example, may be compromised by temperature fluctuations, as for interferometric sensors.

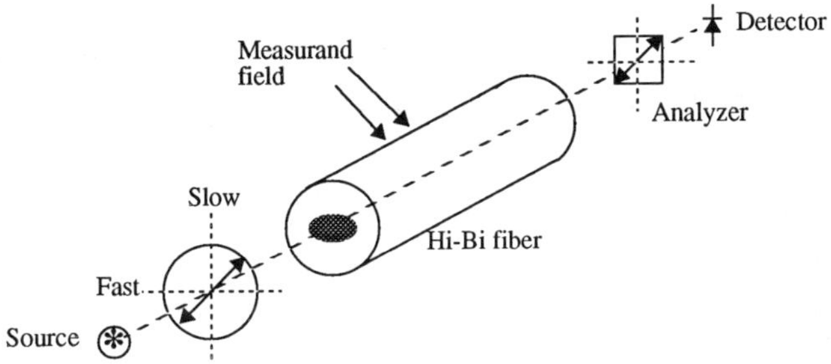

Figure 3: Schematic of a fiber optic polarimetric sensor.

Simultaneous Measurement of Strain and Temperature

Three different techniques to measure and identify the effect of temperature change and applied strain to a length of optical fiber have been described in the literatures. One method is using a length of elliptical core two-moded fiber in which the LP_{01} and LP_{11} modes polarized along the core ellipse minor axis and also the LP_{01} and LP_{11} modes polarized along the core ellipse major axis interfere at the fiber output (Blake *et al.*, 1987). These two interferometric signals could be simultaneously used to measure the temperature change and the axial strain applied to the fiber. Twin core fiber has also been used by considering the fact that temperature and strain sensitivities are wavelength dependent (Meltz *et al.*, 1986). Finally a fiber sensing system has been described where the two polarization modes of a piece of highly birefringent fiber were used to act as two fiber Fabry-Perot cavities (Leilabady *et al.*, 1985). The differential phase between these two independent interferometers was used to obtain polarization information. Since the ratio of phase to polarization sensitivity is different from temperature and strain, by determining interferometric phases and polarization state, the temperature and strain may be found.

Here we describe these techniques and their merits. The latter method will be used to demonstrate the practical limitations of previously reported works. Experimental data for two different types of fiber will be presented which show that the cross-sensitivity effect is significant when large strain or temperature variation is applied to the sensing fiber. This problem will be analyzed theoretically and an experimental technique for the measurement of cross-sensitivity of strain and temperature will be presented.

Elliptical Core Fiber Strain Gauge

An elliptical core two-mode fiber propagates four non-degenerate linearly polarized eigenmodes, two spatial modes each containing two polarization modes. Since the eigenmodes are non-degenerate and hence stable, intermodal interferometry becomes practical. For each polarization, light from the symmetric LP_{01} mode interferes with light from the anti-symmetric LP_{11} mode at the fiber output. The resulting interference pattern is offset from the fiber axis by an angle depending on the relative phase between the LP_{01} and LP_{11} modes. The interference between the LP_{01} and LP_{11} modes is sensitive to any environmental perturbations that induce a differential phase between the modes such that temperature change or applied strain to the fiber. For a detector placed in a position offset from the fiber axis, the light intensity varies cosinusoidally with the induced phase change. Using both eigen polarizations the two independent LP_{01} and LP_{11} modes interferometers can be separated, therefore, the phase of both interferometers can be measured. This phase information leads to simultaneous measurement of strain and temperature (Blake *et al.*, 1987).

Dual Core Fiber Strain Gauge

A dual core fiber is a waveguide formed from a pair of closely spaced single mode cores in a common cladding. However, when one core is illuminated a complete energy exchange can occur from the initially populated core to the other in a half of a beat length. Therefore at the end of the fiber a periodic variation of intensity at the output of each core can be seen. In general both the beat length and the fiber length are affected by temperature change and axial strain to the fiber. The beat length is also a function of the wavelength of the light source used to illuminate the fiber. The phase of the output signal can be obtained from Eq. (5). A pulsed dye laser as a variable wavelength source has been used to simultaneously measure the temperature and strain (Meltz *et al.*, 1986).

Strain Measurement Using Highly Birefringent Fiber

A length of highly birefringent fiber was used in conjunction with suitable polarization components to form two independent interferometers in a reflective configuration (Leilabady *et al.*, 1985). The interference pattern can be observed when light from the input face of the fiber constructively interfere with light reflected back from the distal end. The two interferometers associated with the two eigen modes have difference optical paths because the refractive index of the fast mode is different from that of the slow mode. It has been shown that the phase of these interferometers are functions of temperature and strain in the following manner:

$$\begin{aligned} \Delta\phi_s &= L\,[A_s(T)\,\delta T + B_s(\varepsilon)\,\delta\varepsilon] \\ \Delta\phi_f &= L\,[A_f(T)\,\delta T + B_f(\varepsilon)\,\delta\varepsilon] \end{aligned} \qquad \text{(EQ 7)}$$

where A_s, B_s, A_f and B_f are the temperature and strain sensitivity coefficients for the slow and fast mode of the fiber. Thus, knowing these coefficients one can solve the above simultaneous equations and calculate the strain and temperature change.

All these methods are viable in many practical applications when the sensing fiber is not subject to a large temperature change or a high axial strain. For applications where temperature variation is so large that affects the strain sensitivity or applied strain affects the thermal sensitivity, these methods are not applicable. In the following section we describe a technique which considers the cross-sensitivity effect. This theoretical analysis will be supported by experimental results obtain for two different types of fiber.

Temperature and Strain Measurement: Cross-sensitivity considerations

Consider the configuration shown in Fig. 4. Linearly polarized light from a single mode laser is launched into the sensing fiber via a half-wave plate ($\lambda/2$) and a beam splitter. The half-wave plate is used to rotate the state of polarization of the input beam such that it makes an angle of $\pi/4$ with the fiber eigenaxis, thus equally populates both eigenaxes. Two independent Fabry-Perot cavities are formed using a length of two moded fiber and coherently mixing the light reflected from the input face and the light reflected from distal end of the fiber. The two independent outputs correspond to the two fiber Fabry-Perot cavities can be monitored using photo detectors PD1 and PD2.

The set of Eqs. (7) can be generalized for the configuration shown in Fig. 4 as:

$$\begin{aligned}\phi_1 &= \phi_{10} + L\,[A_1\,(T)\,\delta T + B_1\,(\varepsilon)\,\delta\varepsilon + C_1\,(T,\varepsilon)\,\delta T\delta\varepsilon + \dots] \\ \phi_2 &= \phi_{20} + L\,[A_2\,(T)\,\delta T + B_2\,(\varepsilon)\,\delta\varepsilon + C_2\,(T,\varepsilon)\,\delta T\delta\varepsilon + \dots]\end{aligned} \qquad \text{(EQ 8)}$$

where ϕ_{i0} is the initial phase of the ith interferometer, $A_i(T)$ is the thermal coefficient, $B_i(\varepsilon)$ is the strain coefficient and $C_i(T,\varepsilon)$ is the first order cross-sensitivity coefficient between the temperature and strain effect. For the rest of this work we neglect the higher order effects and use the simplified version of the above equations, i.e,

$$\begin{aligned}\phi_1 - \phi_{10} &= L\,[A_1\,(T)\,\delta T + B_1\,(\varepsilon)\,\delta\varepsilon + C_1\,(T,\varepsilon)\,\delta T\delta\varepsilon] \\ \phi_2 - \phi_{20} &= L\,[A_2\,(T)\,\delta T + B_2\,(\varepsilon)\,\delta\varepsilon + C_2\,(T,\varepsilon)\,\delta T\delta\varepsilon]\end{aligned} \qquad \text{(EQ 9)}$$

We will see that this approximation is valid for a very large temperature variation and high applied strain. It has been shown before (Farahi *et al.*, 1990) that the cross-sensitivity coefficient is also proportional to the length of sensing element, therefore, when a long length of fiber is used as a strain sensor the effect of temperature becomes more important than that of a short sensing element.

Here we present data for two different types of fiber. The experimental arrangements used for both experiments were similar to that shown in Fig. 4. In the first experiment a 7 cm length of the high birefringence thermally stressed (bow-tie) type fiber was enclosed in an electrically heated furnace, and was clamped at one end whilst the other end was fixed into a disc piezo-electric transducer. This configuration enabled us to simultaneously heat and apply axial strain to the common sensing element. The system was thermally isolated using a polystyrene box and the sensing element was also similarly isolated from the remaining fiber.

A pseudo-heterodyne signal processing scheme (Jackson *et al.*, 1982) was used and the phase of each interferometer was measured using a lock-in amplifier. The thermal and strain coefficients of the sensing fiber were measured to be;

$$\begin{aligned}LA_f\,(T) &= 8.681 \pm 0.031 \;\; \text{rad} \;\; {}^{\circ}C^{-1} \\ LA_s\,(T) &= 8.089 \pm 0.023 \;\; \text{rad} \;\; {}^{\circ}C^{-1}\end{aligned} \qquad \text{(EQ 10)}$$

and

$$\begin{aligned}LB_f\,(\varepsilon) &= 10.601 \pm 0.021 \;\; \text{rad} \;\; (\mu m)^{-1} \\ LB_s\,(\varepsilon) &= 10.730 \pm 0.015 \;\; \text{rad} \;\; (\mu m)^{-1}\end{aligned} \qquad \text{(EQ 11)}$$

To determine the cross-sensitivity effect the sensing element was axially stretched in step, and for each step the thermal coefficients were measured. The cross sensitivity coefficients for 7 cm length of sensing element were found to be:

$$LC_f(T, \varepsilon) = -0.143 \pm 7\times10^{-3} \text{ rad } {}^\circ C^{-1} (\mu m)^{-1}$$
$$LC_s(T, \varepsilon) = -0.139 \pm 5\times10^{-3} \text{ rad } {}^\circ C^{-1} (\mu m)^{-1} \qquad \text{(EQ 12)}$$

These experimentally evaluated coefficients and set of Eq.(9) are sufficient to determine δT and $\delta\varepsilon$ for each pair of measured ϕ_f and ϕ_s. Figure (5.a) shows the variation of thermal coefficients against applied axial strain for the bow-tie fiber used in this experiment.

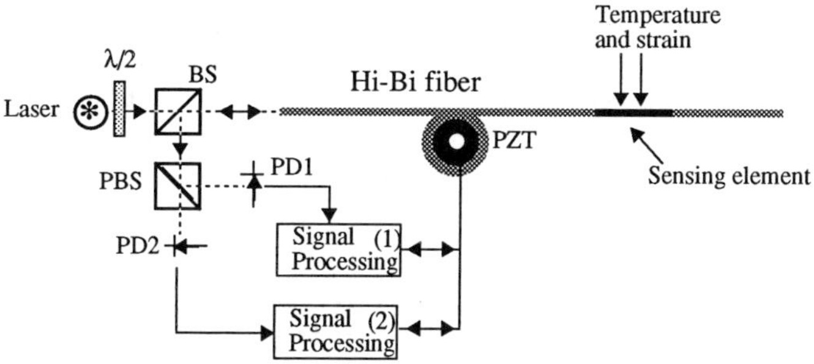

Figure 4: Schematic of the experimental arrangement for simultaneous measurement of strain and temperature using dual interferometry in a length of Hi-Bi fiber.

In the second experiment a 10 cm length of elliptical core fiber was subject to temperature change, axial strain or both. When configuration similar to that shown in Fig. 4 was used four interferometers were formed based upon two polarization states and two spatial modes. During the experiment it was found that the interference signals associated with polarization states of LP_{01} mode are relatively more stable than those of LP_{11} mode. These interference signals were used and for no axial strain, the thermal coefficients were measured to be:

$$LA_f(T) = 10.412 \pm 0.145 \text{ rad } {}^\circ C^{-1}$$
$$LA_s(T) = 9.870 \pm 0.156 \text{ rad } {}^\circ C^{-1} \qquad \text{(EQ 13)}$$

and at a constant temperature the strain coefficients were found to be:

$$LB_f(\varepsilon) = 13.012 \pm 0.092 \text{ rad } (\mu m)^{-1}$$
$$LB_s(\varepsilon) = 13.275 \pm 0.155 \text{ rad } (\mu m)^{-1} \qquad \text{(EQ 14)}$$

Similar method as discussed in the previous experiment was used to measure thermal coefficients as the sensing fiber was axially strained in steps. The results are given in Fig. (5.b). The cross-sensitivity coefficient were calculated to be:

$$\begin{aligned} LC_f(T, \varepsilon) &= -0.211 \pm 9.0\times10^{-3} \ \text{rad} \ \ {}^{\circ}C^{-1}(\mu m)^{-1} \\ LC_s(T, \varepsilon) &= -0.179 \pm 8.2\times10^{-3} \ \text{rad} \ \ {}^{\circ}C^{-1}(\mu m)^{-1} \end{aligned} \qquad \text{(EQ 15)}$$

As may be appreciated from Fig.(5.b), for very high axial strain this fiber has a non-linear behavior.

This system has a limited measurement range as both interferometers outputs have the unambiguous measurement range of 2π rad. This problem can no longer be solved by combined interferometric-polarimetric system, because the interferometric and polarimetric outputs have already been used in the recovery of the two measurands. A dual wavelength technique (Kersey *et al.*, 1986) or alternatively a white light technique in conjunction with dual interferometers (Newson *et al.*, 1988) can be used to extend the unambiguous measurement range.

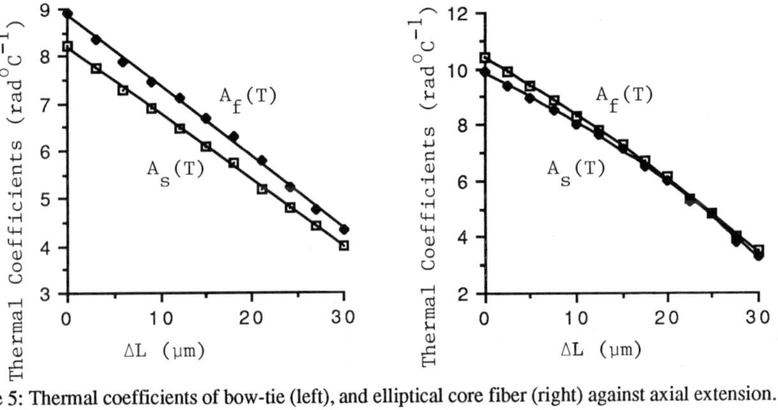

Figure 5: Thermal coefficients of bow-tie (left), and elliptical core fiber (right) against axial extension.

Temperature Compensated Strain Sensor

Dakin and Wade (1984) reported a new type of polarimetric sensor, which has been described as 'compensated'. The sensing element of the device was formed from two sections of highly birefringent fiber, joined with the polarization eigenaxes of the two parts mutually perpendicular. That is, the fast eigenaxis of the first section was aligned with the slow eigenaxis of the second, and *vice versa*. Later this technique was utilized for extending the measurement range of interferometric fiber-optic temperature sensor (Farahi *et al.*, 1988). This method can also be used to design a temperature compensated fiber optic strain sensor.

A multiplexed system containing two strain sensors is shown in Fig. 6. Light from a short coherence light source illuminates an unbalanced Michelson interferometer (local processing interferometer) and the output from this interferometer is injected into the length of highly birefringent fiber such that both axes are equally populated. The interference fringes associated with each sensing interferometers can be observed when the optical path imbalances of the fiber and the local processing interferometer are identical. Each fiber interferometer is made of two equal length segments of fiber which are spliced at $\pi/2$ angle with respect to each other, hence forming two identical interferometers in the absence of any measurand field. Only one segment of the fiber was subject to axial strain and no temperature controller was used. Under these conditions the phase retardation of the interference signals monitored by the two detectors have the following forms:

$$\Delta\phi_1 = \delta\phi_{fT} + \delta\phi_{sT} + \delta\phi_{f\varepsilon}$$
$$\Delta\phi_2 = \delta\phi_{fT} + \delta\phi_{sT} + \delta\phi_{\varepsilon s} \quad \text{(EQ 16)}$$

The difference between these two phases will be independent of the temperature. The stability of a 40 cm fiber interferometer with 20 cm sensing length under a temperature fluctuation of $\sim 5°C$ is shown in Fig. 7.

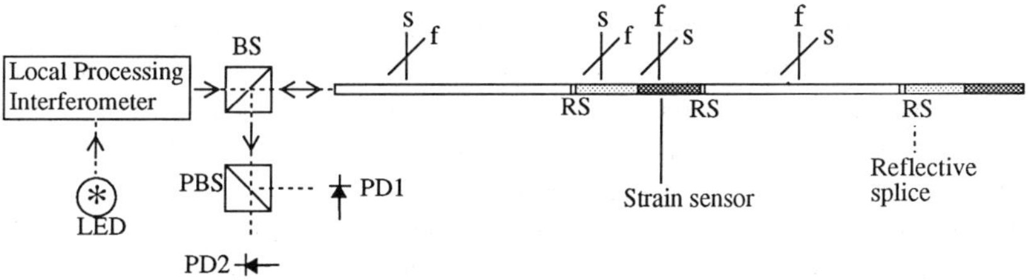

Figure 6: Schematic diagram of a multipexed system consist of two temperature compensated strain sensors.

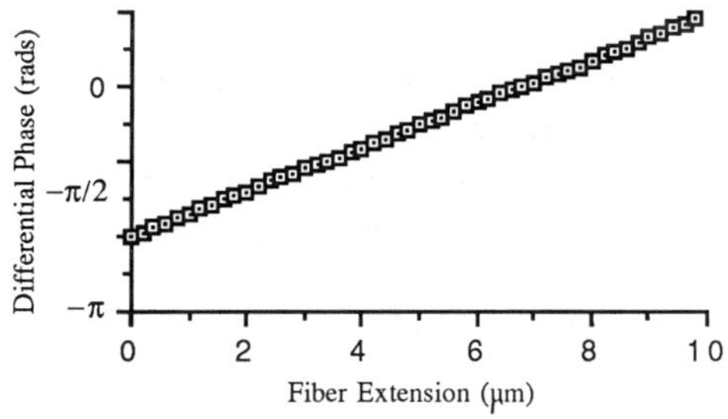

Figure 7: Differential phase as a function of fiber axial extension.

Strain Sensor Using Gratings in Optical Fibers

The formation of holographic gratings in Ge doped optical fibers using UV light was reported by Meltz *et al.* (1989), and later has been used for temperature and strain measurement by many other scientists. The fiber is illuminated from the side by the interference pattern at the output of an interferometer using UV light source. This induces a sinusoidal variation in the index of refraction along the fiber axis in the region of the interfering beam due to a single photon process. To date, grating with reflectivities of up to 90% over 5-10 mm length of fiber have been achieved (Morey *et al.*, 1990). Transmission spectrum of a fiber with Bragg gratings is sensitive to both temperature and strain, since the refractive index of the fiber changes with temperature and strain-optic effects. This is an attractive method since the physical integrity of the fiber remains unaffected during the formation of gratings. It is also possible to use the gratings as fiber internal mirrors and produce fiber optic interferometers.

Discussions and Conclusions

The aim of this article has been to introduce and discuss the techniques for strain and temperature measurement using optical fibers. Fiber optic interferometric, polarimetric and grating sensors have been the subject of interest for many years for smart skin applications. There are now many other methods for measuring these two quantities which have been experimentally demonstrated and only a few of them have been discussed here. Those named 'distributed sensors' have not been discussed and a useful summary of these sensors can be found elsewhere (Culshaw and Dakin, 1989). Fiber optic distributed temperature sensors using liquid core fiber (Hartog and Payne, 1982), using anti-Stokes-Raman spectrum (Dakin *et al.*, 1985), and stimulated Brillouin scattering (Culverhouse *et al.*, 1989) have also been developed. Recently a new non-linear phenomenon in optical fiber called forward stimulated Brillouin scattering has been reported (Russell *et al.*, 1990) which is described by periodic coupling between two spatial modes in optical fibers with the coupling provided by a travelling acoustic wave. Flextural wave is generated through the strain-optic effect, and when the wavelength equals the beat length of the fiber the mode coupling occurs with a down-shift in the frequency of the coupled beam. The beat frequency between the two spatial modes is a function of fiber beat length which is sensitive to temperature and strain. Although all these methods are very attractive for temperature measurement their applications for strain measurement have yet to be shown.

Much of the pioneering work on single mode fiber sensors is now entering a more advance phase of engineering development. The implementation of optical fiber sensors particularly in smart structures and material systems is a good example.

However, there remains considerable scope for research, mainly concerned with their potential to compromise the compressive strength of the composite, to increase the damage vulnerability of the structure, to reduce their fatigue life, and the performance life of the structurally integrated fiber optic sensors themselves. Other areas of interests are multiplexed, multi-measurand sensor systems, and fiber optical signal processing units which can be all embedded within the smart structure.

References

Blake, J.N., Huang, S.Y., and Kim, B.Y. (1987). Elliptical core two-mode fiber strain gauge, *SPIE conf. Proc.*, **838**, 332.

Butter, C.D., and Hocker, G.B. (1978). Fiber optic strain gauge, *Appl. Opt.*, **17**, 2867.

Corke, M., Kersey, A.D., Jackson, D.A., and Jones, J.D.C. (1983). All fiber Michelson thermometer, *Electron. Lett.*, **19**, 471.

Corke, M., Kersey, A.D., Liu, K., and Jackson, D.A. (1984). Remote temperature sensing using polarization preserving fiber, *Electron. Lett.*, **20**, 67.

Culshaw, B., and Dakin, J.P. (1989). *Optical Fiber Sensors: Systems and Applications*, Artech House, Norwood.

Culverhouse, D., Farahi, F., Pannell, C.N., and Jackson, D.A., (1989). Potential of stimulated Brillouin scattering as sensing mechanism for distributed temperature sensors, *Electron. Lett.*, **25**, 913.

Eickhoff, W.E. (1981). Temperature sensing by mode-mode interference in birefringent optical fiber, *Opt. Lett.*, **6**, 204.

Cotter, D. (1983). Stimulated Brillouin scattering in monomode optical fiber, *J. Opt. Commun.*, **4**, 10.

Dakin, J.P., and Wade, C.A. (1984). Compensated polarimetric sensor using polarization-maintaining fiber in a differential configuration, *Electron. Lett.*, **20**, 51.

Dakin, J.P., Pratt, D.J., Bibby, G.W., and Ross, J.N. (1985). Distributed optical fiber Raman temperature sensor using a semiconductor light source and detector, *Electron. Lett.*, **21**, 569.

Dakin, J.P., and Culshaw, B. (1988). *Optical Fiber Sensors: Principles and Components*, Artech House, Norwood.

Farahi, F., Jones, J.D.C., and Jackson, D.A. (1987). High speed thermometry utilizing multiplexed fiber Fabry-Perot interferometers, *SPIE conf. Proc.*, **838**, 216.

Farahi, F., Jones, J.D.C. and Jackson, D.A. (1988). An extended range fiber-optic temperature sensor, *Int. J. Optoelectronics.*, **3**, 237.

Farahi, F., Webb, D.J., Jones, J.D.C., and Jackson, D.A. (1990). Simultaneous measurement of temperature and strain: Cross-sensitivity considerations, *J. Lightwave Tech.*, **8**, 138.

Fields, J.N., Asawa, C.K., Ramer, O.G. and Barnoski, M.K. (1980). Fiber optic pressure sensor, *J. Acous. Soc. America*, **67**, 816.

Hartog, A.H., and Payne, D.N. (1982). Fiber optic temperature distribution sensor, *Proc. IEE Colloq Optical Fiber Sensors*. (London).

Hocker, G.B. (1979). Fiber-optic sensing of pressure and temperature, *Appl. Opt.*, **18**, 1445.

Jackson, D.A., Kersey, A.D., Corke, M., and Jones, J.D.C. (1982). Pseudoheterodyne detection scheme for optical interferometers, *Electron. Lett.*, **18**, 1081.

Kersey, A.D., Dandridge, A., and Burnes, W.K. (1986). Two-wavelength fiber gyroscope with wide dynamic range, *Electron. Lett.*, **22**, 935.

Jones, B.E. (1985). Optical fiber sensors and systems for industry, *J. Phys. E: Sci. Instrum.*, **18**, 770.

Leilabady, P.A., Jones, J.D.C., and Jackson, D.A. (1985). Interferometric strain measurement using optical fibers, *SPIE conf. Proc.*, **586**, 230.

Leilabady, P.A., Jones, J.D.C., and Jackson, D.A. (1986). Combined interferometric-polarimetric fiber optic sensor capable of remote operation, *Opt. Commun.*, **57**, 77.

Meltz, G., Dunphy, J.R., and Leonberger, F.J. (1986). Multi-wavelength twin-core fiber optic sensors, *OFS-4 conf. Proc.*, 67.

Meltz, G., Morey, W.W., and Glenn, W.H. (1989). Formation of Bragg grating in optical fibers by a transverse holographic method, *Opt. Lett.*, **14**, 823.

Morey, W.W., Meltz, G., and Glenn, W.H. (1990). Holographically generated gratings in optical fibers, *Opt. Photonics News*, **1**, No.7, 14.

Newson, T.P., Farahi, F., Jones, J.D.C., and Jackson, D.A. (1988). Combined interferometric and polarimetric fiber optic temperature sensor with a short coherence length source, *Opt. Commun.*, **68**, 161.

Palmer, A.W., Croft, J.R., and Valsier, R. (1983). Potential applications for total internal reflection fiber optic sensors, *SPIE conf. Proc.*, **374**, 206.

Russell, P.St.J., Culverhouse, D., and Farahi, F. (1990). Experimental observation of forward stimulated Brillouin scattering in dual-mode single-core fiber, *Electron. Lett.*, **26**, 1195.

Spillman, W.B., and McMahon, D.H., (1980). Frustrated-total-internal-reflection multimode fiber optic hydrophone, *Appl. Opt.*, **19**, 113.

Valis, T., Hogg, D., and Measures, R.M. (1990). Composite material embedded fiber-optic Fabry-Perot strain rosette, *SPIE conf. Proc.*, **1370**, 154.

Valis, T., Tapanes, E., Liu, K., and Measures, R.M. (1991). Passive-quadrature demodulation localized-Michelson fiber-optical strain sensor embedded in composite materials, *J. Lightwave Tech.*, **9**, 535.

Waite, S.R., Tatam, R.P., and Jackson, D.A. (1988). Use of optical fiber for damage and strain detection in composite materials, *Composite*, **19**, 435.

VI - MEASUREMENT OF PRESSURE

Single- and Few-Mode Fiber-Optic Pressure Sensors

WOJTEK J. BOCK[1] , TINKO A. EFTIMOV[2]

Abstract. The principle of operation of polarimetric single-mode and intermodal interferometric few-mode hydrostatic pressure sensors using highly birefringent optical fibers is described. The experimental responses of the sensor to pressure in the two regimes of operation are presented. The prototype of a fiber-optic pressure cell for stress monitoring utilizing these effects and developed in collaboration with Glötzl GmbH (Germany) is also presented in the paper.

1. Introduction

In many applications in the civil engineering industry such as mining, large dams, construction projects and tunneling, there is an important need to evaluate the response of a structure for safety purposes, including structural integrity and damage assessment in foundations, pillars or anchors and total ground pressure and load evaluation. For static or quasi-static conditions this is usually done by monitoring real stress (or pressure) distribution in concrete, rock or earth. Many specific requirements relative to sensing pressure in these environments can best be met using emerging fiber-optic sensors.

These devices, configured most often in the form of intrinsic sensors where an optical fiber acts as a sensing element and simultaneously as a guiding medium for input/output light signals, offer many advantages over their conventional electrical counterparts. Some of the most important are: inherent immunity to electromagnetic interference (EMI) and to lightning problems, safety in hazardous or explosive environments, high sensitivity, compatibility with industrial data transmission systems based on optical fibers, potentially low cost and miniature dimensions.

For obvious reasons, most of the early research on fiber-optic pressure sensors focused on dynamic applications, especially for the detection of underwater acoustic signals [6]; these devices were mostly configured as interferometric sensors and

[1] Professor, Laboratory of Optoelectronics, Computer Sciences Department, University of Québec at Hull, P. O. Box 1250, Station "B", Hull, Québec, J8X 3X7 CANADA.

[2] on an NSERC Postdoctoral Fellowship from the Fiber Optics and Fiber Optic Sensors Research Laboratory, Institute of Applied Physics, Technical University, Sofia-1156, BULGARIA.

required sophisticated and expensive instrumentation. To determine the real stress in structural components or to measure slow changes of pressure in typical process-control applications, the sensor has to measure static or slowly varying pressure. Most such sensors reported in the literature were designed for relatively low pressures (typically up to 20 MPa) and primarily as modifications of displacement and stress-intensity transducers using multimode technology [6]. These devices have relatively low sensitivity and use mechanically moving parts. Perhaps the most mature fiber-optic pressure sensor of this kind was one recently developed [7] and tested [8] by Voet et al., a modification of a microbending sensor. At the opposite end of the scale, some recently reported phase-sensitive devices are too unstable, often overly sensitive and require signal recovery and processing that is too sophisticated and costly to ever find any industrial application.

This paper discusses development of the theoretical understanding of the principle of operation of hydrostatic pressure sensors based on polarization coupling and intermodal interference occurring in highly birefringent (Hi-Bi) single-mode and two-mode optical fibers. The prototype of a fiber-optic pressure cell for stress monitoring utilizing these effects and developed in collaboration with Glötzl GmbH (Germany) is also presented in the paper.

2. Principle of operation of a fiber-optic polarimetric and intermodal interferometric sensor

2.1. Regimes of Operation and Types of Sensors

An isotropic round optical fiber consists of a core and a cladding of corresponding radii a and b, and refractive indices n_1 and n_2. If light of wavelength λ is launched into the core it will be guided along the fiber due to the total internal reflection which takes place since $n_1 > n_2$. Depending on how the input light is introduced it can propagate following different paths. Since the core and the cladding form a dielectric waveguide, the electromagnetic wave will propagate in definite states known as *modes*. Each mode is characterized by its own *propagation constant* β, *polarization* X or Y, and *intensity distribution* known as *mode pattern*. An important dimensionless quantity characterizing the waveguide properties of the fiber is the V-parameter, defined as [5]: $V = (2\pi/\lambda)a\,(n^2{}_1 - n^2{}_2)^{1/2}$. Depending on the value of the V-parameter one or more modes can exist. For an isotropic round core/cladding fiber the following mode propagation condititions hold. For $V \leq V_c = 2.405$ only the fundamental LP_{01} mode propagates. The mode is two-fold degenerate and can be X- or Y-polarized denoted respectively by $LP_{01}{}^x$ and $LP_{01}{}^y$, both having the same propagation constant β_{01}. For $V_c = 2.405 < V < 4$, a second spatial mode LP_{11} can propagate as well. The second mode is, however, fourfold degenerate and can exist in two spatial orientations *even* and *odd*, each of them being X- or Y-polarized. We thus have $LP_{11}{}^{e,x}$, $LP_{11}{}^{e,y}$, $LP_{11}{}^{o,x}$ and $LP_{11}{}^{o,y}$ polarization modes with the same propagation constant β_{11}. The total number of polarization modes along the fiber is six [5].

For a fiber with given core/cladding refractive indices and a given core radius, the V-parameter and thus the number of modes can be changed only via launching light of a different colour, i.e. wavelength λ. The smaller the wavelength, the greater the V-parameter, and the more modes can propagate. All three characteristics of a mode - its propagation constant β, polarization and mode distribution depend on the V-parameter.

If anisotropy is introduced in the fiber, then the refractive indices along the X and Y axes become different, and hence the V-parameters along X and Y-axes are also different which automatically leads to different propagation constants β^x and β^y.

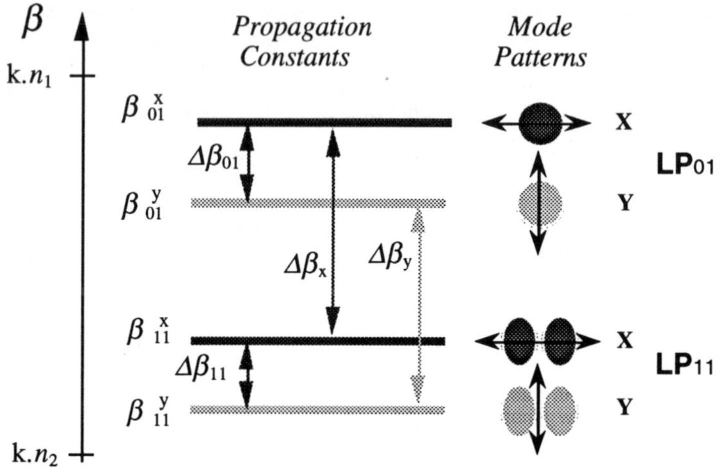

Fig. 1. Propagation constants and mode patterns in Hi-Bi bimodal fibers.

The fiber is then birefringent or highly birefringent (Hi-Bi). Anisotropy can be achieved by making the core elliptical (e-core fiber), or by laterally compressing the core with the help of stress applying parts (bow-tie, panda-type fiber) or elliptical cladding (e-cladding fiber).

When very high birefringence is introduced, the even and the odd LP_{11} modes will have different cut-off wavelengths. This means that over a specific wavelength region, only four polarization modes can propagate along the waveguide: $LP_{01}{}^x$, $LP_{01}{}^y$, and the even $LP_{11}{}^x$ and $LP_{11}{}^y$ whose propagation constants we denote by $\beta_{01}{}^x$, $\beta_{01}{}^y$, $\beta_{11}{}^x$ and $\beta_{11}{}^y$. This effect is clearly observable in e-core and bow-tie Hi-Bi fibers. The schematic diagram of the propagation constants and the corresponding mode patterns is presented in Fig.1.

Single-Mode Regime of Operation. The Polarimetric Sensor.

Single-mode operation occurs when only one spatial mode LP_{01} or LP_{11} is excited at fiber's input. If the fiber is bimodal at the operating wavelength, the single-mode regime can be achieved by adjusting the launching conditions. However, the most common solution is to use a single-mode optical fiber in which only the fundamental LP_{01} mode will propagate. No intermodal interference is observed in this case as the second mode is absent.

If quasimonochromatic light linearly polarized at an angle φ with respect to the fiber's X-axis is launched into the fiber and an analyzer turned to an angle α is placed at the output of the fiber then the optical intensity detected will be

$$I = \frac{1}{2}\left[1 + \cos 2\alpha . \cos 2\varphi + |\gamma| \sin 2\alpha . \sin 2\varphi \ \cos\Phi_0\right] \tag{1}$$

where $\Phi_0 = \Delta\beta_{01}L$ is the phase. When external perturbations are introduced they cause changes in the phase $\Phi_0 = \Delta\beta_{01}L$ of the fundamental LP_{01} mode (or correspondingly $\Phi_1 = \Delta\beta_{11}L$ for the LP_{11}). These will lead to a cosine variation of the observed intensity I measured after the analyzer, a variation that is in fact a polarization interference. The set-up is then a *polarimetric sensor.* The interfering waves in this case are the $LP_{01}{}^x$ and the $LP_{01}{}^y$ polarization modes. With $|\gamma|$ we represent the correlation function between the polarization modes. This is a function of the product of the length of the fiber L, the polarization dispersion of the fiber $\delta\tau$ and the spectral halfwidth of the source $\delta\lambda$.

The visibility V of the observed polarimetric response is:

$$V = \frac{I_{max} - I_{min}}{I_{max} + I_{min}} = |\gamma| \frac{\sin 2\alpha . \sin 2\varphi}{1 + \cos 2\alpha . \cos 2\varphi} \tag{2}$$

An input polarizer (if the light is not linearly polarized) acts as a splitter and the analyzer acts as a recombiner. If we define $k_1 = \sin^2\varphi$ and $k_2 = \sin^2\alpha$ as the power coupling coefficients of the splitter and the recombiner, then the expression for the visibility becomes:

$$V = |\gamma| \frac{4\sqrt{k_1 k_2 (1 - k_1).(1 - k_2)}}{1 + (1 - 2k_1^2).(1 - 2k_2^2)} \tag{3}$$

This expression is analogous to the well-known formula for the visibility in classical two-beam interferometry [6]. So maximum visibility is obtained if $k_1 = k_2 = 0.5$ or equally for $\varphi = \alpha = 45°$. This is when both $LP_{01}{}^x$ and $LP_{01}{}^y$ -polarization modes are equally excited by launching a $\pm 45°$ linearly polarized light. In this case the observed intensity and the visibility become:

$$I = \frac{1}{2}\left[1 + |\gamma| \cos \Phi_0\right] \quad \text{with} \quad V = |\gamma| \tag{4}$$

Evidently, if a monochromatic source is used $\delta\lambda = 0$ and then the visibility will be $|\gamma| = 1$.

Two-Mode Regime of Operation. The Intermodal Interference Sensor.

For a two-mode regime of operation, a bimodal fiber must be used and the interference between either X- or Y-polarized LP_{01} and LP_{11} spatial modes is observed. By letting $\Phi_x = \Delta\beta_x L$ and $\Phi_y = \Delta\beta_y L$, the intensity observed at the output of the fiber excited with X- or Y-polarized quasimonochromatic light can be obtained [4] as:

$$I_X(x,y,z) = \eta_0 f_0^2(x,y) + \eta_1 f_1^2(x,y) + 2\eta_{01} f_0(x,y)\, f_1(x,y)\, cos\left(\Phi_X\right)$$
$$I_Y(x,y,z) = \eta_0 f_0^2(x,y) + \eta_1 f_1^2(x,y) + 2\eta_{01} f_0(x,y)\, f_1(x,y)\, cos\left(\Phi_Y\right) \quad , \qquad (5)$$

In the equations above η_0 and η_1 are the relative optical powers carried by each spatial mode while $|\Gamma_{x,y}|$ are the correlation functions between the two interfering modes. Depending on the detection set-up different expressions for the visibility can be obtained. Assuming the Gaussian approximation [5], the distributions of the two modes are expressed as:

$$f_0(x,y) = \frac{1}{\sqrt{\pi\, \omega_x\, \omega_y}} F(x,y)\,, \qquad F(x,y) = exp\left[-\frac{1}{2}\left(\frac{x^2}{\omega_x^2} + \frac{y^2}{\omega_y^2} \right) \right]$$
$$f_1(x,y) = \sqrt{\frac{2}{\pi\, \omega_x\, \omega_y}}\, \frac{x}{\omega_x} F(x,y) \qquad (6)$$

The following three important subcases can readily be outlined: visibility at a point (pixel), visibility when half the pattern is detected and visibility when an offset single-mode fiber is used. The visibility is then presented in the form $V(z,\eta_1) = |\Gamma_{x,y}| V_0$, where V_0 is a factor depending on the relative excitation coefficient of the second mode as well as on the detection scheme. In all of the cases, maximum visibility is obtained if the source is monochromatic and both modes are equally excited i.e. $\eta_0 = \eta_1 = 0.5$.

It is important to note that a fiber may be used in both single- and two-mode regime by operating it at two wavelengths at the same time. Thus *e*-core, D-shape and bow-tie fibers designed for single-mode operation at 850 nm (laser diode), will be bimodal when operated at He-Ne wavelength 633 nm. Or, a polarimetric set-up can be designed for 850 nm and a polarization sensitive intermodal interferometer for 633 nm. Note, that the input and output polarization conditions in single- and two-mode regimes are different. If the fiber is operated at both wavelengths and all possible polarizations, then three independent cosine responses can be obtained.

2.2. Differential phase shifts

When an external perturbation is applied to a fiber of length L, it will change the phases $\Phi_i = \Delta\beta_i\, L$ (i = 0, x, y) by an amount of $\delta\Phi_i$

$$\delta\Phi_i = \delta\left(\Delta\beta_i\right) L + \Delta\beta_i\, \delta L \qquad (7)$$

If the external perturbation is hydrostatic pressure p, then an increase by δp will cause both $\Delta\beta_i$ and L to change by $\partial(\Delta\beta_i)/\partial p$ and $\partial L/\partial p$. So from (7) we can obtain

$$\frac{\delta\Phi_i}{\delta p} = \frac{\partial\left(\Delta\beta_i\right)}{\partial p} L + \Delta\beta_i \frac{\partial L}{\partial p} = \frac{2\pi}{T_{i,p}} = \Lambda_{i,p} \qquad (8a)$$

The above equation can also be rewritten as:

$$\delta\Phi_i = \left[\frac{\partial\left(\Delta\beta_i\right)}{\partial p} L + \Delta\beta_i \frac{\partial L}{\partial p} \right] \delta p = \frac{2\pi}{T_{i,p}}\, \delta p = \Lambda_{i,p}\, \delta p \qquad (8b)$$

Note that $T_{i,p}$ (i = 0, x, y) has the dimension of pressure while $\Lambda_{i,p}$ has the inverse dimension. These are experimentally measurable parameters and determine the sensitivity of the sensor to a given external perturbation.

The design of a practical high hydrostatic pressure sensor with a predesigned sensitivity as expressed by the parameter $T_{i,p}$, is based on an important relation expressed simply as [1]:

$$T_{i,p}L = C_i.\lambda \tag{9}$$

where C_i is some constant. Thus, a longer fiber will exhibit greater sensitivity and vice versa. Note that if the source is not monochromatic a longer fiber will lead to a lower value of the correlation function $|\gamma|$ which in turn will null the cosine response of the sensor. Therefore, a proper choice of the laser source is essential.

3. Experimental Results and Practical Devices

3.1. Experimental Set-up and Results

The experimental set-up used to determine the pressure response of both polarimetric and intermodal interference sensors is shown in Fig.2. Three different Hi-Bi fibers, single-mode at 850 nm and operated either in single-mode or in bimodal regime at 633 nm, were investigated: bow-tie (HB 800, York), *e* -core and D-shape (Andrew). In case only polarimetric pressure sensor was under study a bow-tie single-mode fiber (HB 600, York) was used. In the set-up shown schematically in Fig. 2, the length of the fiber subjected to external perturbations was always 1m, a length determined by the construction of the universal stress generating apparatus [9]. This device allows three types of perturbations, such as pressure *p*, temperature T and longitudinal strain ε, to be independently imposed on a fiber. The pressure chamber was fed from a

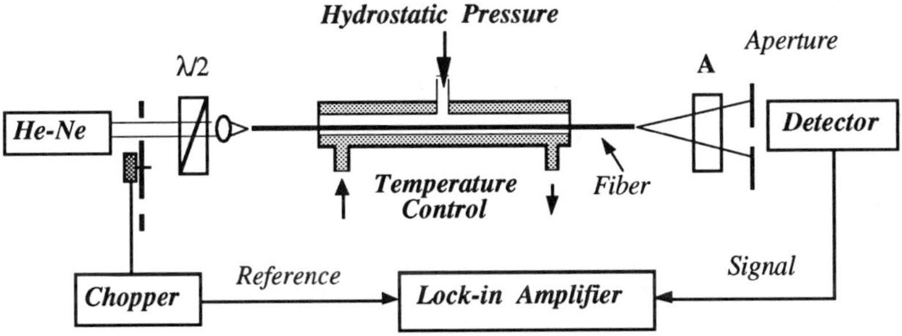

Fig. 2. Experimental set-up.

Harwood DWT-35 deadweight tester, allowing pressure calibration with an accuracy of at least 0.1 % in 100MPa, traceable to NIST. The temperature was controlled by a Haake 3F thermostat and the light source was a linearly polarized He-Ne laser for the single-mode HB 600 and the bimodal HB 800 bow-tie York fibers, as well as for the e-core and D-shape Andrew fibers. In the case HB 600 sensing fiber was used, three versions of a multifiber configuration were developed. The first consists of a low-birefringence (Lo-Bi) lead-in and lead-out fiber spliced to both ends of the sensing single-mode fiber. The second version uses a HB 600 lead-in fiber spliced with its axes at 45° to the sensing fiber's axes. The lead-out fiber is again a Lo-Bi. The third, and very promissing construction is a HB 800 sensing fiber operated at 840 nm and equipped with polarizing 3M fibers spliced at 45° relative to the eigenaxes of the sensing fiber. These polarizing fibers act as lead-in and lead-out distributed polarizers and the whole construction becomes all-fiber, with neither bulk-optic polarizers nor phase plates used in the set-up. The only serious requirement in this case is the source's spectrum which must be narrow (a single-longitudinal laser diode) and very stable. The reason for this is that bow-tie fibers have a very high polarization dispersion $\delta\tau$ = 1.89 ns/km, and thus a source with high spectral halfwidth $\delta\lambda$ reduces the value of the correlation function $|\gamma|$, which will null the cosine response for a longer sensing fiber (L > 0.15 m).

It is crucial to point out that while in the polarimetric set-up the whole optical power is measured, in the interferometric set-up only a part of the pattern is detected in order to register the cosine response.

Fig. 3 shows the polarimetric response to pressure of a HB800 bow-tie fiber operated in single-mode regime at 837 nm. The three curves were taken at different temperatures. From these curves it is found that $T_{01,p}$ L = 950 bar·cm at a wavelength of 837 nm. Note that the visibility is V = 0.91 for a sensing fiber length of about 8 cm due to the high polarization dispersion. The sensing fiber was incorporated in a multifiber assem-bly (shown in the inset) using a 3M polarizing lead-out fiber whose length was ≈ 2 m.

When the same fiber is operated in two-mode regime by using a He-Ne laser emitting at 633 nm, polarization-sensitive intermodal interference is observed. The interferometric responses for a fiber length of 1 m are shown in Fig.4.

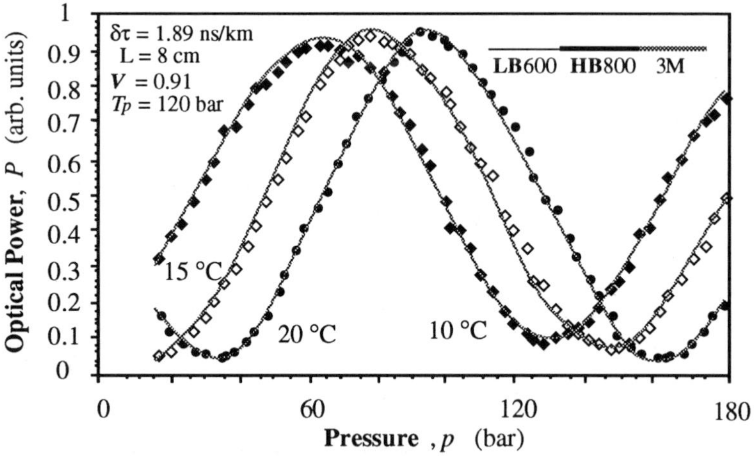

Fig.3. Polarimetric response at different temperatures of a HB800-based single-mode pressure sensor using a Lo-Bi lead-in and a polarizing 3M lead-out fiber in a spliced multifiber assembly. A SHARP laser diode (837nm) was used as a source.

In this case, only a part of the interference pattern is measured in order to detect the cosine responses as illustrated in the Fig.4. Note that, in agreement with the theoretical considerations, the visibility is far from unit. The most important in the interferometric response is the fact that the sensitivities for X- and Y-polarized excitation are clealy different as evidenced by the different periodicities, namely: $T_{x,p}$ L = 10 400 bar·cm and $T_{01,p}$ L = 4700 bar·cm. The polarimetric response of the fundamental mode at 633 nm yielded $T_{01,p}$ L = 487 bar·cm versus 950 bar·cm at 850 nm; in other words the sensitivity decreases with wavelength.

The fact that the two interferometric and the polarimetric responses exhibit such a different sensitivity defines two important advantages of bimodal sensing devices. First, two responses of different periodicities can be used to overcome the problem of ambiguity, making it possible to develop wide-range highly sensitive devices. Secondly, since the response to temperature is also characterized by different sensitivities, the third channel can be used to simultaneously measure temperature. This is an important issue, given that the pressure polarimeter is temperature-dependent as Fig.3 shows.

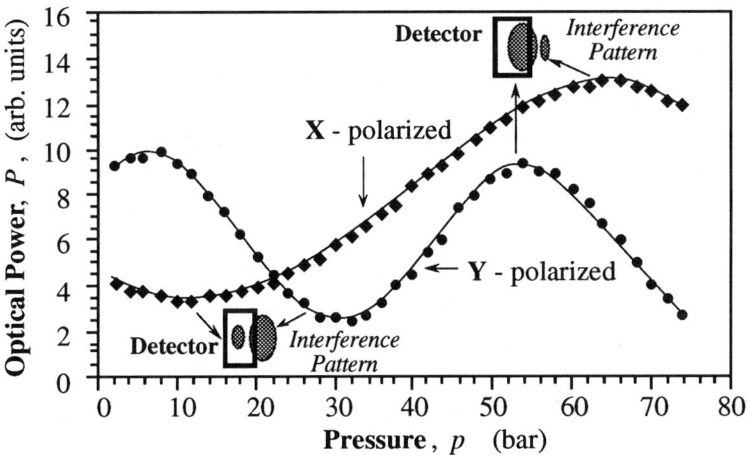

Fig. 4. Polarization-sensitive X and Y intermodal interferometric responses of a 1 m length of HB800 bow-tie fiber when operated in bimodal regime at 633 nm (He-Ne laser).

From the above it is clear that two-mode sensing devices have the potential of both extended range and two-parameter sensing.

We also studied the response to pressure of the e-core and the D-shape Andrew fibers. Here we found the interferometric responses to be closer to each other and twice as sensitive as the polarimetric response of the fundamental mode for which $T_{01,p}$ L = 896 bar.cm (e-core) at 633 nm. However, the inconvenience with e-core and D-shape fibers was the observed nonlinearity of the parameter $T_{i,p}$. This effect is most pronounced for the interferometric responses. Nonlinearities were also observed for temperature changes. From these we can conclude that HB600 and HB 800 bow-tie fibers appear best suited for hydrostatic pressure sensing.

3.2. Fiber-Optic Pressure Cell

The standard technique of stress analysis in large-scale structures with pressure cells involves various configurations of hydraulic measuring systems based on the compensation method, in which stress in the material surrounding the pressure cell is compensated by automatically adjusted pneumatic or hydraulic pressure within the device. This type of system requires installation of great lengths of pressure tubing and creates serious maintenance problems when the number of discrete sensors increases. Dynamic stress measurements are not possible in such a configuration due to the slow response of hydraulic media. In many applications electrical transducers are not acceptable either, since they are susceptible to interference and can be even destroyed by strong magnetic or electric fields. To avoid such problems and to capitalize on the impressive advantages of fiber-optic technology, we proposed to replace the hydraulic measurement system by a direct read-out of the inside of the pressure cell [3] using a polarimetric fiber-optic sensor that we previously developed [2] based on the principle explained earlier in this paper.

The sensor head, shown in Fig.5, is composed of two metal plates which are joined together using a welded sealing to form a thin flat cell filled with a hydrostatic pressure medium.

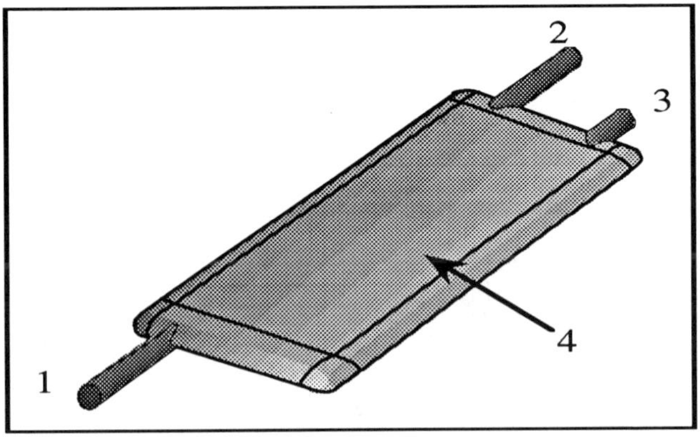

Fig. 5. Construction of the fiber-optic pressure cell: 1,2 - fiber-optic pressure leadthroughs, 3- pressure inlet, 4 - pressure pad.

The dimensions of the flat jack cell are 70 mm x 140 mm x 6 mm, with slight longitudinal deformation in a form of a V-groove to house a polarimetric pressure sensor. The pressure cell is optimized for stress measurements, what involved adaptation of the plate surface/thickness factor to minimize both measurement error and influence of the different moduli of elasticity between the cell and the surrounding material. The whole assembly is then ready to be cast into concrete. The sensor itself is based on the effect of polarization mode coupling which occurs in HiBi polarization maintaining fibers under the influence of hydrostatic stress, as discussed earlier.

The instrumentation system of the all-fiber-optic pressure cell (AFOPC), shown in Fig.6 in a configuration designed to calibrate the sensor against a primary pressure standard, included a semiconductor laser diode LD emitting at 837 nm and directly pigtailed to a York HB 800 fiber (Seastar Model PT-560-830-HX), serving to guide a linearly polarized light parallel to one of the two eigenaxes of the fiber with a polarization extinction ratio of as high as 500/1.

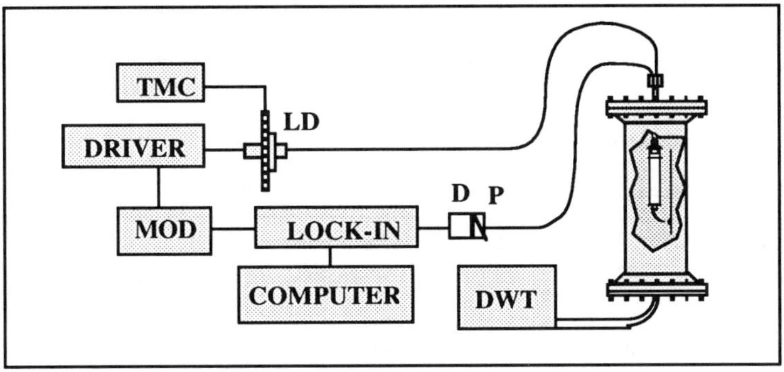

Fig.6. Instrumentation system of AFOPC.

To allow the incoming light to equally excite both polarization modes in the sensing element, the laser pigtail has to be fusion-spliced at 45° to the birefringence axes of the sensing fiber. High accuracy of azimuthal alignment is required in this case in order to avoid instabilities and diminution of device sensitivity, as explained in the preceding paragraph. In this case the system is "all-fiber" and does not require any additional polarizers or phase retarders. A modulated output signal obtained by electronic modulation of the thermoelectrically controlled semiconductor laser was delivered to the silicon photodetector (D). The analyzer (A) was later replaced by a 3M polarizing fiber in the output part of the set-up that allowed analog conversion of the pressure-induced phase shift into intensity-pressure characteristics. An all-fiber manual device analogue to the bulk-optics Soleil-Babinet compensator could also be added to the system in order to allow easy adjustment of the zero pressure point and to permit selection of that part of the sensor characteristics to be actually utilized in conversion of

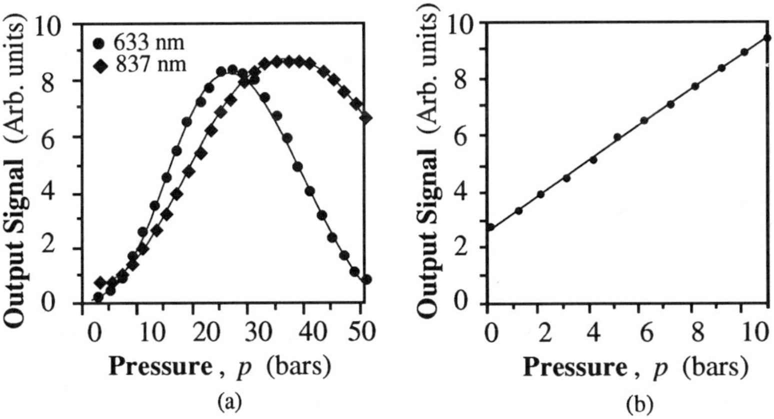

Fig.7. Output signal of the AFOPC.

the measurand. The final stage of the analog signal recovery system was provided by a dual phase lock-in "on a card" directly installed inside an IBM-AT compatible computer driven by specially designed data-acquisition software. For calibration purposes, the AFOPC was placed inside a large pressure chamber filled with a liquid medium and fed directly from a Harwood DWT-35 deadweight tester. The range of pressures generated inside the system was designed as required for typical stress-measurement applications and did not exceed 70 bar, with a reading accuracy of at least 0.1% traceable to NIST.

Shown in Fig.7a are two characteristics of the AFOPC taken at two different wavelengths, 633 nm and 837 nm, while Fig.7b illustrates how the same sensor can be tailored to measure the pressure range of 10 bar by applying an additional external phase shift and using only the middle part of an increasing branch of its sine-like polarization-coupling characteristic. As can be predicted from equation (9), the output signal may also be easily adjusted for less than its half cycle for the maximum range of pressure by choosing a different length of the sensing fiber. This is obviously a much simpler procedure than using sophisticated and costly fringe-counting signal processing systems as is the case with interferometric sensors.

4. Conclusions.

The direct measurement of internal stress in concrete structures has important advantages over the method of stress evaluation based on deformation of concrete monitored with a system of strain gauges. In this paper we have discussed the use of single- or few-mode fiber-optic pressure sensors based on Hi-Bi optical fibers with a view of this particular application. The technique is EMI-free, compatible with optical transmission systems and has the potential of two-parameter sensing; the sensor can be easily designed for a specific range of pressure with maximum sensitivity significantly greater than conventional transducers. In situ tests of a temperature desensitized device are in progress.

Acknowledgements. The support of NSERC, University of Quebec and Glötzl GmbH is acknowledged by the authors.

References

1. Bock, W. J., and Domanski, A.W., "High Hydrostatic Pressure Effects in Highly Birefringent Optical Fibers", J. Lightwave Techn., Vol. 7, pp. 1279 - 1283, 1989.

2. Bock W.J., Wolinski T.R., A.Barwicz, "Development of a polarimetric optical fiber sensor for electronic measurement of high pressure", IEEE Trans.Instrum.Meas., vol. 39, pp. 715-721, 1990.

3. Bock W.J., Voet M.R.H., Beaulieu M., Wolinski T.R., Chen J., "Prototype Fiber-Optic Pressure Cell for Stress Monitoring", IEEE Trans. Instrum. Meas., Vol. 41, 6, 1992 (to appear in December 1992).

4. Eftimov, T.A., "Resultant mode pattern and polarization in a LP_{01},LP_{02} two-mode linearly birefringent optical fibre", Opt. & Quant. Electron., Vol. 23, pp. 1143-1160, 1991.

5. Snyder, A.W., and Love, J.D., "Optical Waveguide Theory ", Ed. London, New York: Chapman and Hall, 1983.

6. Udd, E., Editor, 'Fiber Optic Sensors - an introduction for engineers and scientists", John Wiley & Sons, New York, 1991.

7. Voet, M.R.H., and Barel, A.R.F., "Performances of the Glötzl fibre optic pore water pressure transducer in a line neutral sensing system", Proc. 7th Optical Fiber Sensors Conference, Sydney NSW, Australia, pp. 159-162, 1990.

8. Voet, M.R.H., and Barel, A.R.F., "Line-neutral fibre-optic sensing combining time division and frequency division multiplexing techniques", Proc. ECO3, The Hague, March 1990.

9. Wisniewski, R., and Bock, W., "Universal stress generating device for investigations of transmission properties of optical fibers under mechanical perturbations", Rev. Sci. Instrum., Vol. 62, pp. 2955 - 2958, 1991.

VII - MEASUREMENT OF MAGNETIC FIELD

Fiber Optic Sensing of Magnetic Field Gradients in Near and Far Field

Raymond D. Rempt[1]

Abstract

The detection of magnetic field gradients by optical common mode rejection of the field interferometrically is reviewed. Configurations are presented for detection of far field gradients as well as near field gradients, with performance examples for each. Various applications are presented, with an evaluation of the future of this technology.

Introduction

Common mode rejection techniques are very well applied to the detection of various types of spatial gradients, since they preclude the requirement of making absolute measurements, The field whose gradient is of interest, is common rejected at two points of detection, allowing a direct determination of only the difference of the field values between the points, which is the gradient. This is preferable to the performance of two absolute measurements, since doing so involves subtracting the two measurements, which includes an uncertainty larger than either of the uncertainties of the individual measurements themselves. With the advent of the fiber optic interferometric sensors [T. Giallorenzi, et al, 1982], this common mode rejection may be done directly by optical means. This was first done by Koo and Sigel

---

[1]Principle Engineer, Mechanical Physics, The Boeing Company, P.O. Box 3999, Seattle, WA 98124-2499

[1983], and later by others, [Dube, et al, 1987], [Rempt 1989]. All of these efforts were focussed on detecting the magnetic field gradient in the far field. Rempt [1991 & 1993], first proposed detection of near field magnetic gradients. After describing basic fiber optic interferometry, we will first address the far field case and its development, and then examine the near field case.

Basic Fiber Optic Interferometry

Referring to figure 1, the so called "high performance" fiber optic interferometric sensors are shown in a series of stages. First, the generic Mach-Zehnder interferometer, in which coherent light is injected into a fiber, and split into two optical paths by means of an optical coupler. The coupler allows the light to retain its coherent property, and the two fibers constitute two optical paths, which when recombined at a second optical coupler produce interference fringes. If one of the two fibers changes length more or less than the other, the fringes shift accordingly. The shift in fringes, or phase shift across the interferometer, is a measure of how much the difference in optical path between the two fibers has changed. The output of the interferometer appears as a voltage from the PIN diode detectors at the end of each of the optical fibers in the output coupler. Carefully constructed interferometers of this type are capable of detecting phase shifts of somewhat less than a microradian, hence their categorization as high performance sensors.

If one of the two fibers is coupled to a field of interest,which in our case is a magnetic field, and the other is not, then the interference phase shift at the output of the interferometer is an indication of how much the magnetic field at the first fiber (called the measurement fiber), has changed its length compared to the non influenced fiber (called the reference fiber). In the case of the magnetic field, this coupling is accomplished by means of a magnetostrictive material that strains in the presence of the field. This material imparts a strain to the measurement fiber only, and since the output phase shift of the interferometer is the difference in the optical path length changes of the two fibers, it is then related to the strength of the field experienced by the magnetostrictive material.

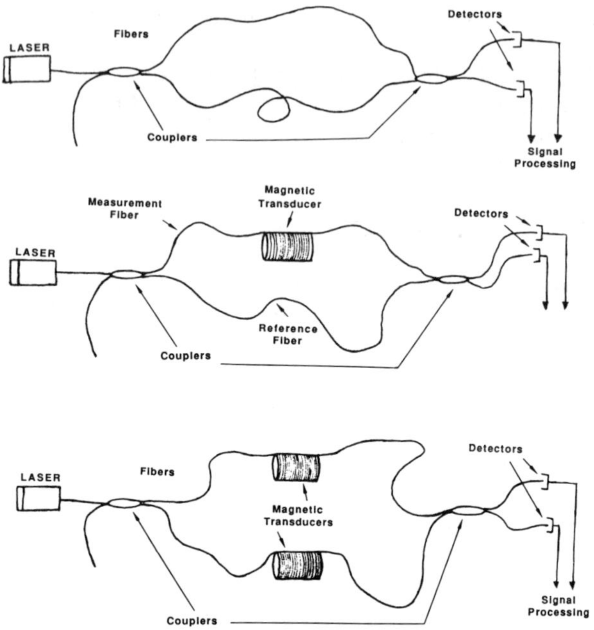

Figure 1 - Mach-Zehuder Interferometer Examples: Upper - Generic; Center - Field Sensor; Lower - Field Gradient Sensor

If both of the fibers are coupled to the field as shown in the third part of the figure, then the output phase shift is an indication of the difference of the two fields across the distance spanning the two of them. This difference is proportional to the magnetic field gradient. Note that the subtraction is automatic, via common mode rejection, and so the gradient is detected directly by optical means, without having to make any absolute measurements.

Far Field Gradiometry

Detecting the field gradient rather than the field is important for two reasons. First, the noise associated with the earth's field is much lower in the gradient than in the field itself, permitting a greater sensitivity. Second, the gradient is less susceptible to orientational changes of the sensor. As most far field magnetic detection schemes want to increase detection range and/or sensitivity as much as possible, the value of the gradient becomes apparent. In addition to being less noisy and orientationally sensitive, the gradient allows far better localization of the "target" being detected, whether it be an underground pipe, an archaeological site, a shipwreck on the ocean floor, or a submarine.

Several investigators have worked in developing this technology, but no off-the-shelf systems currently exist. The first were Koo and Sigel [1983], followed by Dube, et al, [1987], and Rempt [1989]. The first two used planar transducers, and Rempt used cylindrical transducers, permitting much more fiber to be affixed to the magnetostrictive material. With a separation of one foot between the transducers, and transducers approximately 2 inches in diameter and 4 inches long, Rempt [1989] reported minimum detectable gradients of a few milligamma per foot at 1 hertz. An example of a cylindrical transducer gradiometer is shown in figure 2.

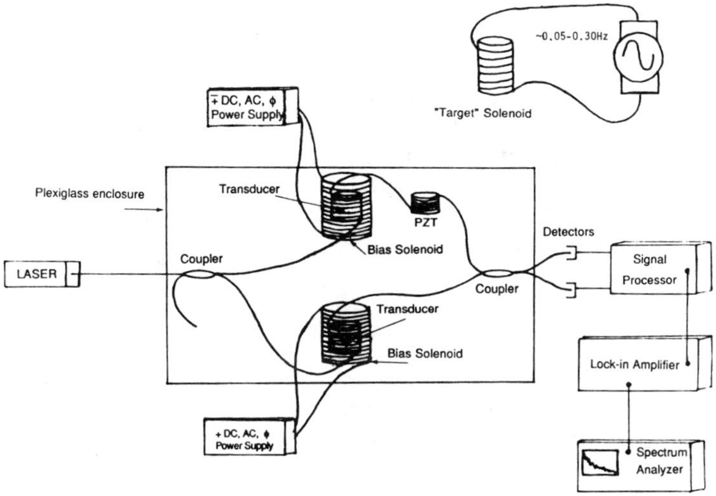

Figure 2 - Cylindrical Transducer Gradiometer

Calibration

Far field gradient sensors may be calibrated by placing a magnetometer (such as a fluxgate) at the position of each of the transducers and observing the signal resulting from a coil placed slightly closer to one of the transducers than to the other. These measurements may be made on a convenient scale of, say, 10's or 100's of gamma and then extrapolated down. The extrapolation is fully valid as long as the scale of the magnetometer remains the same for all readings. This requires that the calibration coil be so oriented that its field at one transducer is approximately 50% larger than its field at the other. The calibration coil can be excited at about a 0.5 hertz rate, so that interfering noise can be disregarded. The difference in the values at the transducers divided by the separation distance can then be directly expressed as a calibrated gradient signal whose magnitued is dependent only on the current through the coil.

Operation

Referring to figure 2, each transducer is immersed in a DC and AC field provided by its bias solenoid. The elongation of the magnetostrictive material, and hence the strain imparted to the fiber in each transducer, is proportional to the square of the field that the magnetostrictive material experiences. The material itself is called METGLAS® which is a trade name registered to the Allied Corporation. The path difference between the two interferometer arms is given by

$$\Delta L = L_1 K_1 H_1^2 - L_2 K_2 H_2^2 \tag{1}$$

Where the K's are material constants and the L's are the lengths of the fiber in each arm which are bonded to the magnetostrictive material. As pointed out by Koo & Sigel, H_1 may be represented as the sum of several fields:

$$H_1 = H_1^{dc} + H_{1E} + H_{ext}^{dc} + H_1^{ac} \tag{2}$$

where H_{1E} is the earth's field at transducer 1, H_1^{dc} the applied dc field from the solenoid, H_1^{ac} the applied ac field from the solenoid, and H_{ext}^{dc} the external field of interest.

At transducer 2, H_2 is as follows:

$$H_2 = H_2^{dc} + H_{2E} + H_{ext}^{dc} + \Delta H_{ext}^{dc} + H_2^{ac} \tag{3}$$

where similar notation applies, with the additional expression of the external field of interest being incremented by ΔH_{ext}^{dc} which represents the gradient.

The applied AC "dither" fields are at a suitably selected frequency which is provided by a single oscillator to both solenoids. Squaring H_1 and H_2, inserting into equation (1), and collecting only terms corresponding to the frequency ω yields:

$$\begin{aligned} \Delta L|\omega &= 2L_1K_1H_1^{ac}\left(H_1^{dc} + H_{1E}\right) - 2L_2K_2H_2^{ac}\left(H_2^{dc} + H_{2E}\right) \\ &+ 2\left(L_1K_1H_1^{ac} - L_2K_2H_2^{AC}\right)H_{ext}^{dc} - 2L_2K_2H_2^{ac}\Delta H_{ext}^{dc} \end{aligned} \tag{4}$$

As pointed out by Koo and Sigel, the following conditions may be set experimentally:

$$\begin{aligned} &H_1^{dc} + H_{1E} = 0, \\ &H_2^{dc} + H_{2E} = 0 \end{aligned} \tag{5}$$

This is done by adjusting the applied DC field in each solenoid to minimize the interferometer output at ω.

Also, the dither amplitudes of one or both of the solenoids may be adjusted such that

$$L_1K_1H_1^{ac} - L_2K_2H_2^{ac} = 0 \tag{6}$$

When conditions (5) and (6) are satisfied, then equation (4) simplifies to

$$\Delta L|\omega = -2L_2 K_2 H_2^{ac} \Delta H_{ext}^{dc} \qquad (7)$$

So the interferometer output at the dither frequency is proportional to the gradient when nulling conditions (5) and balance condition (6) are established.

Balance is demonstrated by bringing a bar magnet past the sensor with its moment parallel to the transducer axes of sensitivity, and the path parallel to the line joining the two transducers. When balanced, the sensor response should be a positive and a negative bump as shown in figure 3. As described above, a remote "target" coil may then be turned on at a very low frequency of a few tenths of a hertz to a few hertz, as shown on the right side of figure 3. The output spectrum of the interferometer with the calibrated signal of 1.2 gamma per foot pk-pk is shown in figure 4. The noise level is a few milligamma per foot at 1 hertz. These data are typical for a state-of-the-art gradiometer.

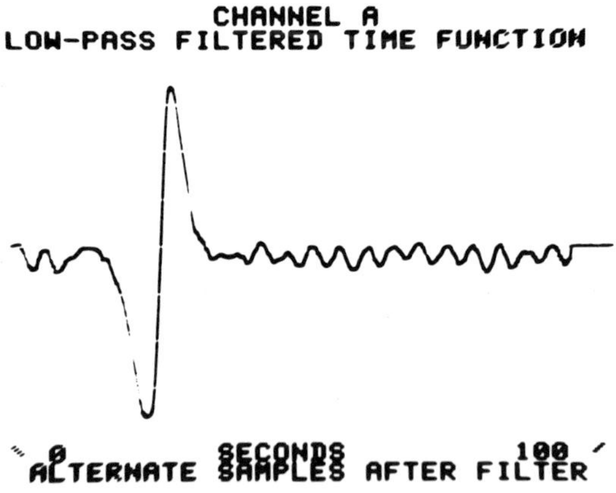

Figure 3 - Far Field Gradiometer Output for Bar Magnet Followed by a 1.2 γ/ft pk-pk Signal

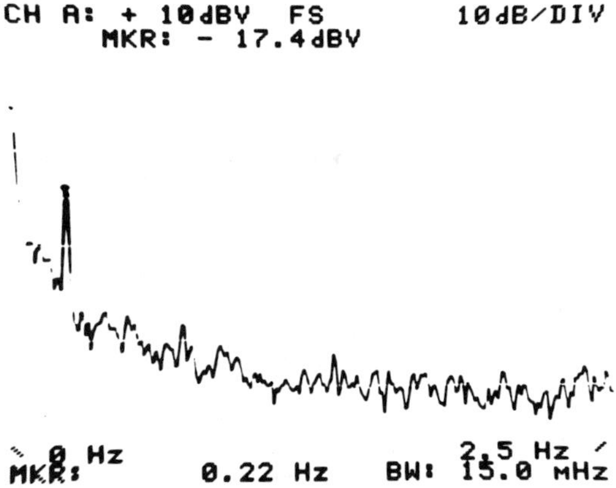

Figure 4 - Output Spectrum of 1.2 γ/ft pk-pk Signal

Near Field Gradiometry

When the target being detected is very near to the sensor, as in corrosion currents at or just under the surface of an airplane skin, or biomagnetic signals, near field gradiometry is applicable as an analysis tool. For these types of applications, it has been customary to use the sensor as a gradiometer operating in a magnetometer mode. That is to say, the gradiometer is balanced to reject noise and other ambient magnetic signals, but only one of the two transducers is near enough to the signal source or target to detect it. Thus, that transducer is acting basically as a magnetometer, while the sensor as a whole enjoys the common mode rejection of field that are spatially present over distances of the order of the sensor dimensions or larger. For the fiber optic sensors, since each of the magnetostrictive transducers is made of materials of relatively high permeability, it is possible that one of the transducers will shield the other from the signal of interest in certain configurations.

These considerations suggest the configuration of the sensor with both of the transducers as near to the signal source as possible. The field components parallel with the surface of the test object are thus those that are sensed, and the differentiation coordinates are also restricted to lie in the plane of the test object, since the transducers cannot be stacked one on top of the other.

In order to get the transducers as close as possible to the test article surface, they must be constructed very thin. Planar METGLAS® coupons are affixed to many passes of fiber and immersed in the field of a very thin bias solenoid. With care, it is possible to construct these transducers so that the METGLAS® is less than 0.030 inch from the surface being tested.

Two possible configurations of the near field gradiometer exist. These are the "side-by-side" and the "end-to-end" configurations. The first permits detection of the gradient of the field component along one horizontal direction with respect to the other horizontal coordinate, as shown in figure 5. With a fluxgate magnetometer, we calibrated the gradient due to a current flowing through a "pennant" shaped foil which was confirmed to be linear with the axial coordinate near the central portion of the foil. With a signal of 0.15 gamma/cm rms, the noise level was about 17 dB down as shown in figure 6A, or about 21 milligamma/cm rms at 7 Hz. The peak at about 4.8 Hz is due to an undamped mechanical vibration.

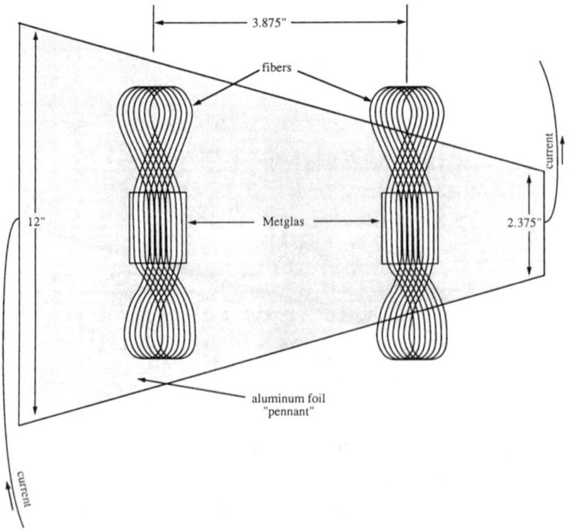

Figure 5 - Calibration Setup for Side-By-Side Near Field Gradiometer

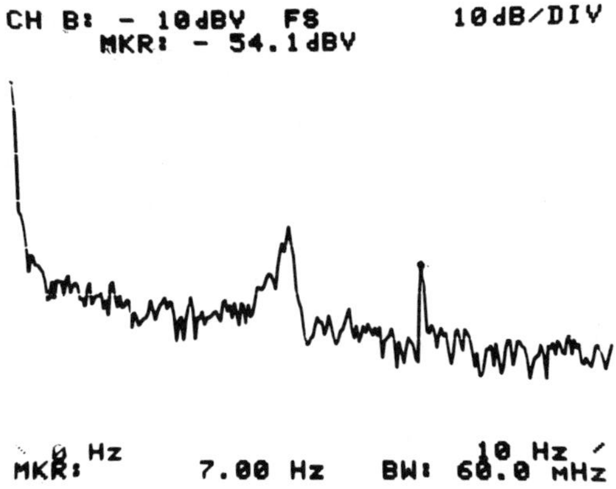

Figure 6A - Output Spectrum of Side-By-Side Gradiometer with 0.15 γ/cm rms Signal

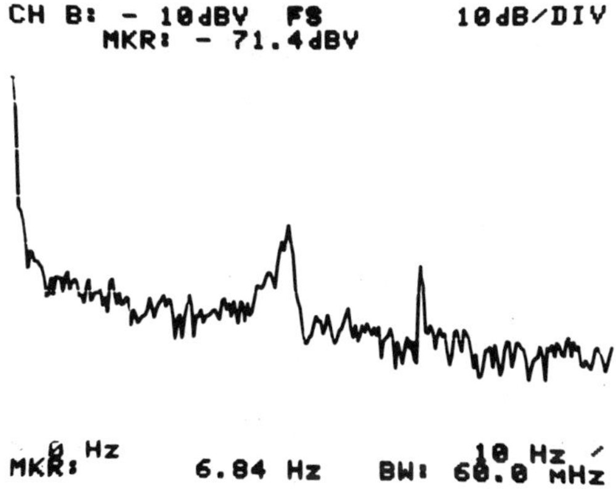

Figure 6B - Noise Level Estimation for 0.15 γ/cm rms Signal on Side-By-Side Gradiometer

We also configured the transducers "end-to-end" such that detection of the gradient of the horizontal component of the field with respect to the same coordinate was permitted, as shown in figure 7. The wire interconnecting the two foil sheets was located a long distance from them. The calibration was again done with a fluxgate magnetometer. The noise spectrum from a 0.13 gamma/cm rms signal at 7 Hz is shown in figure 8. The noise level is 24.8 dB down, or about 7 milligamma/cm rms at 7 Hz. The peak at about 4.8 Hz is again due to a mechanical vibration.

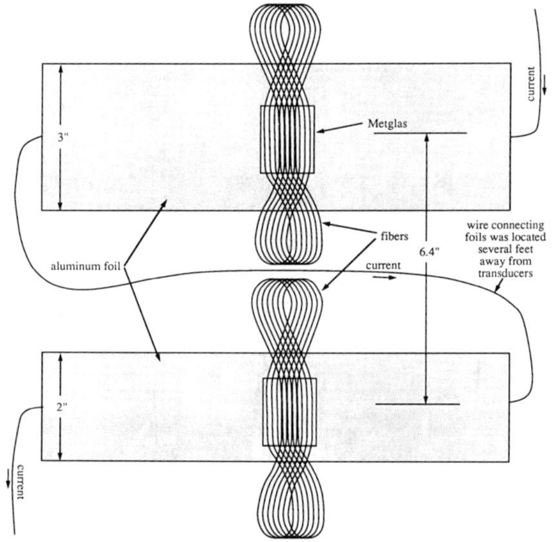

Figure 7 - Calibration Setup for End-to-End Near Field Gradiometer

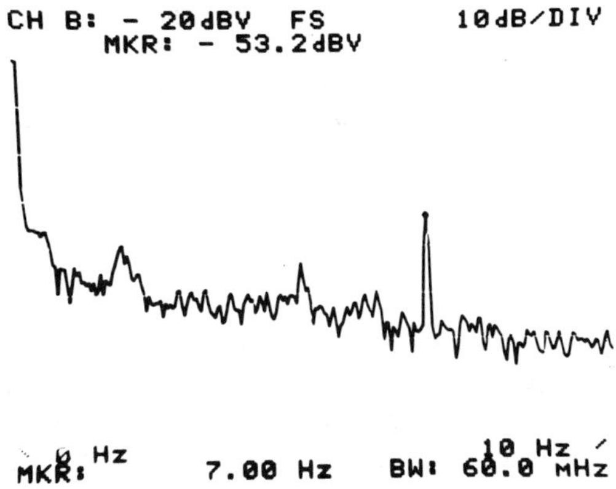

Figure 8A - Output spectrum for End-to-End Gradiometer with 0.13 γ/cm rms Signal

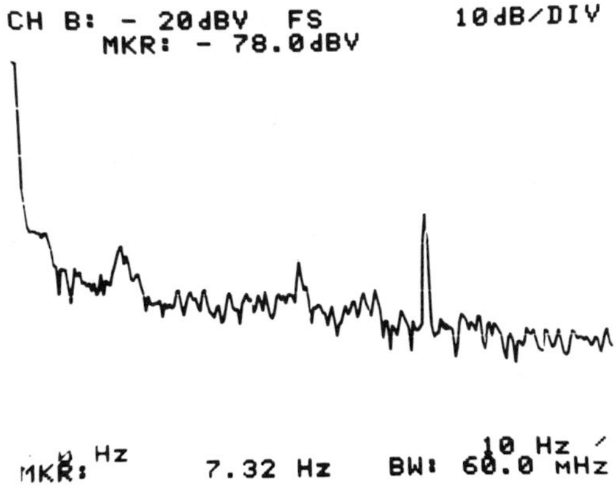

Figure 8B - Noise Level Estimation For 0.13 γ/cm rms Signal on End-to-End Gradiometer

These near field applications show some promise for inspection of articles for sub surface corrosion, as well as for circuit board inspection. In addition, there is potential for use in mapping of bio magnetic signals. The cardiac magnetic signal has been successfully detected with a near field fiber optic gradiometer [Rempt 1993].

Conclusions

Magnetic field common mode rejection by optical interferometry has been demonstrated to be viable for detection of small magnetic field gradients. In the far field, as the market develops, there may be some niches for petroleum exploration, archeological site location, or military applications. For the near field, there are a number of applications that presently are attractive for this technology, including circuit board inspection, and bio magnetic imaging.

References

C. M. Dube, S. Thordarson, H. K. Wanser, Proc. SPIE 883 p 17, 1987

T. G. Giallorenzi, et al, IEEE JQE Vol. QE-18 No. 4, p 626, 1982

K. P. Koo, G. H. Sigel, Jr., J. Lightwave Tech. Vol. LT-17 p 509, 1983

R. D. Rempt, Photonics Tech. Lett. Vol. No. 5, p 105, 1989

R. D. Rempt, Review of Progreess in Quantitative NDE No. 11A, p 1175, 1992

R. D. Rempt, Photonics Tech Lett. Vol. 5 to appear in Jan. 1993

VIII - LOCALIZED DETECTION AND SMART STRUCTURES

Spatially Multiplexed Optical Fiber Time Domain Sensors for Civil Engineering Applications

Bernd D. Zimmermann[1], Richard O. Claus[2]

Abstract

We report on the performance of an Optical Fiber Time Domain (OFTD) strain monitoring system for civil engineering applications. The feasibility of such a system has been demonstrated with multiplexed fiber optic sensor strings embedded in composite anchoring tendons. As many as 10 in-line segments are monitored using a commercially available high-resolution Optical Time Domain Reflectometer (OTDR). Hardware and software configurations for the system used in these trials are presented.

Introduction

Optical fiber sensors have been developed for the instrumentation of advanced systems and structures (Knowles, 1991). However, most applications have been confined to the military sector, and have only recently been identified in the commercial sector (Claus, 1991). These applications include the use of optical fiber sensors for civil engineering structures. Various industrial and academic organizations have started promoting this rapidly growing technology for the implementation in bridges, buildings, pipelines, dams, anchoring systems, etc. (Hecht, 1992). The fiber optic (FO)-based monitoring systems measure (and control) parameters such as strain, temperature, pressure, displacement, vibration, and chemical processes. These measurements are performed using intensity-, phase-, time-, spectrum-, and polarization-based methods to convert unprocessed sensor data into the measurands of interest (Culshaw, 1984).

Optical fiber time domain (OFTD) methods are particularly well suited for systems which require interrogation of a large number of locations. Multiplexed sensor networks are accomplished with serial and parallel fiber arrays using in-line optical reflectors and fiber optic splitters (Zimmermann, 1992, 1991, 1990), (Yaniv, 1993). The first commercial application of the OFTD strain sensor has been implemented with anchoring systems. Sensor strings are embedded in composite tendons which are used to anchor natural or man-made rock formations.

[1] President, FIMOD Corporation, P.O. Box 11192, Blacksburg, VA 24062
[2] Director, Fiber & Electro-Optics Research Center, VPI&SU, Blacksburg, VA 24061

The requirements of these applications are such that conventional strain gages cannot be used. Unlike strain gages, OFTD sensors are immune to electro-magnetic interference (EMI), resistant to corrosion, easily embeddable in composite rods, compatible with in-line sensor multiplexing formats, and inherently redundant for critical applications. The performance of this system is demonstrated with a commercially available high-resolution Optical Time Domain Reflectometer (OTDR) and customized software.

Background

The OFTD sensor approach utilizes an Opto-Electronics Inc. MF20/TDR30 picosecond-resolution OTDR system (see Figure 1). This system consists of a signal processor (1), a power supply (2), a signal delay generator (3), a sampling unit (4), a photo-detector (5), and a laser (6). The individual system components, as well as the FO sensor with its in-line optical reflectors are shown in Figure 2. The principle of operation is described next.

The laser (6) emits a very narrow (< 200 ps pulse-width), highly stable optical pulse at a wavelength of approximately 850 nm and a repetition rate of 30 kHz. The output of the laser (6) is routed to a FO splitter through a FO cable. The FO splitter has two output ports. One port is used to generate a reference return signal, the other is connected to the FO sensor. The sensor consists of an optical fiber which is physically segmented with partially reflective markers ("reflectors") such that integral strain between any two adjacent reflectors can be monitored. This is accomplished by measuring the relative time delay of the respective reflections. These signals return to the OFTD monitor through the FO splitter to the photo-detector (5). The optical signals are converted into electrical signals and amplified. The amplified electrical signals are then sampled using the built-in sampling module, averaged using the signal processor (1), transferred via a GPIB data bus to a Personal Computer (PC), converted into strain data, and finally displayed on the PC-screen. An oscilloscope is also used to display the sampled signals and to easily locate all sensor reflections.

Theory

The reflections of the FO sensor shown in Figure 2 can be used to determine the absolute length of each of the sensor segments by considering the time delay, Δt_i and Δt_{i+1} of any two adjacent reflectors. The change in length Δl_i of the segment confined by these two reflectors is given by:

$$\Delta l_i = (\Delta t_{i+1} - \Delta t_i)\,\mathbf{K} \quad , \tag{1}$$

where $\mathbf{K}$ is a calibration factor in units of μm/ps that can be expressed as:

$$\mathbf{K} = \frac{c}{2n}\left(\frac{1}{1+\mathbf{a}}\right) . \tag{2}$$

In equation (2) above, **a** denotes the fiber photo-elastic coefficient, c is the speed of light in vacuum (3.0E08 m/s), and n is the group refractive index of the optical fiber (~ 1.46). The calibration factor **K** is measured experimentally for the fiber type that is used. A 100/140 μm core/cladding outer diameter fiber, coated with a thin polyimide jacket to 165 μm, was shown to be a suitable sensor fiber candidate. This fiber was tested to have a value for **K** of approximately 130 μm/ps, which in terms corresponds to a +/- 260 μm measurement resolution for Δl_i (this value is based on a 1 ps resolution OTDR system such as the MF20/TDR30). The number of in-line segments which can be monitored is limited theoretically only by the excess loss of each of the reflectors. The minimum spacing between reflectors is determined by the width of the optical pulse of the OTDR system. Although 10 cm reflector spacings have been used, 1.0 m spacings are more typical.

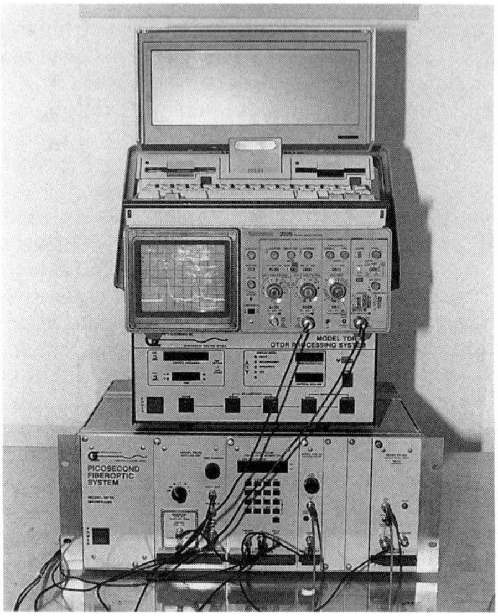

Figure 1 - Photograph of the OFTD Monitoring System.

Experiments

A sensor system (hardware and software) for anchoring systems was developed. Although this system was very application-specific, it is envisioned that implementation of this system (or a very similar version) in other applications is possible. The system consisted of the monitoring hardware shown previously in Figures 1 and 2.

The 100/140/165 μm polyimide coated fiber was segmented using 10 in-line reflectors. The segments varied in length from 0.5 to > 5.0 m. The typical sensor overall length was approximately 20.0 m with access fibers typically no longer than 80.0 m. For redundancy purposes, both ends of the sensor were accessible in case of accidental damage. Since operation from either end is conceivable, each sensor string

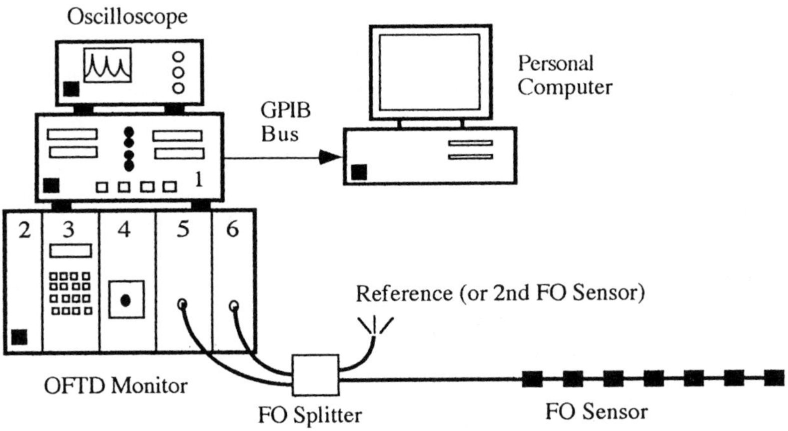

Figure 2 - Schematic of the OFTD Monitoring System Components.

was characterized in the forward (FWD) and backward (BWD) measurement mode. After measuring the characteristic Δt-versus-Δl behavior for the 100/140/165 μm fiber (see Figure 3), each sensor was tested for reflected power as a function of reflector number in the FWD and BWD measurement mode (see Figure 4). The average excess loss per reflector, calculated from an exponential curve fit, ranges from 0.40 to 0.60 dB. The correlation factor (R^2) between the the exponential curve fit and the measured data was typically > 0.96. Preliminary specifications for the segmented OFTD sensor are summarized in Table 1.

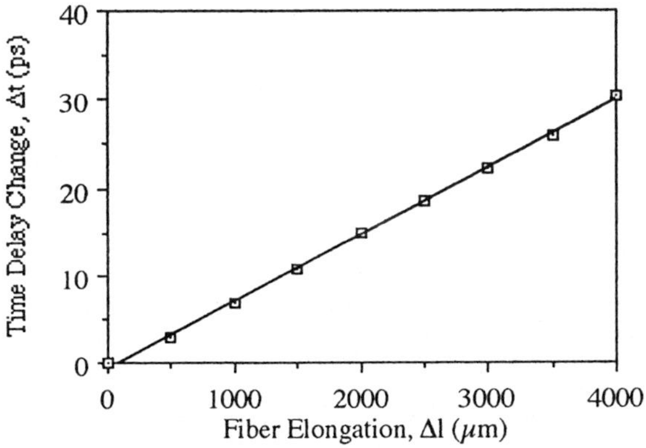

Figure 3 - OFTD Sensor Characterization Results: Δt vs Δl.

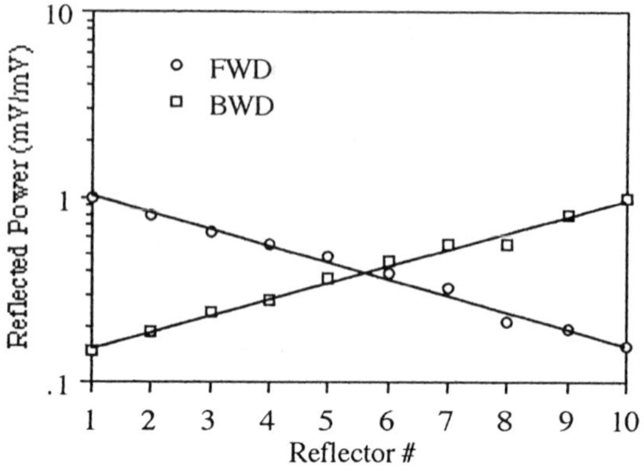

Figure 4 - OFTD Sensor Test Results: Reflected Power vs Reflector #.

Fiber Type	100/140/165 μm, Polyimide Coated
Resolution	+/- 260 μm
Min. Reflector Spacing	10cm
Reflector Dimensions	1.0 mm Dia. x 1.0 cm Length
Reflector Tensile Strength	> 10 N (long term)
Reflector Excess Loss	0.40 - 0.60 dB
Reflectivity	7.0 % per Reflector
Connectors	Optional (ST, SMA, FC)

Table 1 - Preliminary OFTD Sensor Specifications.

A software package (OTDR Rev. 1.1) written in Microsoft Quick-C was developed to interface the OTDR monitoring system with an IBM-compatible computer. The software is based on National Instruments' standard General Purpose Interface Bus (GPIB) protocol. The software functions, starting with the input of all sensor parameters under no-load (zero-strain) conditions and ending with the output/display of strain data under load (in-service) conditions, are shown in the software flowchart (see Figure 5). In addition to the real-time display of all strain and elongation values, data can also be stored in ASCII format on the hard-disk of the computer for off-line signal analysis. Despite the simplicity of the menu-driven format of the software, an operating manual with fully detailed instructions on each of the software functions is delivered to the customer.

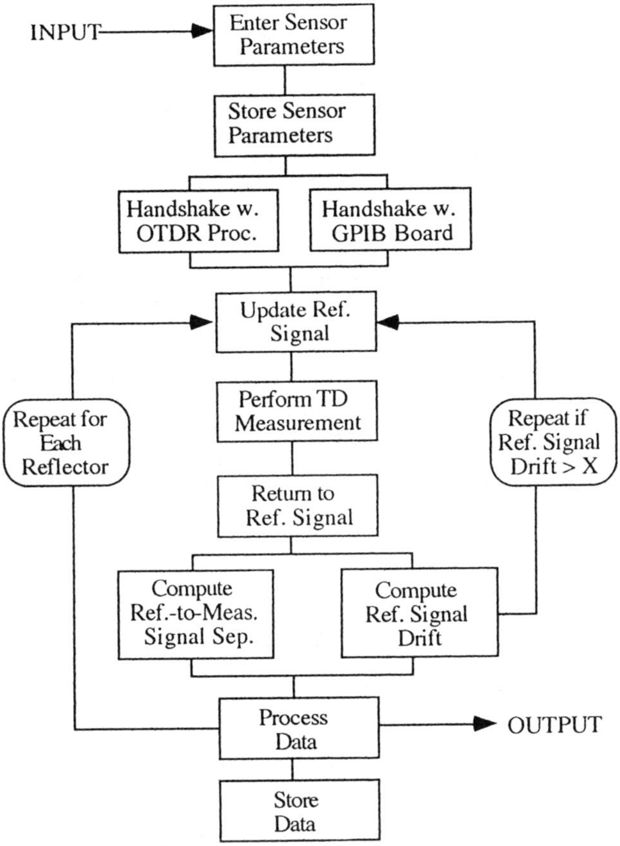

Figure 5 - Flowchart of the OFTD Sensor Software.

Applications

The segmented OFTD strain sensor is well suited for large, semi-static structural applications. These include strain and fatigue monitoring on and within bridges, dams, pipelines, buildings, and aircraft. A number of field trials have been conducted to demonstrate the capabilities of the system. We have reported on the performance of both surface-attached sensors tested on the wings of an F-15 fighter aircraft, as well as embedded sensors monitored within advanced thermoset graphite/epoxy composite laminates.

More recent efforts focus on the application of the OFTD sensor for anchoring systems. These systems typically consist of a number of tensioning cables (such as steel or composite tendons) which are anchored through a grouting material to stabilize and prevent movement of large man-made and natural structures (e.g., concrete walls, dams, and rock formations). In one particular case, tests are being conducted to determine the feasibility of the OFTD sensor system to a.) monitor pretensioning loads during installation and b.) measure long-term elongation of tendons embedded within hydroelectric dams. The typical implementation of a concrete wall anchoring system is shown in Figure 6. The segmented OFTD sensor is embedded within a composite tendon which is in direct contact with the "structure". The elongation and strain in any segment along the length of the tendon from the concrete wall to the grouting area can be determined The monitoring system is inherently redundant since measurements can be made from either end of the FO sensor. In case of damage to the sensor anywhere along its length (due to excesive loading, for example), information in the segments beyond the point of damage can still be performed by accessing its opposite end. This feature is extremely important for many critical structural applications.

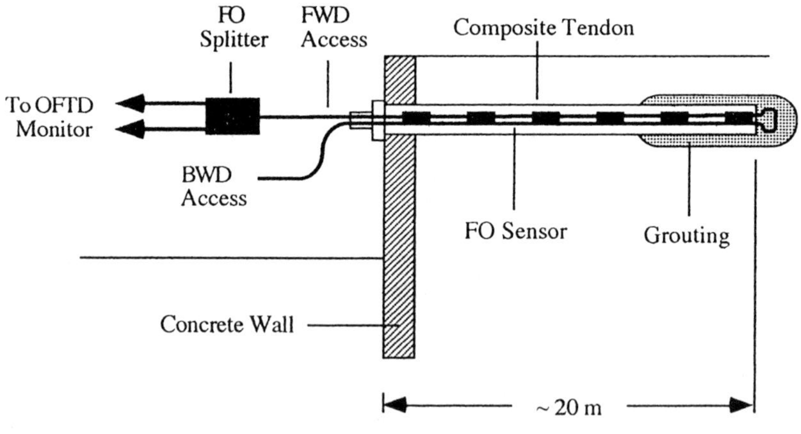

Figure 6 - OFTD Sensor for Concrete Wall Anchoring Systems.

Summary

The feasibility of an Optical Fiber Time Domain (OFTD) strain sensor system for commercial applications has been demonstrated. The system allows monitoring of strain and elongation within large, semi-static structures. Hardware and software components have been integrated to result in a stand-alone, easy-to-use, computer-driven system. More recent trials in the field of civil engineering have been conducted with anchoring systems for natural and man-made structures. These include concrete walls, rock formations, and hydroelectric dams. Post-installation results for these trials are anticipated by the middle of 1993. Additional applications to monitor the dynamic behavior of bridges, pipelines, and aircraft are envisioned.

References

Claus, R.O., 1991, "Optical Fiber Sensor-Based Smart Materials and Structures," Conf. Proceedings, Blacksburg, VA April 3-4, 1991.

Culshaw, B., 1984, Optical Fibre Sensing and Signal Processing, Peter Peregrinus Ltd., London, UK, pp. 62-72.

Hecht, J., 1992, "Fiber Sensors Built into Univ. of Vermont Structure," from Lasers & Optronics, p. 10, Nov. 1991.

Knowles, G.J., "Active Materials and Adaptive Structures," Proc. of the ADPA/AIAA/ASME/SPIE Conf. on Active Materials and Adaptive Structures, Alexandria, VA, Nov. 4-8, 1991.

Yaniv, G., Zimmermann. B.D., Lou, K-A, 1993, " Development of an Optical Fiber Time Domain Sensor for Monitoring Static and Fatigue Strains in Composite Laminates," to be presented at the 1993 SPIE North American Conference on Smart Structures and Materials, Albuquerque, NM, Jan. 31-Feb. 4, 1993.

Zimmermann, B.D., Lessing, R., 1992, "Optical Fiber Time Domain Strain Sensors - An Integrated System for Commercial Applications," Proc. of 1992 Conf. of Fiber Optic Sensor-Based Smart Materials and Structures, Blacksburg, VA, April 15-16, 1992.

Zimmermann, B.D, Claus, R.O, 1991, "Multi-Segment OTDR Based Fiber Strain Sensors for Dynamic Structures," Proc. of 1991 SPIE Conf. on Distributed and Multiplexed Fiber Optic Sensors, Boston, MA, Sep. 4-5, 1991.

Zimmermann, B.D., Claus, R.O., Kapp, D.A., Murphy, K.A., 1990, "Fiber Optic Sensors Using High Resolution Optical Time Domain Instrumentation Systems," J. of Lightwave Tech., Sep. 1990.

Shape Memory Alloy Actuators Energized Optically Through Optical Fibers for Smart Civil Structures

Dr. Barry G. Grossman[1] and Jorge E. Franke[2]

Abstract

One type of actuator that is capable of producing large displacements and large forces for use in civil engineering applications is the shape memory alloy (SMA) actuator. We have demonstrated that these actuators can be effectively energized using optical signals transmitted through optical fibers. This would allow actuators to be embedded into structures along with optical fibers, and quite possibly be multiplexed onto a single optical fiber along with a number of fiberoptic sensors.

In this paper we report on our initial investigations that were performed to determine the feasibility of the optical activation technique and provide additional insight into the properties of shape memory alloys. We demonstrate that the strain produced is repeatable and shows little sign of fatigue after hundreds of cycles. We attached SMA actuators, much like muscles, on both sides of a cantilever beam and demonstrated that we could deflect the beam in both directions as well as reduce the thermal relaxation time. We show that the rise and fall times for activation are dominated by thermal effects.

Introduction

Research on smart aerospace structures has been ongoing for over ten years (Udd, 1988) . In the past few years, there has been a growing interest in applying the technologies developed for aerospace applications to civil engineering structures such as highways, bridges, buildings, etc. (Robinson, 1992). These applications impact our everyday life and someday may allow mechanical health monitoring of these and

[1]Associate Professor of Electrical Engineering, Florida Institute of Technology, 150 University; Blvd., Melbourne, FL 32901
[2]Formerly at Florida Institute of Technology, currently a Research Engineer at INTELLIKEY Corp., Melbourne, FL 32901

other structures and possibly save lives. Smart structure research has centered, in great part on fiberoptic sensor technology because optical fibers can be readily embedded into composite materials, including concrete and lend themselves to remote readout of strain, temperature and other measurands with possibly dozens of sensors on a single optical fiber kilometers in length. Along with fiberoptic sensors, various types of actuators have also been developed. Figure 1 shows the concept of a smart structure.

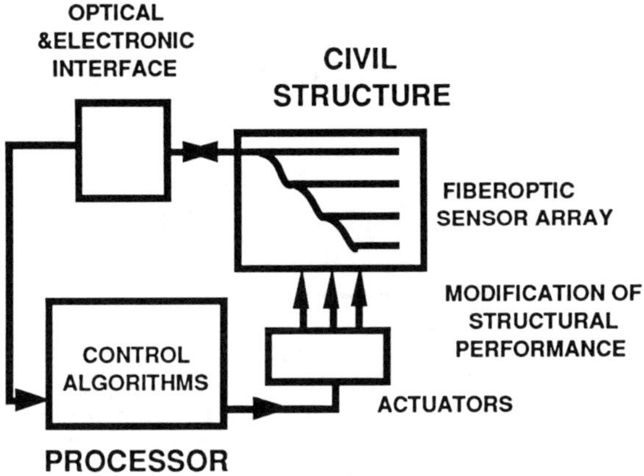

Figure 1. Smart structure concept

During construction, sensors and possibly actuators can be embedded into the structure. The sensor optical signal outputs are first converted into electrical signals and then processed to extract the strain, pressure, or other measurand information. These inputs are fed to a processor which can be a conventional digital computer or an artificial neural network(Grossman, 1989,1993). The processor output can be either an indication of the mechanical heath of the structure, such as the presence and location of cracks, vibration, or other faults, or generate control signals to embedded actuators which then could be used to modify the structural characteristics to prevent or counteract damage or damp out vibrations.

Typically, arrays of both sensors and actuators will be required to monitor and control large structures. This will require sensors and actuators that can be easily multiplexed, perhaps onto one or a few fibers. This would minimize the installation task during construction. Although many types of fiberoptic sensors have long been able to be multiplexed, electrically energized actuators are more difficult to multiplex. Unlike most actuators, shape memory alloy actuators (SMA) are activated through a temperature induced transition in crystal structure. Researchers have typically produced this temperature change through ohmic losses in the actuators (Baz 1989; Schetky 1984,) when large currents are passed through them. This loss of electrical energy is converted to thermal energy which then raises the temperature, resulting in actuation.

Electrical techniques, however, can be inefficient and difficult to multiplex over long distances. We have demonstrated a technique that uses absorption of optical energy to change the actuator state. This is shown in Figure 2. Light from a semiconductor laser diode is coupled into an optical fiber that is embedded into concrete or other material

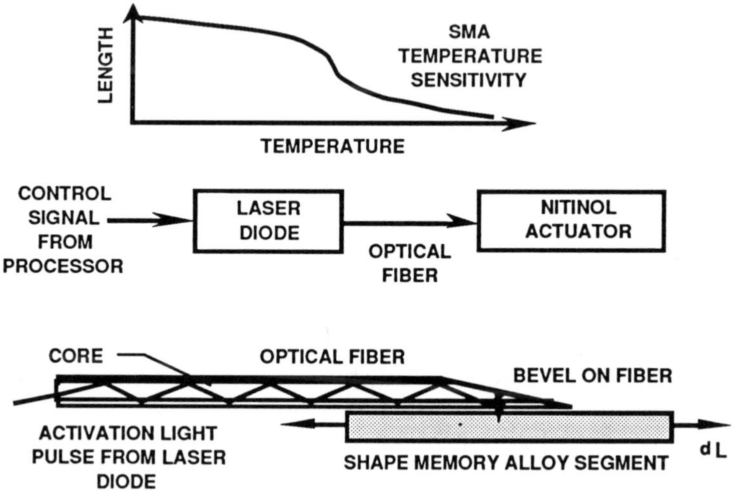

Figure 2. SMA Actuators energized optically through embedded optical fibers

along with the SMA actuators. At the actuator, the light is absorbed by the SMA which is heated past its transition temperature. After an initial electrical to optical conversion loss in the laser, the system is very efficient, with little loss in the optical fiber and a high thermal conversion efficiency when used with a highly absorptive optical coating on the SMA. Thus, the actuator can be kilometers away from the laser source and existing wavelength multiplexing techniques can be used to selectively energize specific actuators.

The remainder of the paper briefly discusses the theory of operation of shape memory alloy actuators and presents results and conclusions for preliminary experiments we have performed.

Theory

The crystal structure of a shape memory alloy (SMA) depends on the composition of the alloy, the temperature, and the internal stress. For our research, a SMA composed of nickel and titanium (Ni-Ti) was used. The composition of these two alloys consists of 50 atomic percent of nickel and 50 atomic percent of titanium. The austenite or parent phase crystal structure of the SMA corresponds to the high temperature phase. This phase has the characteristic of having a face centered cubic structure in which the atoms are arranged to form a cubic lattice with one atom in the center of each face. As the martensite phase is induced by lowering the temperature, the crystal structure transforms into a body centered cubic structure. Martensite plates are formed by groups of atoms having the same stacking sequence(Vernon, 1975). This sequence of atoms can be modified by dislocations of atoms, defects in the material, or slip, thus generating a new stacking sequence. Thermoelastic behavior has been described as the formation or growth of martensitic plates as temperature is gradually decreased below M_s, the temperature at which martensite starts to form. Existing martensite plates will reduce in size as the temperature is increased.

Pseudoelasticity is a "rubber like" behavior that can be observed under certain conditions when an applied stress is removed. This phenomenon involves an elastic and reversible process which results from growth of a martensitic particle. The term "pseudo" is used since strain and stress are not linear. Shape memory alloys have the characteristic of being able to achieve martensite transformation in two ways: thermally and stress induced. Austenite to martensite transformations can be observed to occur thermally due to thermoelastic mechanisms. On the other hand, a martensite particle can be forced to grow or nucleate by an applied stress. The latter is considered to be stress induced martensite. The characteristic of pseudoelasticity in a SMA is in part attributed to the growth or shrinkage of martensite particles as external tensile stress is applied or removed, respectively (Delaey, 1974).

Another important characteristic is that the amount of martensite growth is not only dependent on the magnitude of the applied stress, but is also dependent on the temperature at which the stress is applied. Consider an alloy that is subjected to different stress values while strain is measured at a constant temperature, T_1. First let T_1 be greater than A_f, the temperature at which the austenite phase is completed. For this specific temperature range ($T_1 > A_f$), the stress-strain curve has the pseudoelastic characteristic(Wasilewski 1975) for a loading and unloading of stress as shown in figure 3.

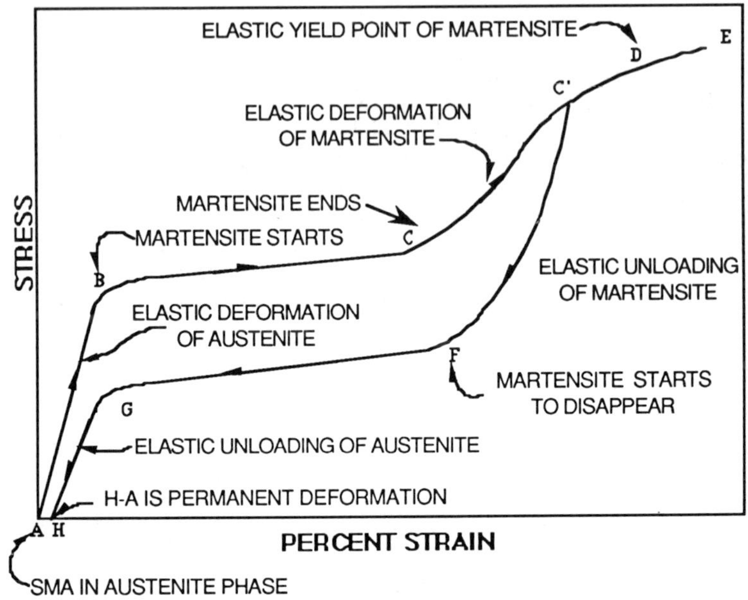

Figure 3. Typical stress-strain relationship of SMA materials

In figure 3, the segment of the curve AB represents the effect of an elastic deformation of the austenite phase due to the increase of applied stress. Point B is the point where the formation of the first stress induced martensite plate starts to form. The martensitic transformation is complete when point C is reached. During the segment BC, the slope of the segment represents the ease of transformation. Furthermore, this segment of the stress-strain curve is where the maximum variation of strain is observed with relatively small increases in stress. Stressing beyond point C, the martensite plates are being elastically deformed up to point D. Point D indicates the plastic yield point of the martensite. If the material is subjected to an increase in stress beyond point D, fracture may occur. Assume that the elastic yield point is never reached. Instead, point C' is reached to further ensure that the material will recover from its martensitic plastic deformation. If the stress is released at point C', the strain is recovered in different stages. In the unloading process, segment C'F shows the elastic unloading of the martensite plates. At point F the reverse transformation starts, ie., martensite plates start to shrink. The reverse transformation is completed and the austenite phase is reached at point G. From point G to H the unloading of the elastic deformation of the austenite phase is observed. The process of loading and unloading a SMA causes a permanent deformation. This deformation is illustrated by the difference in strain corresponding to points A and H.

Another temperature range that is of interest is around M_f, the temperature at which the martensite transformation is completed. If stress is applied at a temperature below or equal to M_f, the already existing martensite particles will no longer suffer any change in growth. Therefore, the pseudoelastic effect will not be observed. At this same temperature upon release of the stress, no recovered strain due to the pseudoelastic effect will be observed. The reason is that once the martensite particle is fully grown, it is not possible to reverse its transformation when the specimen is kept at a temperature where thermoelastic reversion is not possible. However, some recovery strain may be observed which is attributed to the recovery of the plastic deformation of the martensite plates.

The upper part of Figure 3 (points A to D) represents the temperature range below A_f, the range of temperature where the shape memory effect is observed. Shape memory effect can be observed by deforming the material at one temperature; when it is heated to a higher temperature, it tends to recover its original shape. The shape memory effect and pseudoelasticity are related because the non-recoverable strain after loading and unloading is a function of testing temperature. The strain that is not recovered by the pseudoelastic effect will mostly be recovered by heating. In figure 3 the segment AH is the non-recoverable strain. This strain can not be recovered by further heating the material.

Figure 4 shows the relation between pseudoelasticity and shape memory effect. The upper curve in figure 4 represents the deformation stage as stress is applied at a testing temperature less than A_s (but not lower than M_f). The upper part of the graph illustrates the pseudoelastic effect(Delaey 1974). Segment AB is the elastic deformation of the austenite phase; point B is where the formation of the first martensite plate occurs. Point C is where martensitic formation is completed and segment CC' is where Martensite plates are being elastically deformed. Point C' is below the elastic yield point of the martensite phase, and point D corresponds to the elastic strain recovery. The non-recoverable strain at point D is much bigger than that in figure 3.

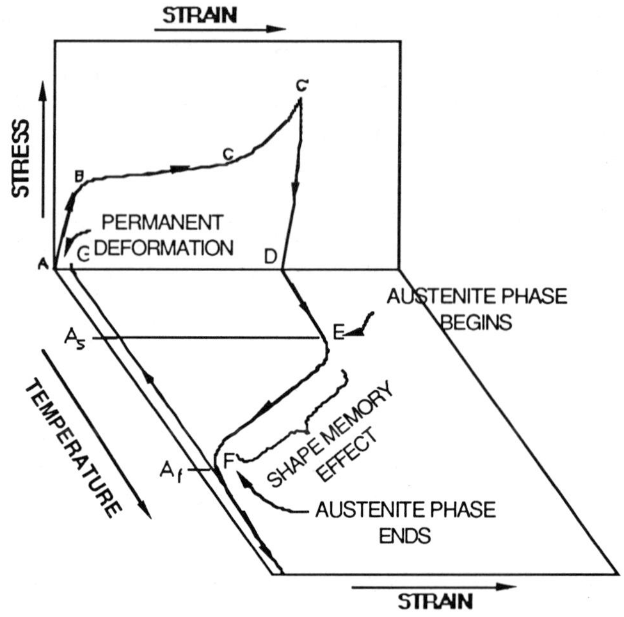

Figure 4. Pseudoelasticity and shape memory effect

The lower part of the figure 4 represents the shape memory strain recovery as a function of temperature. As temperature is increased from the testing temperature to A_s, no change occurs in strain. The segment from points E to F corresponds to the increase in temperature from A_s to A_f where the martensite phase progressively reverts to austenite. The non recoverable strain is the segment A-G. If the temperature is now decreased below A_s, the path followed is shown by segment from point F to G. Stressing the material will yield to elastic deformation of the parent phase as shown before (segment AB). The path difference between the first and consecutive cycles is caused mainly by the initial non-recoverable strain (segment AG in figure 4). Therefore, no significant change in path should exist after the first cycle is completed.

Experimental Investigation

After training the SMA samples, the first investigation performed was to determine the repeatability of the SMA when it is used as a mechanical actuator. A shape memory alloy wire of 0.012in. (12mils) in diameter was used. Two stress levels were used in this experiment to study the change in performance as stress was varied. The variation of length of a trained sample, when the temperature was varied from room temperature to a temperature higher than the transition temperature, T_c, was measured. The variation in length .was expressed in percent strain which was measured as a function of number of cycles. Each cycle represents the formation of austenite and then reversion to martensite caused by a cyclic variation of temperature. To increase the temperature of the SMA, laser power (wavelength ~ 1.06um) transmitted through an optical fiber was used to illuminate the sample. A temperature increase was produced due to absorption of the laser energy. Each cycle was achieved by switching the optical power on and off.

In this experiment, two different stress levels were used. First a stress level of 9.725Ksi was used. The optical power of the laser was set to 0.4W at the output of the optical fiber. The optical power was left on for 5 seconds. In this time interval, the SMA alloy reached a constant value of strain. The maximum temperature measured was 110°C. , well above the transition temperature of 45°C. After the optical power was turned off, the strain in the SMA gradually returned to zero. When the strain reached zero, the laser was again turned on and the experimental procedure of reading the maximum strain was repeated. This procedure was first repeated 255 times with an applied stress of 9.725 Ksi. The stress level was then increased to 19.45 Ksi for the next 180 cycles. Finally, the stress level was decreased back to 9.725 Ksi and data taken for the next 410 cycles. Results are plotted in figure 5.

Figure 5 shows that the recoverable strain is dependent on the amount of external stress. This experiment also proved that the strain recovered on each cycle was repeatable. It can also be seen from the figure that when the stress is changed, the SMA undergoes a transitional "learning" stage before reaching a constant strain. At these stress levels no memory fatigue was observed. It appears from these results that SMA actuators can be used for many cycles and in many applications, particularly those that have less stringent requirements for exact actuation displacements and recovery stress. Many civil engineering applications would fall into this category.

Our next goal was to investigate the response times of a trained SMA sample when activated by a pulse of optical power. Figure 6 illustrates the variation in length of the SMA wire as a function of time for the same SMA sample used before (1cm. long and 12 mils in diameter). The variation in length of the SMA sample is shown in figure 6 as displacement (measured in mm). The displacement represents the elongation of the SMA actuator.

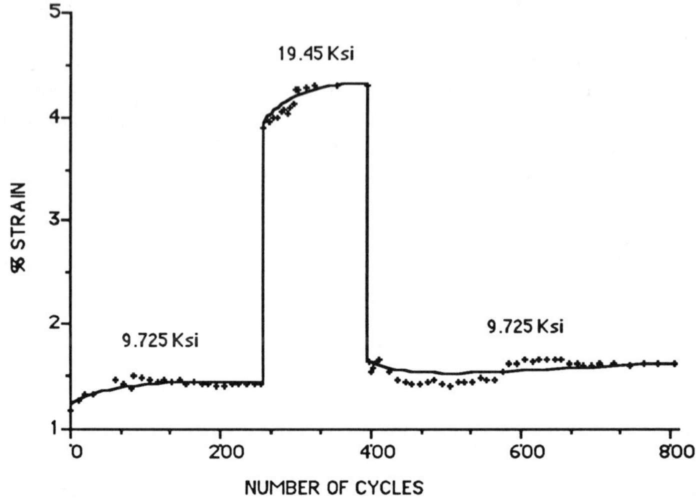

Figure 5. Measured repeatability of shape memory effect

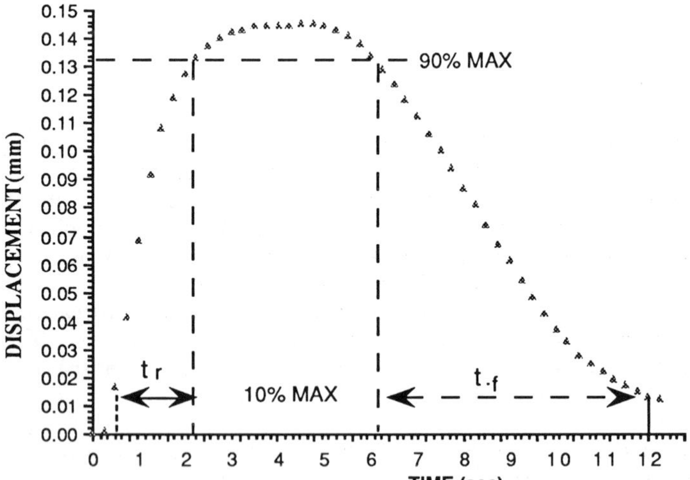

Figure 6.Variation in length of a trained SMA sample vs. time when activated by 0.4W of optical power for 5 seconds.

At time t=0, 0.4W of optical power was applied to the SMA. The optical power was turned off after 5 seconds, once steady state had been reached. The total area of the SMA illuminated by the output of the optical fiber was 0.1135 mm^2. The steady state temperature was measured to be 110°C at the center of the SMA wire. The rise and fall times can be determined from figure 6. Rise time is defined as the time duration from 10 percent to 90 percent of the maximum displacement. In similar fashion, fall time is defined as the time duration from 90 percent to 10 percent of the maximum displacement. The maximum displacement observed was 0.145 mm at time t = 4.5 seconds. The rise time was found to be 1.75 seconds and the fall time was 5.8 seconds. To fully characterize the behavior of displacement as a function of time, delay time and the storage time can be determined. Delay time is the time it takes for the displacement to reach 10 percent of the maximum after the optical power was turned on. This value was found to be 0.45 seconds. Similarly, the storage time is the time that the displacement takes to fall to 90 percent after the optical power was turned off. At 6.2 seconds, the displacement reached 90 percent of its maximum value after the optical power was turned off at 5 seconds; thus the storage time was 1.2 seconds. The experiment demonstrated that a trained SMA sample could be be optically energized through an optical fiber(Grossman,1990). The variation in length as a function of time indicated that thermal dissipation dominated the time response of the actuator, since the SMA transition between states should be on the order of milliseconds or less.

The goal of the next experiment was to use the optically energized SMA to deflect a fiberglass cantilever beam and to measure this deflection as a function of time. This experiment demonstrated that the relaxation time obtained when using only one actuator on the upper surface can be decreased by the use of another actuator on the lower surface. The lower actuator, when energized, forced the beam downwards. The same size actuators were used on both surfaces (0.4 inches long and 12 mils diameter). Their placement is shown in figure 7.

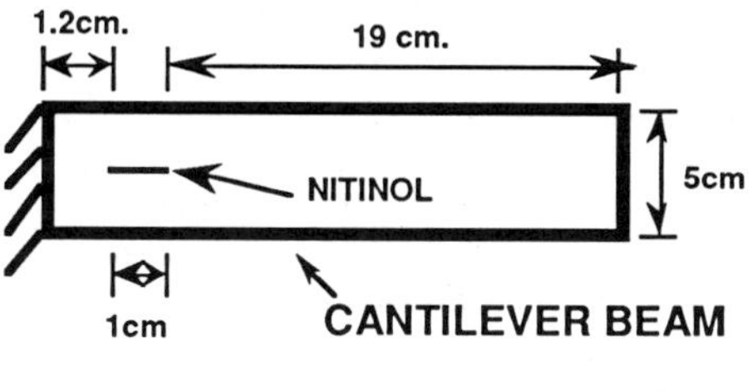

Figure 7. Actuator attachment location on the cantilever beam(bottom actuator is directly below upper SMA

The time that it takes the beam to return to its undeflected position after removing the optical power depends mostly by the time it takes to the actuator to cool down. The second actuator, placed at the bottom of the beam, when actuated will force the beam to deflect in the downwards direction. The results obtained are shown in figure 8.

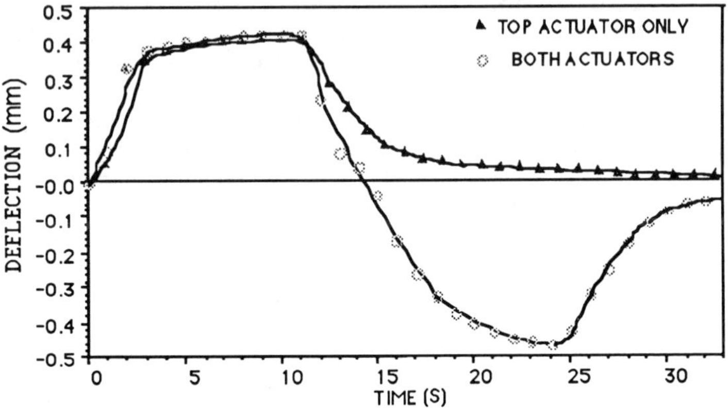

Figure 8. Deflection of cantilever beam with one and then two actuators

At time t=0 the upper actuator was activated with 10mW of optical power (labeled in the graph as TOP ACTUATOR ONLY). The power was turned off at t=11 seconds. During this interval (t=0 to t=11 seconds) the optical power of the lower actuator was off. From the graph it can be determined that the fall time was 9.25 seconds. The other curve in the same figure (labeled as BOTH ACTUATORS) represents the deflection as a function of time when both actuators have been used as follows. At t=0 seconds, only the upper actuator was energized. At 11 seconds, the optical power of the upper actuator was turned off and the optical power in the lower one was turned on. The optical power of the lower actuator was turned off at 24 seconds. When two actuators were used as described in the experiment, the the deflection fall time of the beam was reduced. The lower actuator forced the beam to deflect downwards. When only the top actuator was used, the fall time of the beam depended on the time that it took the actuator to cool to room temperature. Therefore, this experiment showed that the cantilever beam after being deflected upwards, then returned to its undeflected position faster by using two actuators, one on each surface as well as demonstrating that multiple actuators can be used to generate a variety of motions in a structure.

The next experiment involved monitoring the strain at the lower surface of the fiberglass cantilever beam while the SMA actuator, at the upper surface, was being optically energized. The amount of strain induced by the actuator attached to the upper surface is equal in magnitude but opposite in polarity to that measured at the lower surface of the cantilever beam. The results of this experiment correlate the amount of deflection and the amount of strain induced by the SMA actuator. A constant optical power of 5mW was used to actuate the SMA. The actuator was epoxied to one surface and the strain gage the opposite side. The optical power was turned on at time t=0, and off after 40 seconds. At 40 seconds the strain had reached a steady state value. The deflection at the end of the beam was measured along with the strain values. The results are graphed in figure 9. The left axis is the strain and the right axis is the corresponding deflection of the end of the beam.

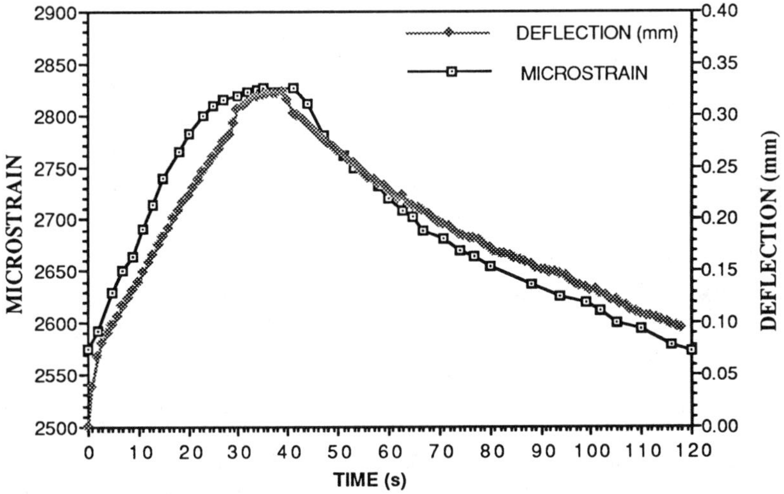

Figure 9. Strain and deflection versus time for an optically energized actuator epoxied to the upper surface of a fiberglass cantilever beam.

Figure 9 shows that the amount of strain induced by the actuator roughly follows the amount of deflection of the cantilever beam. For a maximum induced strain of 2830 microstrain, a maximum beam deflection of 0.33mm was observed. This experiment demonstrated that the SMA actuator was able to induce a reasonable amount of strain in the structure, and that the deflection was approximately proportional to the strain, as expected.

The objective of final experiment was to verify that the time responses being measured were primarily thermal in origin and not a limitation due to the transformation process. The fraction of the SMA length above the transition temperature was measured as a function of time. This result is compared to the rise time of the deflection of the cantilever beam as shown in figure 10.

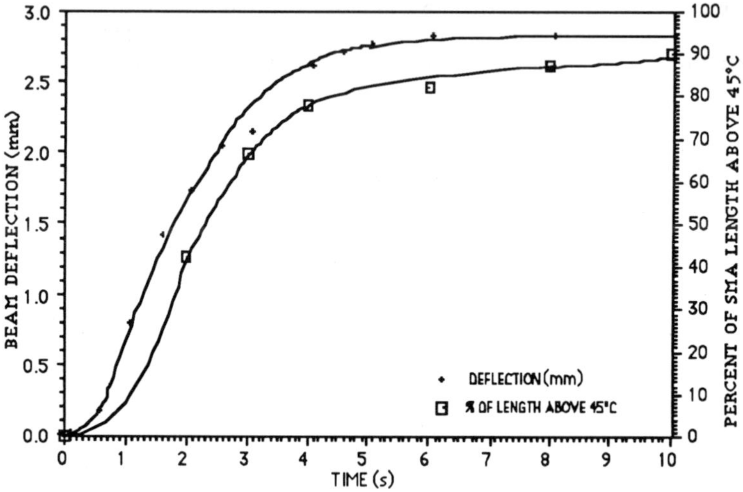

Figure 10. Percent of the total length of the SMA above its transition temperature and deflection measured as a function of time

For maximum change in length, the whole SMA sample must reach transition temperature. Therefore, the length of the SMA should be at a temperature higher than or equal to transition temperature of 45°C . From figure 10, the rise time measured for the fraction above transition temperature was 3.55 seconds. The rise time of the deflection of the beam was 3.425 seconds. These results are equal within 3%. Therefore, one can conclude that the rise time of the deflection of the beam was almost completely dominated by the time it takes the whole actuator to reach transition temperature.

Conclusions

We have demonstrated the feasibility of an optical technique for energizing shape memory alloy actuators. In the future this could allow multiple fiberoptic sensors and SMA actuators embedded in a structure to be wavelength division multiplexed onto a single fiber with control over each individual actuator. It also allows the power source to be located kilometers away from the actuators, if desired, thus providing significant flexibility in the placement and control of the system. We have performed preliminary experiments using the optical technique to investigate repeatability, time constants, and dual actuator control. We have found that for many civil engineering applications, the performance would be quite satisfactory. With optimized illumination techniques millisecond operation should be possible.

References

Baz, Iman,K., and McCoy, J., "The Dynamic and Thermal Characteristics of Shape Memory Actuators", Proceedings OE/Fiber'89, September, 1989.

Delaey, L., Krishnan, R.V., Tas, H., "Thermoelasticity, Pseudoelasticity and the Memory Effects Associated with Martensitic Tranformations", J. of Materials & Science, Vol. 9, pgs. 1521-1535, 1974.

Grossman,B.G., Thursby, M.H., Alavie, T., and Yoo, K.S., "Smart Structures Incorporating Artificial Neural Networks, Fiber-Optic Senors, and Solid-State Actuators", Proceedings of SPIE OE/FIBERS '89, September 1989.

Grossman, B.G., Franke, J., Thursby, M., et al, "Fiber-Optic Sensors and Smart Structures Research at Florida Institute of Technology", Proceedings of SPIE OE/FIBERS '89, September, 1989.

Grossman,B.G., Caimi, F., Alavie, T., Franke, J., et al, "Smart Structures and Fiberoptic Sensor Research at Florida Institute of Technology-1990", Proceedings of SPIE OE/FIBERS '90, September, 1990.

Grossman, B., and Thursby, M., "Fiberoptic Sensor Signal Processing using Neural Networks", chapter written for Fiberoptic Smart Structures , Edited by E. Udd, Wiley and Sons, NY.., to be published in 1993.

Miyazaki, S., Kimura, S., Takei, F., Miura, T., et al, "Shape Memory Effect and Pseudoelasticity in a Ti-Ni Single Crystal", Scripta METALLURGICA, Vol. 17, pp. 1057-1062, 1983.

Robinson,R., "Smart Structures", Civil Engineering, November, 1992.

Schetky,L.,"Shape Memory Effect Alloys for Robotic Devices", Robotics Age, July, 1984.

Udd, E., "Embedded Sensors Make Structures Smart", Laser Focus Magazine, may, 1988.

Vernon, M.W., Lovell, M.C., Avery, A.J., Physical Properties of Materials, Wiley Inc., 1975.

Wasilewski, R.J., The Shape Memory Effect in TiNi: One Aspect of Stress-Assisted Martensitic Transformation, Plenum Press, New York and London, 1975.

Distributed Fiber Optic Sensors for Civil Structures using OTDR

Keith H. Wanser[1] , Michael Haselhuhn[2] , Michael Lafond[2], and Jeffrey Williams[2]

Abstract

Theory and experiments for various distributed sensing applications in Civil Structure monitoring are discussed. In particular, the theory of incoherent OTDR, backscatter sensor resolution, microbending sensors, and new results on fiber kink loss and serial sensor interaction effects are presented. New applications, methods, packaging techniques, and experimental results for OTDR addressed microbending sensors are presented for; fiber embedded metal lap joints, pipeline strain/deflection monitoring and prototype sensor fabrication, and localized crack detection on metallic and other structures.

Introduction

There are several different types of optical reflectometry employing fiber optics, and numerous research publications in the literature describing them. It is not our purpose here to review the various techniques. An excellent entry point to the literature in the field may be found in references 1-4. Initially, the various reflectometry methods were developed to serve the needs of the telecommunications industry. As such, they were not intended to probe relatively short structures, (of the order of 1 m to 1 km) with high spatial resolution (between 1 cm and 1 m).

On the other hand, there have been several studies on the use of millimeter and extremely high resolution (micron) reflectometry for use in integrated optical systems. These systems tend not to operate on Rayleigh backscatter, but rather on discrete reflections. Other studies have employed a variety of different reflectometry techniques based on Rayleigh (or Raman) backscattering. Some of the more important methods are; coherent optical time domain reflectometry (COTDR), complementary correlation OTDR (CCOTDR), optical frequency domain reflectometry (OFDR), both incoherent and coherent (IOFDR and COFDR), optical coherence domain reflectometry (OCDR), Raman OTDR (ROTDR) and polarization OTDR (POTDR). Most of these techniques tend to employ cumbersome and

---

1 Associate Professor of Physics, Department of Physics, California State University Fullerton, Fullerton CA 92634

2 Department of Physics, California State University Fullerton, Fullerton CA 92634

expensive laboratory apparatus, such as mode-locked picosecond dye lasers, or multigigahertz spectrum analyzers, so that practical implementation of fieldable equipment at a reasonable cost is not feasible in the near term. Possible exceptions to this are CCOTDR and IOFDR, although systems employing these methods are not in wide use. An extensive bibliography of these reflectometry methods can be found in reference 3.

In the present work, we will discuss the most widely used and commercially available reflectometry method, which is incoherent optical time domain reflectometry (IOTDR, or just OTDR). In this method, a short pulse of light is launched into the fiber through one port of a fiber optic beamsplitter. The backscattered and reflected light from the fiber is monitored on the other port of the beamsplitter as a function of time, with a high speed photodetector and sampling electronics. For short structure sensing applications, the emphasis is on graded index multimode fiber at the ~ 850 nm and possibly ~ 670 nm wavelength ranges for improved backscattering coefficient, as well as the greater launched power available with high power laser diode sources at 850 nm. The use of relatively large core diameter multimode fiber (typically 50 or 62.5 μm), as opposed to the considerably smaller core diameter of single mode fiber (typically 4 - 9 μm), allows for a much larger optical power to be launched, in addition to a larger backscatter capture fraction, and thus improved signal to noise ratio, before the threshold for nonlinear optical processes, such as stimulated Raman scattering, is reached. These nonlinear processes limit the amount of optical power that can be launched down the fiber without severe pulse distortion effects. A disadvantage of multimode fiber is that intermodal dispersion causes a resolution limitation due to pulse spreading, compared to longer range, lower resolution systems employing multimode fiber and longer wavelength sources (1300 nm), or single mode fibers. This slight disadvantage is not serious for most sensor applications.

Although IOTDR measurements are relatively slow (typical data acquisition times for Rayleigh backscatter signals using commercially available equipment range from about 20 seconds to several minutes, depending on the measurement accuracy and spatial resolution required), on the time scale of many structural monitoring requirements, this is essentially real time. Most commercially available OTDRs also have a reduced accuracy, "real time" mode, which updates about every 1-2 seconds or so, and is often useful for quick diagnostics and some process monitoring applications. Several techniques for improving the acquisition time of OTDR based systems have been suggested and demonstrated. As yet, there is no real time, commercially available reflectometry instrument capable of monitoring Rayleigh backscatter at rates suitable for structural vibration monitoring (typically up to several hundred Hertz). The difficulty in doing so with IOTDR will be seen in the following discussion. Other techniques, such as IOFDR[4], show promise of being able to address the higher data acquisition rates required for the structural vibration monitoring problem using sensors based on Rayleigh backscatter. Many of the sensor techniques presented here are suitable for structural vibration monitoring using other types of optical reflectometry.

Incoherent Optical Time Domain Reflectometry (IOTDR) Model

Consider the case of incoherent optical time domain reflectometry as first demonstrated by Barnoski and Jensen[5] in 1976 and elaborated by Personick[6] in 1977. IOTDR is simply an optical radar technique that allows both the magnitude and location of fiber perturbations to be determined. It is analogous to scattering of radar by clouds when we look at ground radar weather maps on television. Optical backscattering from propagation medium inhomogenieties is dramatically illustrated by turning on the bright lights of an automobile when driving through fog. Since there are many ways of affecting the local fiber loss, reflection, and scattering properties, IOTDR lends itself to many different sensing applications, and is a very powerful and general technique, much as ordinary radar is. Since typical commercial OTDR instruments can easily measure 10 - 20 km lengths of fiber, depending on the total attenuation, the potential for remote sensing applications and large sensor arrays is very good. In addition, recent developments in optical fiber amplifiers have extended the range of the OTDR technique into the hundreds of km.[7], and thus many new possibilities for remote sensing applications will become available as this technology matures.

In the following discussion, the IOTDR model we consider consists of Rayleigh backscattering from the entire length of fiber, and discrete reflections from the front and rear of the fiber. Additional discrete reflections and transmission losses from the fiber due to splices, fiber parameter discontinuities, and/or Bragg type reflections can be included in a straightforward fashion, but are not discussed in this section. For short structure monitoring situations that do not require millimeter resolution, it is appropriate to describe the backscattering response neglecting the effects of modal and chromatic dispersion, mode coupling, polarization, and source coherence effects. Given the above assumptions, the Rayleigh backscatter contribution to the IOTDR return signal for the general situation of nonuniform fiber attenuation and arbitrary time dependence of the laser source power can be modeled by the equation[4],

$$P_B(0,t) = \int_0^L dx\, P_o\left(t - \frac{2x}{v}\right) S(x)\alpha_s(x)\exp\left[-2\int_0^x dy\,\alpha(y)\right] \tag{1}$$

Where $P_B(0, t)$ is the returned Rayleigh backscattered power at the fiber input end at time t, x is the distance from the input end of the fiber, $\alpha(x)$ is the local total power attenuation coefficient, including scattering, material absorption, and macro and microbending loss, $\alpha_S(x)$ is the local scattering attenuation coefficient, $S(x)$ is the local backscattering trapping factor, v is the fiber group velocity (assumed constant for simplicity), L is the total length of the fiber, and $P_o(t)$ is the time dependent optical power launched into the fiber at the input end.

The backscattering trapping factor S, which is the fraction of scattered power that is trapped in the backward propagating direction, is given for multimode fibers by[4]

$$S = \frac{3 g NA^2}{8(g+1) n_{co}^2} \tag{2}$$

Where NA is the fiber numerical aperture[8], n_{co} is the fiber core refractive index (on axis), and g is the index profile parameter ($g = 2$ for parabolic graded index fiber and $g = \infty$ for step index fibers). For a typical multimode fiber having an NA of 0.20 and core refractive index of 1.457, the trapping factor S is 4.71×10^{-3} for graded index and 7.07×10^{-3} for step index fiber. Thus for typical commercial multimode fibers, about one half percent of the Rayleigh scattered light intensity is captured (into guided mode power) by the optical fiber in the backward direction.

In order to understand how the basic factors come into play in the OTDR backscatter response, consider first the simple case of a square optical input pulse of the form

$$P_o(t) = P_o\, \theta(t)\, \theta(T-t) \ , \tag{3}$$

where P_o is the power of the pulse and $\theta(t)$ is the Heaviside step function. Equation 3 is a square pulse of length T that turns on at $t = 0$ and turns off at $t = T$. Substituting eqn. 3 into eqn. 1 yields

$$P_B(t) = \frac{P_o \alpha_s S}{2\alpha}(e^{\alpha v T} - 1) e^{-\alpha v t}\, \theta(t) \theta(T + \frac{2L}{v} - t) \ . \tag{4}$$

For the usual cases of interest, $\alpha\, v\, T << 1$ is well satisfied. As an example, for a 2 dB/km fiber attenuation ($\alpha = 4.6 \times 10^{-4}$ m^{-1}), a (long) pulse length of 10^{-7} sec., and $v \cong 2 \times 10^8$ m/sec., $\alpha\, v\, T = 9.2 \times 10^{-3}$. In the $\alpha\, v\, T << 1$ limit eqn. 4 becomes

$$P_B(t) = \frac{P_o}{2} \alpha_s S v T e^{-\alpha v t}\, \theta(t) \theta(T + \frac{2L}{v} - t) \ . \tag{5}$$

This last result is the fiber backscatter impulse response function, since a delta function input pulse gives the same result as eqn. 5. The impulse response function is fundamental to a quantitative understanding of the factors governing the IOTDR, and eqn. 5 is sometimes called the OTDR maker's equation, in analogy to the lens maker's equation of geometrical optics.

The most important feature of eqn. 5 is the exponential return signal as a function of time from a uniform fiber attenuation. The exponential decay time or slope of the backscattered power on a semilog plot (dB scale) is directly proportional to the total attenuation coefficient α. The backscatter signal strength is linearly proportional to the scattering coefficient α_S, the local trapping factor S, and the pulse length T. The backscatter returns begin as soon as the pulse does, and end at a time corresponding to the round trip time of the leading edge of the pulse plus the pulse width. This has the significant limitation that with ordinary IOTDR, one must wait for this length of time before sending out another pulse. This imposes a constraint on the speed and accuracy of data acquisition for structural vibration applications, although not for most structural monitoring applications.

Figure 1 shows a classic OTDR trace from a commercially available 50 μm core diameter graded index fiber at a wavelength of 1300 nm. Note the extremely linear decay of the signal, on the dB scale, corresponding to uniform attenuation. Time t is converted to distance x in the OTDR by the relation $t = 2\ x/v$. A least square fit to the data between 100 m and 1100 m yields an attenuation coefficient of 0.631 ± 0.002 dB/km, which is essentially the fundamental Rayleigh scattering limit of the fiber due to frozen in density and dopant concentration fluctuations[9]. In addition to the uniform decay of the signal, at the end of the fiber (1295 m) there is a sharp spike. This large signal is from the discrete reflection off of the fiber end known as the Fresnel reflection. As an example, using eqn. 5, the effective Rayleigh backscatter reflection coefficient for this fiber with a pulse width of 10^{-7} sec. is 7.0×10^{-6}, as compared to the 3.5×10^{-2} Fresnel end reflection coefficient from a well cleaved, air-fiber interface. The fact that discrete Fresnel reflections are usually orders of magnitude larger than Rayleigh backscatter signals can be used to great advantage in many fiber optic distributed sensing system applications.

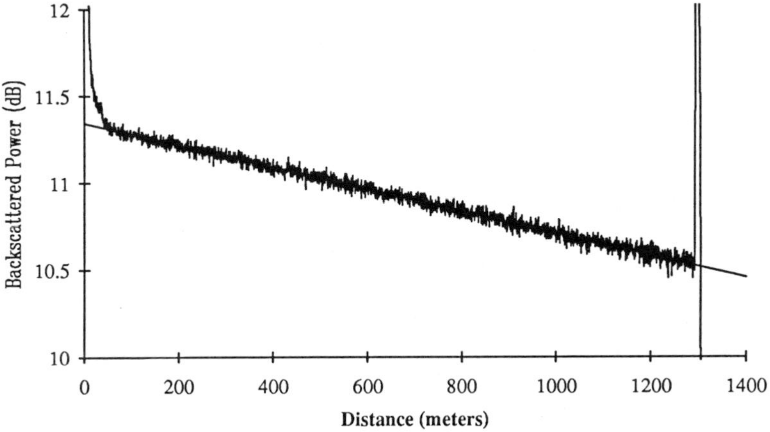

Figure 1. Classic OTDR trace of multimode graded index fiber at 1300 nm wavelength showing uniform fiber attenuation and least square fit line. Least square fit attenuation coefficient $\alpha = 0.631 \pm 0.002$ dB/km.

In addition to the Fresnel end reflection, strong front end reflections are caused by the OTDR connector and internal coupler, and result in a so called "dead zone" at the beginning of the OTDR trace due to the finite receiver response time. In this dead zone, the receiver electronics must recover to the extremely low level of

Rayleigh backscatter power, as compared to the Fresnel front end reflection (this usually requires several receiver response time constants). In addition to the dead zone from the receiver, spatial transient effects are caused by increased backscatter from cladding and radiation modes excited by the fiber light launching from the laser. These spatial transients must decay in order to reach the equilibrium mode distribution[10]. The effects of front end reflection and spatial transients can be seen in fig. 1 as a large decaying pulse at the beginning of the trace out to about 50 meters. The length of the dead zone and spatial transient region varies with the OTDR, and with the pulse length, fiber type, and resolution parameters. Typical values of the dead zone with commercially available OTDRs range from a few meters to 100 meters. It is important to have sufficient lead-in fiber length from the OTDR to a sensor in order to get beyond the deadzone and spatial transient region, otherwise OTDR sensor response measurements can be misleading. Some OTDR manufacturers put 50 to 100 meters of fiber inside the OTDR to eliminate this problem.

In order to study the sensor resolution properties of the IOTDR, we now consider a three segment fiber of total length L, with segment lengths L_1, L_2, and L_3. Each fiber segment has its own set of fiber parameters; namely length, total attenuation, scattering attenuation, *NA*, and backscattering capture factor. The middle section of the fiber will generally be denoted as the "sensor", and its length, L_2, will be called the sensor "gauge length". Thus we write for the returned, backscattered and reflected power $P_B(0, t)$ at the fiber input end at time t,

$$P_B(0,t) = P_{B1}(t) + P_{B2}(t) + P_{B3}(t) + P_R(t) \tag{6}$$

where we have written separately the contributions to the backscattered power from the three consecutive fiber segments 1, 2 and 3 as P_{B1}, P_{B2}, and P_{B3} respectively, and P_R is the power from the discrete front and end reflections.

For the case of a Gaussian pulse waveform at the fiber input end of the form,

$$P_o(t) = P_o \exp\left[-(t/T)^2\right] \tag{7}$$

where P_o is the peak power and T is related to the pulse full width at half maximum height (*FWHM*) by,

$$FWHM = 2T\sqrt{\ln 2} \tag{8}$$

we obtain the following results using eqns. 1 and 7,

$$P_{B1}(t) = \frac{\sqrt{\pi}}{4} P_o \, vT \, \alpha_{s1} \, S_1 \exp\left[-\alpha_1 vt + (\alpha_1 vT/2)^2\right] x$$
$$x \left\{ erf\left[\frac{2L_1}{vT} + \frac{\alpha_1 vT}{2} - \frac{t}{T}\right] - erf\left[\frac{\alpha_1 vT}{2} - \frac{t}{T}\right]\right\} \tag{9}$$

$$P_{B2}(t)=\frac{\sqrt{\pi}}{4}P_o\,vT\,\alpha_{s2}\,S_2\exp\left[-2(\alpha_1-\alpha_2)L_1-\alpha_2 vt+(\alpha_2 vT/2)^2\right]x$$
$$x\left\{erf\left[\frac{2(L_1+L_2)}{vT}+\frac{\alpha_2 vT}{2}-\frac{t}{T}\right]-erf\left[\frac{2L_1}{vT}+\frac{\alpha_2 vT}{2}-\frac{t}{T}\right]\right\} \quad (10)$$

$$P_{B3}(t)=\frac{\sqrt{\pi}}{4}P_o\,vT\,\alpha_{s3}\,S_3\exp\left[-2(\alpha_1-\alpha_3)L_1-2(\alpha_2-\alpha_3)L_2-\alpha_3 vt+(\alpha_3 vT/2)^2\right]x$$
$$x\left\{erf\left[\frac{2L}{vT}+\frac{\alpha_3 vT}{2}-\frac{t}{T}\right]-erf\left[\frac{2(L_1+L_2)}{vT}+\frac{\alpha_3 vT}{2}-\frac{t}{T}\right]\right\}, \quad (11)$$

where erf is the error function[11]. The discrete reflection power $P_R(t)$ consists of two terms,

$$P_R(t)=P_F(t)+P_E(t) \quad . \quad (12)$$

The front reflection is modeled by $P_F(t)$, and the end reflection by $P_E(t)$ where

$$P_F(t)=R_F\,P_o\exp\left[-(t-\frac{2L_F}{v})^2/T^2\right] \quad (13)$$

and

$$P_E(t)=R_E\,P_o\exp\left[-2(\alpha_1 L_1+\alpha_2 L_2+\alpha_3 L_3)-(t-\frac{2L}{v})^2/T^2\right] \quad . \quad (14)$$

The additional parameters are the end reflection coefficient R_E, the effective front reflection coefficient R_F, and the effective distance L_F of the front reflector.

Application of 3 fiber segment IOTDR model to specific cases

A computer program utilizing the three segment optical fiber model has been developed that allows calculation of the absolute backscattered power as a function of time, both the total power and the time dependent power from a particular fiber segment. Shown in fig. 2 is a computer simulation run for a short structure measurement situation of interest, using the three segment fiber model discussed in the previous section. The run parameters described in the figure are appropriate to one manufacturer's millimeter resolution OTDR specifications of 100 psec. pulses with 20 dBm peak launch power, and we have chosen a one inch long, 1 dB one way loss region located 31.8 meters from the fiber front end. The lower line in the figure is the backscattered power from only the second fiber segment. Note that it is only identical to the total backscattered power for a relatively short time, compared to the time of significant returns from this fiber segment. The total integrated energy from segment 2 of the fiber is 5.31 x 10^{-19} Joules, corresponding to 2.24 backscattered photons per pulse from the sensing region of the fiber. Assuming Poisson source statistics, this requires 1.189 x 10^5 photons to be acquired to achieve a measurement accuracy of 0.01 dB, or 8.41 x 10^4 traces to be averaged to achieve the desired quantum limited signal to noise ratio under the best possible circumstances. The data acquisition rate is limited by the length of the structure so that only one scan can be acquired during the round trip transit time of the pulse. This yields a minimum data acquisition time of 0.817 seconds for a short structure requirement of 1 km, having a spatial resolution of one inch, and measurement accuracy of 0.01 dB. Clearly, even in this favorable circumstance where the best possible photon counting sensitivity has been assumed[12], we see that the measurement bandwidth (i. e. Nyquist frequency) is less than 1 Hz, which is less than desirable for many real time structural monitoring applications. Acquisition times become longer for sensors with more loss occurring before the sensing region, as is encountered in many embedded sensor applications. Relaxation of the measurement accuracy requirement by a factor of 10 to 0.1 dB allows a factor of 100 fewer photons to be acquired, thus reducing the acquisition time by a factor of 100. This would give an acceptable bandwidth of ~100 Hz for some low frequency structural monitoring applications. It may thus be necessary in some applications to sacrifice measurement accuracy in order to obtain higher bandwidth, a common systems tradeoff. Use of an analog detector system yields considerably less sensitivity than the photon counting system[13].

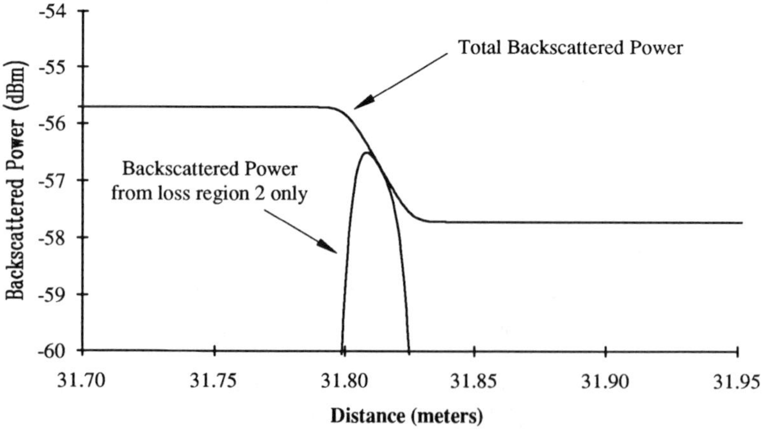

Figure 2. Single pulse IOTDR backscattered power from a segment of multimode step index fiber with a one inch long, 1 dB one way loss region. Parameters for the run are: Gaussian pulse FWHM = 100 psec., 20 dBm peak launch power, fiber NA = 0.204, 1.5 dB/km scattering attenuation, 2.48 dB/km total attenuation (except in the high loss region), L_1 = 31.8 m, n = 1.457, 840 nm. wavelength.

For comparison purposes, another simulation run is shown in fig. 3. The only differences between this run and fig. 2 are that the laser source power has been reduced to 10 dBm, and the Gaussian pulse FWHM is now 4 nsec instead of 100 psec. This corresponds to another commercially available OTDR. Note the extreme broadening of the return signal in the vicinity of the loss region to at least one meter, even though the high loss attenuation region is only 1 inch long. This broadening of the sensing region is independent of the receiver electronics, which are assumed to have a high enough bandwidth that they contribute no loss of signal temporal information. This situation prevails whenever the pulse length is longer than the resolution cell desired. Although the loss information across the sensing region is correct, given two points far enough separated, the slope of the trace has no obvious relation to the local value of the attenuation. That this is the case is further supported by fig. 4. This is the same simulation run as fig. 3, but plotted on a different scale in order to display the Rayleigh backscattering contribution to the total return power from fiber segment 2. As can be seen from the figure, the backscattering contribution from fiber segment 2 is always more than 12.45 dB below the total return signal. This is in contrast to fig. 2, where the contribution from fiber segment 2 contributes essentially the entire return signal for a short period of time. Although the total integrated energy from segment 2 of the fiber is now 2.12×10^{-18} Joules, corresponding to 8.95 backscattered photons per pulse, this energy is spread over a much longer time, and in order to obtain the same resolution as in the first example, one would have to time gate the receiver and thus throw away the additional photons coming from times outside the resolution cell. Clearly this situation will cause difficulty for manufacturer's attempting to use relatively long pulses and trying to signal process in software in order to extract the Rayleigh backscattering contribution from a section of fiber shorter than the pulse length. The preceding simple examples show the usefulness of a computer simulation to assess potential approaches and evaluate fundamental constraints to high resolution OTDR sensing.

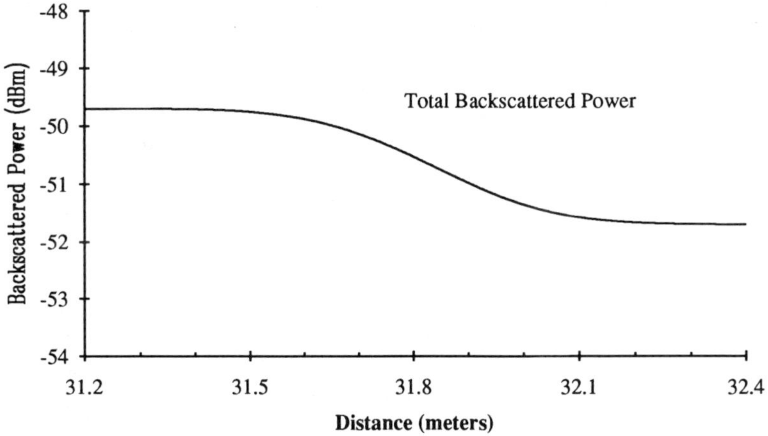

Figure 3. Single Pulse IOTDR backscattered power from a segment of multimode step index fiber with a one inch long, 1 dB one way loss region. Fiber parameters the same as in fig. 2. Source parameters are: Gaussian pulse FWHM = 4 nsec., 10 dBm peak launch power.

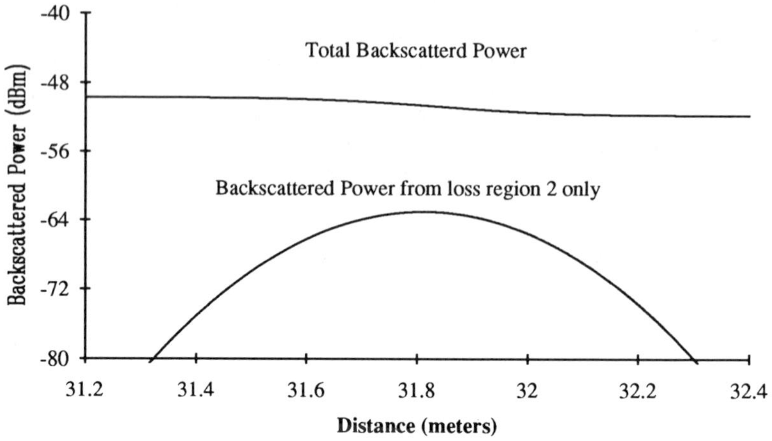

Figure 4. Same simulation run as shown in fig. 3 but plotted on expanded vertical scale to display Rayleigh backscatter signal from fiber segment 2. Note that the Rayleigh backscatter from this fiber segment is always at least 12.45 dB below the total returned power, as discussed in the text.

Serial sensor multiplexing

The simplest type of arrangement for addressing multiple sensors with OTDR is serial sensor multiplexing. In this arrangement, the sensing regions of fiber are arranged consecutively along the length of a single fiber. An example of the OTDR response from such an arrangement is shown in fig. 5. Since the sensing regions are delayed in time due to their spatial separation, it is possible to isolate the response from each sensor or sensing gauge length. Advantages of this arrangement are a single continuous fiber, simplicity in signal processing, and the order of sensing can be arranged according to the spatial location of the fiber relative to the structure of interest. Another advantage is that the sensors can be addressed from either end of the fiber, which allows for flexibility and some redundancy in the case of fiber breakage.

Although the simple serial multiplexing scheme is the easiest to implement, it also has disadvantages. A typical OTDR has a backscatter dynamic range of about 20 dB, depending on the spatial resolution, pulse length, etc. This means that when using backscatter to address loss transducer fiber sensors, one has a total backscatter sensing range of only about a factor of 100. The fact that the sensors are all serial, coupled with the dynamic range limitation of the OTDR, puts a constraint on the number of serial sensors, and the dynamic range of each one before reaching the noise floor. As an example, in a typical measurement situation, the OTDR can make a measurement with 0.01 dB resolution. If we design an individual sensor to have a maximum loss of 2.56 dB (which is a transmission factor 0.5546), then the individual sensor will have 8 bits of resolution. A sensing system with 8 of these sensors arranged with serial multiplexing would require an OTDR backscatter dynamic range of 20.48 dB.

Other disadvantages of serial sensor multiplexing involve sensor interaction effects due to fiber modal depopulation[3] (see section on kink loss) and the fact that fiber breakage can cause dramatic loss of information from the sensing array if it is addressed from one fiber end only. Fortunately additional sensor multiplexing techniques that reduce or eliminate these problems have been developed.

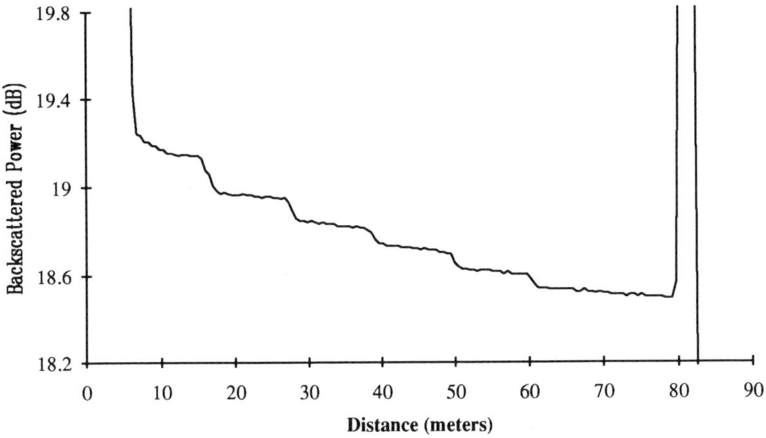

Figure 5. Simple serial sensor multiplexing of five passes of optical fiber embedded in a 1.5 inch wide bonded aluminum lap joint (see figs. 6 and 7). Note the five loss steps showing resolution of all five 12 inch long sensing fiber passes. Each fiber sensing pass has a ten meter fiber delay in between.

Bonded Metal Lap Joint Application

Smart structures utilizing fiber optic sensing techniques is a rapidly developing technology area with a growing number of applications[14]. The major focus of previous work has been on fibers embedded in composite materials, such as graphite epoxy composites and concrete, or else fibers attached to structures in cases where embedment is not possible or desirable. It is widely believed that smart skin technology cannot be used with conventional metal skins or structures[15].

The present work demonstrates that embedment of optical fibers in metallic structures is feasible under certain circumstances, such as the important case of bonded metal lap joints. Recent experiments and development work in our laboratories on the measurement of stress in fiber optic embedded aluminum lap joints have demonstrated the feasibility of using OTDR technology as an in situ NDE method for this application. The technique has application to measurement of stress in metal lap joints in aircraft[16] and other structures, such as bridges and ships. Methods to resolve the stress distribution across a lap joint have been developed to detect the onset of failure.

Udd et al[14] have studied an optical fiber embedded in a thermoplastic composite material bonded lap joint, using a simple optical transmission loss measurement system and a single pass of fiber. The change in optical transmission in their work was no more than 8% for the cured lap joint, even at the failure tension of 35,000 pounds. This is an undesirable situation from a measurement sensitivity and dynamic range standpoint. In addition, results on only a single acrylate coated embedded fiber were reported, with no resolution of stress across the lap joint.

In the present work, we have utilized OTDR to interrogate multiple sensing passes of fiber within a bonded aluminum lap joint. Figure 6 shows the layout of the fibers used and fig. 7 shows a picture of an actual test specimen. The use of OTDR allows for simultaneous measurements to be obtained on all five passes of fiber with a single continuous strand of fiber. The delay loops on each end of the lap joint, shown in fig. 6, were employed to allow resolution of each sensing pass, with an OTDR having a spatial resolution of only 1 meter, in spite of the fact that the lap joint was only 12" long. This delay loop technique can be employed in numerous optical fiber sensing applications to obtain localized information on a length scale much shorter than the OTDR spatial resolution determined by pulse length and receiver electronics.

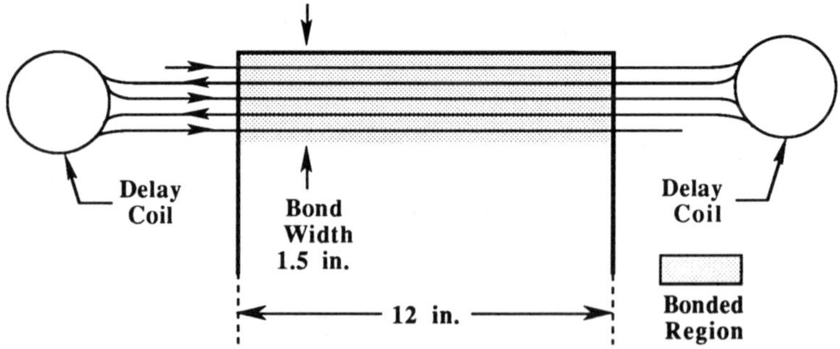

Figure 6. Layout of fiber embedded bonded aluminum lap joint. Note five sensing fiber passes and fiber delay loops.

The OTDR was used to monitor the fiber embedment process, from layup thru the post curing phase. This was found to be essential to insure good quality results and to prevent excess fiber attenuation in the absence of loading. The OTDR was used in real time mode during layup to adjust the lap joint bond and insure no excess fiber attenuation from slight fiber pinching or excess stress in one part of the joint compared to another. This technique also helped insure uniformity of the bond. The test specimens were cured in an oven at 55°C for 12-15 hours. After the cure process, OTDR scans of each specimen were taken to establish a baseline trace prior to application of stress, and to observe if any changes took place during the cure, as shown in fig. 5. Our initial prototype specimens had slight fiber attenuation effects due to layup. As we gained more experience with the process, it was possible to control the amount of built in fiber attenuation due to layup and to fabricate specimens with essentially no static or frozen in optical attenuation. This can be extremely useful, since in many cases it is desirable to pre-bias fiber optic microbend sensors, in order to obtain more linear response characteristics and to tune the sensitivity and dynamic range.

Bonded aluminum lap joints 12 inch square by by 0.25 inch thick with an overlap of 1.5 inch were fabricated using Hysol EA 9309.3 NA adhesive, with five passes of a single continuous optical fiber embedded in the joint. The specimans were tested in extension from 1000 to 35,000 lbs. using OTDR. The strain variation across the joint could be resolved by monitoring the individual loss of each pass of fiber, as shown in fig. 5, as well as the total loss across all five sensing passes. Our first results using acrylate and polyimide coated fibers showed almost no strain sensitivity, in agreement with the extremely low sensitivity results reported by Udd et al[14]. A microbending sensitizer consisting of fine silica-based particulates was then added to the adhesive material in subsequent tests. Testing showed little degradation (less than 25%) of the bond strength due to the combined effect of embedded fibers and particulates. However, the particulates greatly enhance the sensitivity of the embedded fibers and allow their sensitivity to be tuned. Results for the total loss across all five sensing passes (determined from the pre-sensor to Fresnel peak 2 point loss variation) for one of our polyimide coated fiber (147 μm coating outer diameter) test specimens are shown in Figure 8. As can be seen from the figure, the sensor has

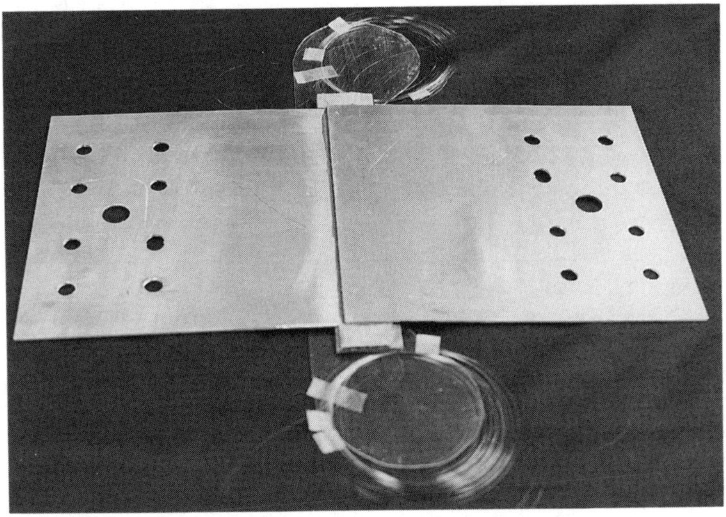

Figure 7. Fiber embedded bonded lap joint test specimen.

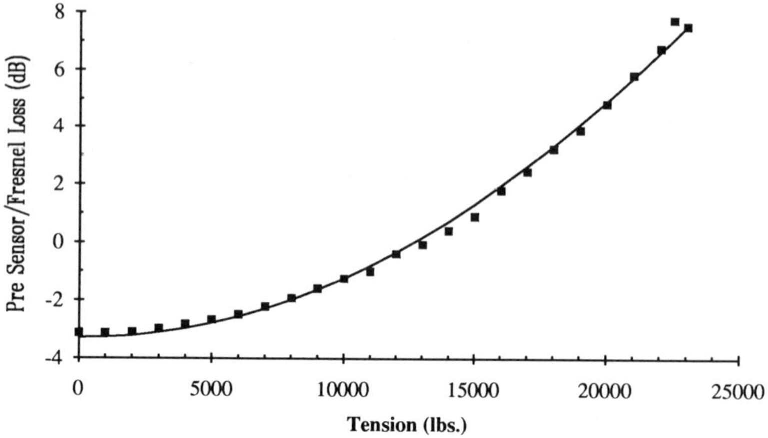

Figure 8. Results for the total loss across all five sensing fiber passes (determined from the pre-sensor to Fresnel peak 2 point loss variation) for polyimide coated fiber bonded lap joint (147 μm coating outer diameter). The curve is a pure quadratic least square fit to the data, (i.e. $y = m x^2 + b$).

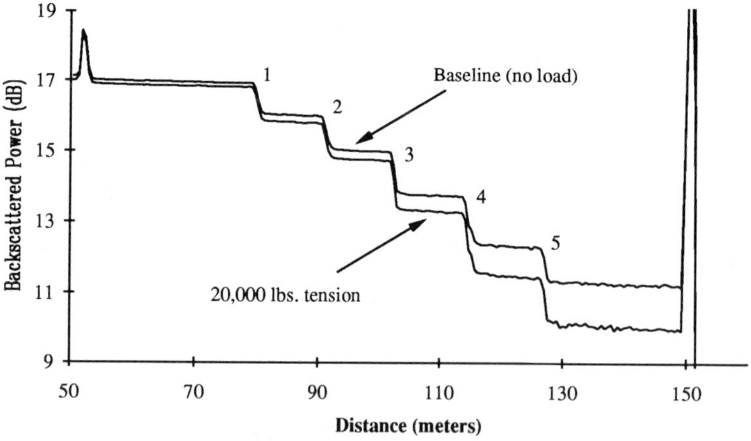

Figure 9. OTDR response of optical fiber embedded aluminum lap joint showing sensing response from all five sensing fiber passes under tension. The loss steps at zero load were purposely introduced to "prebias" the sensors. Note the effect of the cumulative response from each sensing fiber pass. The numbers on the fiber sensing passes refer to the layout inset in fig. 10.

a well defined response with good measurement sensitivity that varies by a factor of 12.6 (11 dB optical power) in the load range of 0 to 23,000 lbs, with a nearly ideal quadratic response curve. We have found that the technique of controlling embedded microbend sensor response with particulate additives is an important method for tailoring sensor response in a variety of applications.

Figure 9 compares the OTDR response obtained from the lap joint embedded fiber sensors with and without tension loading. Note the increased response of the two outermost fiber sensor passes 4 and 5, consistent with the data and inset in fig. 10. In fig. 10 the loss across each fiber sensing pass is plotted as a function of tension. The data clearly show the increased and nearly identical sensor response variation of the outermost fiber sensor passes 4 and 5. All five fibers were equally spaced 0.25 inches apart and the outermost fibers were 0.25 inches from the edge of each plate of the 1.5 inch wide lap joint. The inner fiber passes 1, 2, and 3 showed much smaller response, associated with reduced strain in this region.

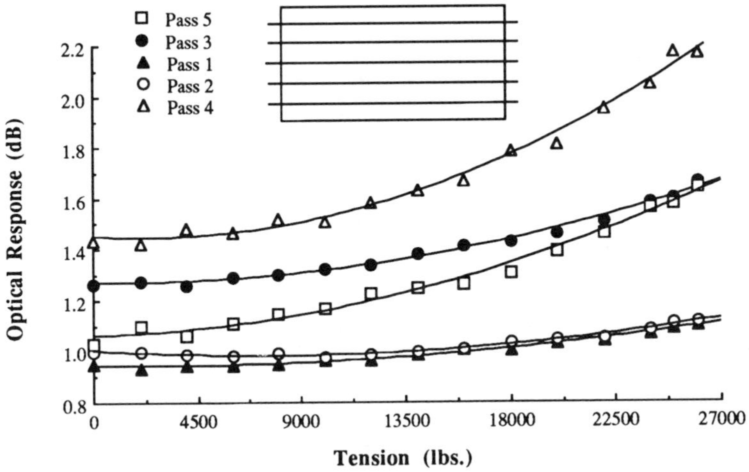

Figure 10. OTDR response of individual fiber passes embedded in a bonded aluminum lap joint. Note the difference in response corresponding to variation of strain across the joint, especially the increased response variation of the outermost sensing passes 4 and 5. Refer to the legend and inset for location of fiber passes.

Pipeline Monitoring Applications

There are several applications for pipeline strain and deflection monitoring using distributed optical fiber sensors. Other pipeline monitoring applications include crack and leak detection. Some of the more important pipelines include oil and natural gas pipelines[17], high temperature steam pipelines in electric utility plants[18,19], and large underwater sewage pipelines[20].

Pipeline failures can have disasterous consequences. People have been killed by high pressure steam pipe failures in power plants. The recent (February 1992) San Diego, California underwater sewer pipe failure[20] was an example of an economic, health, and environmental disaster that could have been avoided had a distributed fiber optic pipeline monitoring system been in place. This pipeline disaster caused approximately 180 million gallons a day of partially treated sewage to be dumped into the ocean less than three-quarters of a mile off shore of Point Loma in shallow water. Joints dislodged in 19 of the 25 foot long sections of 9 foot diameter concrete pipe, each weighing 30 tons. The disaster was blamed on a settling ocean floor and resultant pipeline sag. The repair costs alone were estimated at $10,000,000. In addition, there was a large economic loss to the local fishing industry, tourism, and severe environmental and health hazards. A similar underwater sewage pipeline extending from Hilo, Hawaii broke in 1987.

We have developed optical fiber strain and deflection sensors for monitoring pipelines and other structures. Figure 11 shows front, top, and side views for the pipe loading test fixture that was fabricated for this work. The size of the fixture is appropriate to a 1/5 scale model of a selected header pipe in a fossil-fuel power plant boiler. Figure 12 shows a photograph of the test fixture with a pipe and attached optical fiber sensors under test. A 25 ton actuator with load cell is below the pipe to measure the transverse force, and a Linear Variable Differential Transducer (LVDT) is above the pipe to measure and calibrate the transverse pipe deflection. The four fiber optic sensors shown in fig. 12 are mounted as shown in fig. 13 (a.). Transverse deflections of as little as 0.005 in. could be resolved with the fiber optic sensors in this arrangement, corresponding to longitudinal strains as small as 0.001%. The four sensors are on a single continuous strand of fiber and are monitored simultaneously using OTDR.

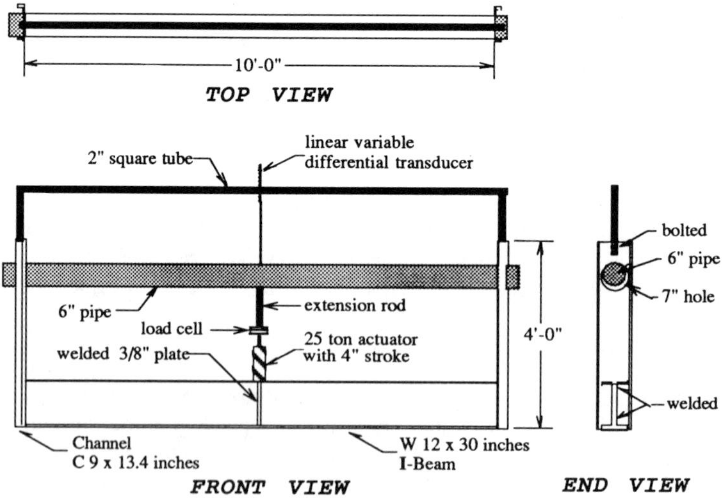

Figure 11. Diagram of pipe loading test frame for testing fiber optic pipeline sensors.

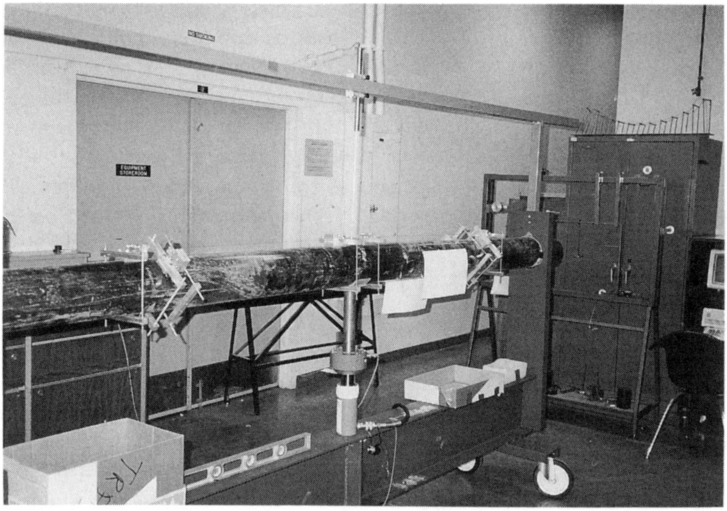

Figure 12. Pipe loading test frame with four fiber optic strain sensors under test.

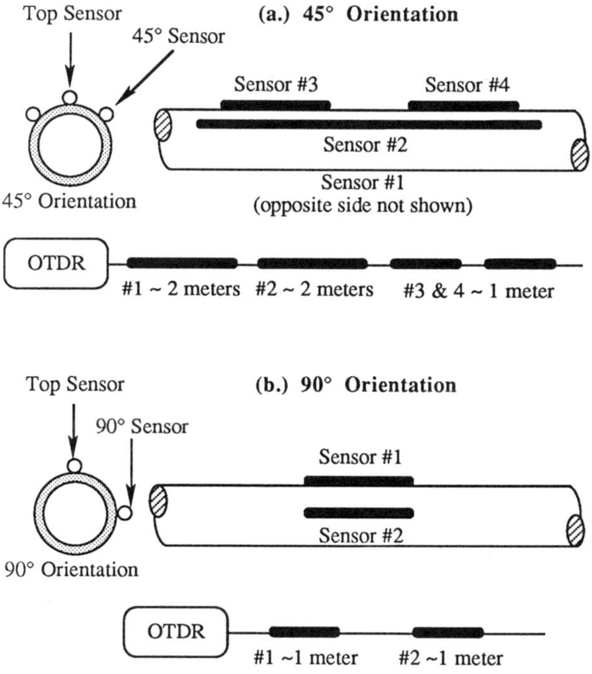

Figure 13. Pipe testing configurations. (a.) four sensors (b.) two sensors

Figure 14 shows the response of the fiber optic strain sensors in the four sensor configuration of fig. 13 (a.). The fiber optic sensors were mounted on an 11 foot long, 6.625 inch diameter, schedule 40 steel pipe which was tested in three point loading. The 45° orientation sensors were employed to test the response of sensors that were not aligned with either the neutral axis of the pipe or the line of maximum strain. The purpose of this was to obtain data for a proposed pipeline installation with sensors placed at 120° intervals around the circumference of the pipe. Tests were performed which verified the expected orientational response of the fiber optic sensors by rotating the pipe 45° and 90° from the configuration shown in fig. 13 (a.) and applying a load to the pipe.

The microbend sensors in these tests consisted of polyimide coated 50/125/152 μm (core/cladding/buffer diameter), graded index multimode fiber. The fiber was surrounded by two identical dummy optical fibers (for packaging redundancy and also for buffers) and two 0.010 inch diameter stainless steel wires packaged inside a stainless steel braid. Special fixtures were designed to pull on the braid, without producing longitudinal strain in the optical fiber, which then induced microbending of the optical fiber in the manner of a "Chinese finger cloth" (i.e. longitudinal strain of the braid produces transverse microbend deflections of the optical fiber). This packaging method has been refined and found to yield good sensitivity, as well as protection to the optical fibers, with little or no hysteresis. The sensors were attached by mounting brackets at each end, as shown in fig. 12, thus defining a sensing or gauge length.

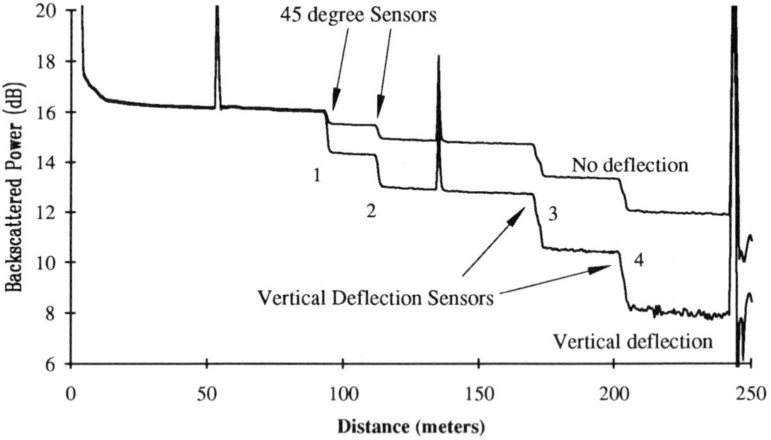

Figure 14. Response of four fiber optic strain sensors tested in vertical deflection, mounted as shown in fig. 13 (a.). Vertical pipe deflection in lower trace was 0.654 inches corresponding to a longitudinal strain of 0.127%, and a deflection force of 19,300 lbs. Note the response from the vertical sensors as well as the response from the 45 degree deflection sensors, as expected. The pulses near 55 meters and 140 meters are reflections from splices.

Figure 15 shows the OTDR response of two ~1 meter long fiber optic sensors mounted on the steel pipe test fixture in the configuration of fig. 13 (b.). This sensor configuration allows horizontal and vertical deflection to be determined simultaneously. As can be seen from the figure, there is excellent isolation between the horizontal sensor (mounted along the neutral axis of the pipe) and the vertical deflection sensor. Longitudinal strains as small as 0.001% could be resolved with these sensors. Figure 16 shows a plot of the response of both sensors in this configuration and further demonstrates the isolation between horizontal and vertical responses, as well as the essentially hysteresis free operation of the sensors. As in the case of the previous tests (fig. 14), experiments were performed with the pipe rotated by 45° and 90° from the configuration shown in fig. 13 (b.). These tests verified that the response of sensors 1 and 2 were nearly identical when they were at the top dead center (TDC) orientation. Furthermore, when sensor 2 was at 90°, it also had negligible response, since it was then along the neutral axis of the pipe.

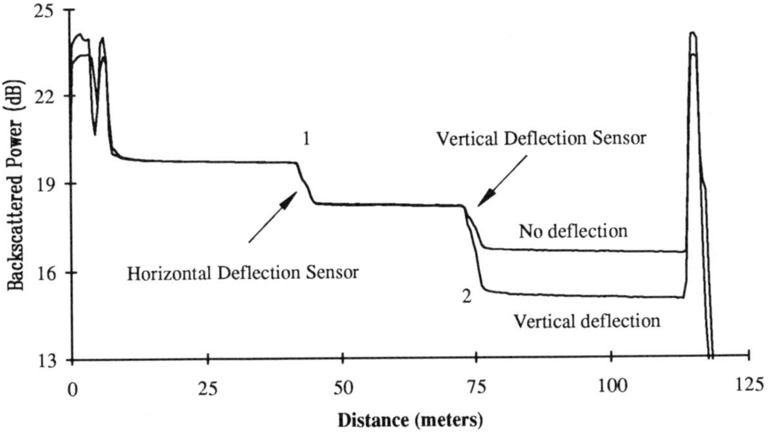

Figure 15. Response of two fiber optic strain sensors tested in vertical deflection, mounted as shown in fig. 13 (b.). Vertical pipe deflection in the lower trace was 0.553 inches corresponding to a longitudinal strain of 0.107% and a deflection force of 11,600 lbs. The horizontal deflection sensor was along the neutral axis of the pipe in this test. Note the response from the vertical sensor and no response from the horizontal sensor, as expected.

Microbending, Kink Loss, Sensor Interaction Effects and Localized Crack Detection

There is an extensive literature on the subject of optical fiber microbending[3,8,9,21,22]. The majority of the theoretical work has used perturbation theory or coupled mode theory, and been applied to single mode or few mode fibers. The basic result from the coupled mode and perturbation theories is that for multimode parabolic graded index fibers, there is a critical spatial periodicity[22] for coupling light out of the propagating core modes into the lossy cladding and radiation modes. When the fiber is transversely deformed at this critical distortion wavelength, (which ranges from 1.0 - 1.4 mm. for typical commercially available graded index fibers), a large microbending loss occurs. In addition, perturbation theory predicts a quadratic response of the loss of a microbending sensor to the magnitude of the external mechanical strain, as we find experimentally in many cases (see fig. 8).

In spite of all the previous theoretical work on microbending sensors, the theories have several shortcomings when applied to practical sensors[22]. Perturbation theories are unable to account for losses of more than a few tenths of a dB in a single sensor. In addition, the coupled mode theory approaches are usually only applied to the case of fibers with just a few modes, usually 2. Typical graded index multimode fibers support propagation of a few hundred modes. In addition, almost all of the prior work has not accounted for the wavefront tilt effect, which has been shown to be important for bend loss. Another deficiency with current theories is that they neglect coherent effects which lead to sensor modal noise, and mode depopulation and excitation effects which lead to serial sensor interactions[3]. Due to these and other shortcomings, we have developed a new approach to microbend sensor transducer analysis.

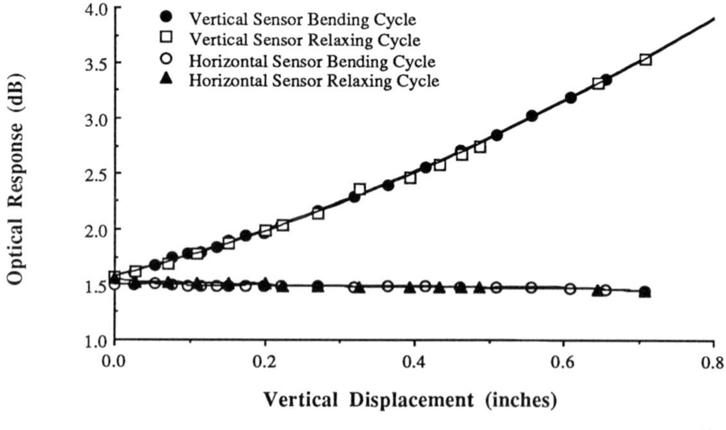

Fig. 16. Optical response for 2 sensor configuration of fig. 13 (b.) and fig. 15. Note the excellent isolation between the vertical sensor and horizontal sensor response and the essentially hysteresis free operation.

The most fundamental fiber microbend is a single kink, since an arbitrary fiber bend profile can be finite element modeled by a sequence of successive kinks. Suprisingly, this simple kink has not been treated properly in prior works on multimode fibers. Figure 17 shows the first exact, all mode wave optics calculation for the transmission loss across a single kink in a highly multimode fiber as a function of the kink angle, including the effects of modal noise. Although the curve looks very simple, a large amount of computer time is required to perform the calculation, since the initial mode phases were averaged over 1000 random configurations for each angle.

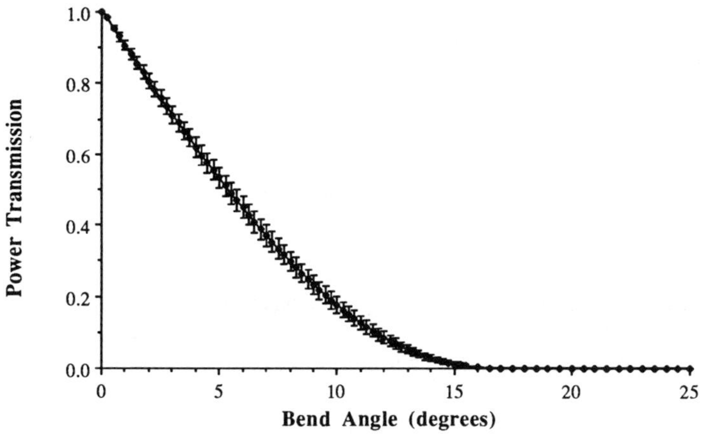

Figure 17. All mode wave optics calculation of the power transmission coefficient thru a single kink in a 50 μm core diameter parabolic graded index fiber. 850 nm wavelength, fiber NA = 0.22, 210 bound modes in the fiber. The error bars denote the standard deviation of the transmission coefficient due to coherent modal noise effects, when averaged over 1000 random initial phase configurations.

Under certain circumstances, we have noticed serial microbend sensor interaction effects[3]. These effects are often noticeable when the loss across a given sensor "upstream" from a sensor is more than a few tenths of a dB. This sensor interaction effect causes the loss of a sensor "downstream" from a given sensor to be affected by the loss of a sensor "upstream". One of the more peculiar features of this interaction is that it is often erratic and unpredictable. We have performed series of systematic experiments to understand these effects, and have concluded that they are due to modal depopulation and excitation of sensors upstream from a given sensor, as well as coherent mode beating effects. Figure 18 shows the extension of our calculations presented in fig. 17 to the case of a double kink, which is the most basic model for two microbend sensors. Previous theories of microbend sensors do not account for or explain these effects. The large, quasiperiodic, and chaotic nature of the sensor interaction effects in our calculations are able to explain the experimental observations for the first time. Detailed explanation of these results will be presented elsewhere.

As a practical method for minimizing sensor interaction effects after passing thru a microbend sensor with a few dB of loss, for the near term, we have found that the easiest way is to insure equilibrium mode distribution is reached by employing sufficient sensor separation (typically several meters to be on the safe side). Further experimental studies and computer simulation of these kink loss and sensor interaction effects are in progress in our laboratories.

As a final application consider the detection of localized cracks by attaching an optical fiber to a metal (or other material) structure in a region of high stress concentration where cracks are likely or known to occur. Examples of such areas are numerous[23,24], and this is an important application for a fiber optic distributed sensing system with the capability of a large number of sensing regions. Statements like "a bridge may fail because a crack develops in a steel girder"[23] and "the best systems would have to find small structural flaws like a hairline crack in a tension flange"[23], point out the need for a distributed fiber sensor system with millimeter or less crack resolution detection capability.

The first localized crack detection study using optical fibers and OTDR was performed by Vishlizki[24] and Roberts, who pioneered the orientation angle approach. The method essentially induces a double kink type of deformation of the fiber axis. We have extended their work by fabrication of a crack simulation test fixture, shown in fig. 19, which allows more rapid, repeatable, and controlled experiments to be performed. The micropositioners allow movement both parallel and perpendicular to the line of separation. The computer modeling results shown in figs. 17 and 18 allow such crack detection sensors to be designed with excellent sensitivity. Shown in fig. 20 is a test of a crack detection sensor. In this test, the kink sites were 0.5 cm apart and the orientation angle was 45°. Acrylate coated 50/125/242 μm (core/cladding/coating diameter), graded index fiber was used.

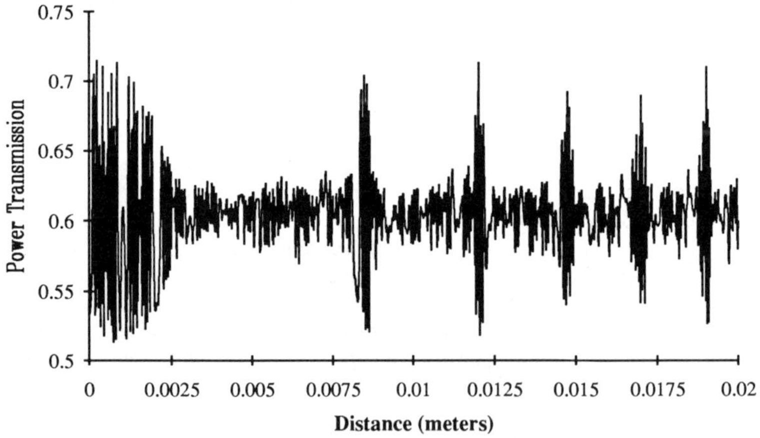

Figure 18. All mode wave optics calculations of the power transmission coefficient through the second bend of a double kink as a function of the distance between the kinks. Fiber parameters the same as in fig. 17. Kink angles are both 3°. Kink 2 average transmission = 0.604, standard deviation = 0.0293; kink 1 average transmission = 0.7113, standard deviation = 0.0246. The quasiperiodic oscillations clearly show the effect of sensor interactions due to modal excitation and depopulation (i.e. the power loss through the second kink depends on the loss and distance from the first kink).

Note the excellent crack detection sensitivity, with a conservative alarm threshold of 0.07 mm. Further refinements of the packaging and attachment of such sensors will allow numerous crack sensors to be rapidly and inexpensively deployed as strain gauges in many structural monitoring applications.

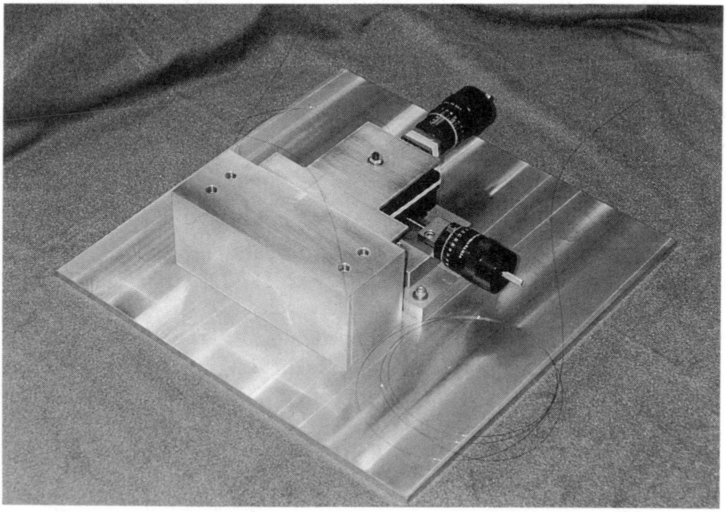

Figure 19. Crack simulation test fixture showing optical fiber oriented at 45°.

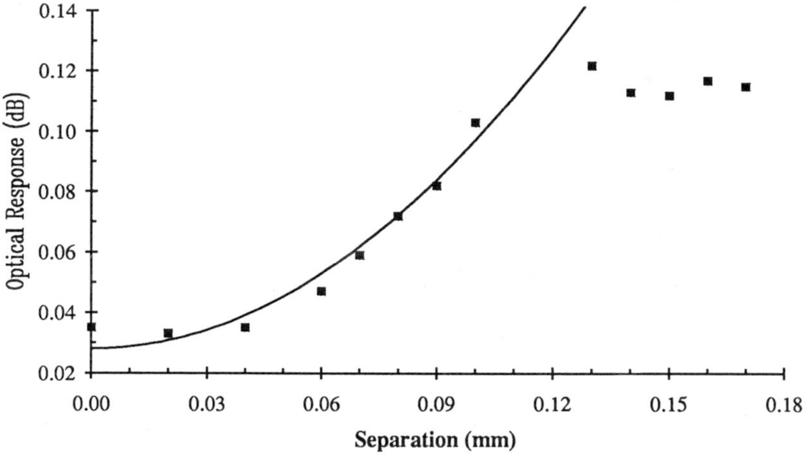

Figure 20. Localized crack detection using configuration and setup shown in fig. 19. Note the threshold for crack detection at about 0.07 mm. In this test the fiber attachment loosened at 0.13 mm. Quadratic line is guide to the eye only.

Acknowledgments:

Support for much of the work reported here was provided by G2 Systems Corporation, the California Department of Commerce, Nova Corporation, and Southern California Edison. The contributions of Pacific Braid and Cable to prototype sensor fabrication are gratefully acknowledged. We wish to thank Professor Roger Roberts and Mr. Yoram Vishlizki of Lehigh University for access to their work on localized crack detection. We thank Mr. Jim Keich of the CSUF Civil Engineering Department for design and fabrication of the pipe strain test fixture and for technical assistance with various aspects of this work.

References

1.) Tateda, M. and Horiguchi, T., "Advances in Optical Time-Domain Reflectometry", *J. Lightwave Technol.*, **LT-7**, pp. 1217-1223, August (1989).

2.) Dakin, J. P., "Multiplexed and distributed optical fibre sensor systems", in *The Distributed Fiber Optic Sensing Handbook*, Dakin, J. P., ed., IFS Publications, UK pp. 3-20 (1990).

3.) Wanser, K. H., Kingsley, S. A. and Griffiths, R. W., "High Spatial Resolution Measurement Terminal for Monitoring Short Structures Using Fiber Optics", *Proc. SPIE*, **1170**, *Fiber Optic Smart Structures and Skins II*, pp. 416-434 (1989).

4.) Kingsley, S. A., "Advances in distributed fibre optic detection and ranging", in *The Distributed Fiber Optic Sensing Handbook*, Dakin, J. P., ed., IFS Publications, UK pp. 45-55 (1990).

5.) Barnoski, M. K. and Jensen, S. M., "Fiber Waveguides: A novel technique for investigating attenuation characteristics", *Appl. Opt.*, vol. 15, p. 2112-2115 (1976).

6.) Personick, S. D., "Photon probe–An optical-fiber time-domain reflectometer", *Bell Syst. Tech. J.*, vol. 56, pp. 355-366 (1977).

7.) Blank, L. C., and Cox, J. D., "Optical Time Domain Reflectometry on Optical Amplifier Systems", *J. Lightwave Technol.*, **7**, pp. 1549 - 1555, October (1989).

8.) Snyder, A. W., and Love, J. D., "Optical Waveguide Theory", p. 27-28, Chapman and Hall (London, 1983).

9.) Olshansky, R., "Propagation in glass optical waveguides", *Rev. Mod. Phys.*, **51**, pp. 341-367, April (1979).

10.) Reference 8, pp. 154, 205, 487, and 514.

11.) Abramowitz, M. and Stegun, I. A., "Handbook of Mathematical Functions", p. 297, (Dover, New York, 9th printing).

12.) Healey, P., "Optical time-domain reflectometry–A performance comparison of the analog and photon counting techniques", *Opt. Quantum Electron.*, **16**, pp. 267-276 (1984).

13.) Garside, B. K., "Photon counting optical time domain reflectometer for distributed sensing applications", *Proc. SPIE*, **1169**, paper 1169-11 (1989).

14.) "Microbending fiber optic sensors for smart structures", E. Udd, J. P. Theriault, A. Markus, and Y. Bar-Cohen, Proc. SPIE, **1170**, *Fiber Optic Smart Structures and Skins II*, pp. 478-482, Boston (1989).

15.) Editor, "Big opportunities, big dislocations foreseen from smart skins", *SPIE OE Reports*, **78**, p. 9, June (1990).

16.) Brown, A. S., "Industry seeks tonic for aging aircraft", *Aerospace America*, **30**, pp. 24-28, 39, March (1992).

17.) Hodgson, B., "The Pipeline: Alaska's Troubled Colossus", *National Geographic*, pp. 684 - 717, November (1976).

18.) Weiss, J., Esselman, W., and Lee, R., "Assess fiberoptic sensors for key powerplant measurements", *Power*, pp. 55-58, Oct. (1990).

19.) Griffiths, R. W., "Fiber Optic Monitoring in the Electric Utility Power Plant Environment", *EPRI Conference Proceedings, Optical Sensing in Utility Applications*, San Francisco, May (1991).

20.) Granberry, M., "San Diego Sewage Spill Grows", *Los Angeles Times*, p. A1, Feb. 6, 1992. See series of articles on this pipeline rupture, p. A1, Feb. 7; p. A1, Feb. 8; p. A3, Feb. 9; Feb. 10; and Feb. 11; Feb. 19.

21.) Marcuse, D., "Losses and Impulse Response of a Parabolic Index Fiber With Random Bends", *Bell Syst. Tech. J.*, **52**, pp. 1423-1437, (1973).

22.) T. G. Giallorenzi et al, "Optical Fiber Sensor Technology", *IEEE J. Quant. Electron.*, **QE-18**, pp. 626-665, April (1982).

23.) Tarricone, P., "Bridges Under Surveillance", *Civil Engineering*, pp. 48-51, May (1990).

24.) Vishlizki, Y., "Crack Detection with Fiber Optic Sensor: A Parametric Study of the Orientation Angle Approach", M. S. Thesis, Lehigh University, (1992).

SUBJECT INDEX
Page number refers to first page of paper

AUTHOR INDEX
Page number fefers to first page of paper